DEVELOPMENTS IN **ENVIRONMENTAL ECONOMICS** **7**

THE INSTITUTIONAL ECONOMICS OF MARKET-BASED CLIMATE POLICY

DEVELOPMENTS IN **ENVIRONMENTAL ECONOMICS 7**

Titles in this series:

1. Economics of Environmental Conservation
   by C.A. Tisdell

2. Macroeconomic Analysis of Environmental Policy
   by E.C. van Ierland

3. Macro-Environmental Policy: Principles and Design
   by G. Huppes

4. Macro-Environmental Economics: Theories,
   Models and Applications to Climate Change,
   International Trade and Acidification
   edited by E.C. van Ierland

5. The Management of Municipal Solid Waste in Europe.
   Economic, Technological and Environmental Perspectives
   edited by A. Quadrio Curzio, L. Properetti and R. Zoboli

6. Marine Ecologonomics.
   The Ecology and Economics of Marine Natural Resources Management
   by A.V. Souvorov

7. The Institutional Economics of Market-Based Climate Policy
   by E. Woerdman

DEVELOPMENTS IN **ENVIRONMENTAL ECONOMICS 7**

THE INSTITUTIONAL ECONOMICS OF MARKET-BASED CLIMATE POLICY

EDWIN WOERDMAN

University of Groningen, Groningen, The Netherlands

2004

ELSEVIER

Amsterdam – Boston – Heidelberg – London – New York – Oxford
Paris – San Diego – San Francisco – Singapore – Sydney – Tokyo

ELSEVIER B.V.
Sara Burgerhartstraat 25
P.O. Box 211, 1000 AE
Amsterdam, The Netherlands

ELSEVIER Inc.
525 B Street, Suite 1900
San Diego, CA 92101-4495
USA

ELSEVIER Ltd
The Boulevard, Langford Lane
Kidlington, Oxford OX5 1GB
UK

ELSEVIER Ltd
84 Theobalds Road
London WC1X 8RR
UK

© 2004 Elsevier B.V. All rights reserved.

This work is protected under copyright by Elsevier B.V., and the following terms and conditions apply to its use:

Photocopying
Single photocopies of single chapters may be made for personal use as allowed by national copyright laws. Permission of the Publisher and payment of a fee is required for all other photocopying, including multiple or systematic copying, copying for advertising or promotional purposes, resale, and all forms of document delivery. Special rates are available for educational institutions that wish to make photocopies for non-profit educational classroom use.

Permissions may be sought directly from Elsevier's Rights Department in Oxford, UK: phone (+44) 1865 843830, fax (+44) 1865 853333, e-mail: permissions@elsevier.com. Requests may also be completed on-line via the Elsevier homepage (http://www.elsevier.com/locate/permissions).

In the USA, users may clear permissions and make payments through the Copyright Clearance Center, Inc., 222 Rosewood Drive, Danvers, MA 01923, USA; phone: (+1) (978) 7508400, fax: (+1) (978) 7504744, and in the UK through the Copyright Licensing Agency Rapid Clearance Service (CLARCS), 90 Tottenham Court Road, London W1P 0LP, UK; phone: (+44) 20 7631 5555; fax: (+44) 20 7631 5500. Other countries may have a local reprographic rights agency for payments.

Derivative Works
Tables of contents may be reproduced for internal circulation, but permission of the Publisher is required for external resale or distribution of such material. Permission of the Publisher is required for all other derivative works, including compilations and translations.

Electronic Storage or Usage
Permission of the Publisher is required to store or use electronically any material contained in this work, including any chapter or part of a chapter.

Except as outlined above, no part of this work may be reproduced, stored in a retrieval system or transmitted in any form or by any means, electronic, mechanical, photocopying, recording or otherwise, without prior written permission of the Publisher.
Address permissions requests to: Elsevier's Rights Department, at the fax and e-mail addresses noted above.

Notice
No responsibility is assumed by the Publisher for any injury and/or damage to persons or property as a matter of products liability, negligence or otherwise, or from any use or operation of any methods, products, instructions or ideas contained in the material herein. Because of rapid advances in the medical sciences, in particular, independent verification of diagnoses and drug dosages should be made.

First edition 2004

Library of Congress Cataloging in Publication Data
A catalog record is available from the Library of Congress.

British Library Cataloguing in Publication Data
A catalogue record is available from the British Library.

ISBN: 0-444-51573-9

⊚ The paper used in this publication meets the requirements of ANSI/NISO Z39.48-1992 (Permanence of Paper).
Printed in The Netherlands.

Contents

Foreword	xi
About the Author	xiii
Chapter 1. Introduction	1
1.1. Introduction	1
1.2. Climate Change, the Kyoto Protocol and Beyond	4
1.3. Market-Based Climate Policy, Public Goods and Property Rights	8
1.4. The Kyoto Mechanisms, Institutional Features and Competitive Advantages	11
1.5. The Emerging International Greenhouse Gas Market	16
1.6. Objective and Approach of the Book	19
1.7. Overview of the Book	23
1.7.1. Part I: Institutional Economics	23
1.7.2. Part II: New Institutional Economics	23
1.7.3. Part III: Institutional Law and Economics	24
1.7.4. Part IV: Neo-Institutional Economics	24
1.7.5. Part V: Conclusion	24

Part I. Institutional Economics

Chapter 2. Design and Implementation of Market-Based Climate Policy	27
2.1. Introduction	27
2.2. Tradeable Emission Rights and the Private Sector	28
2.2.1. Domestic Permit Trading Design	32
2.2.2. Downstream Permit Trading with Upstream Monitoring	36
2.3. Project-Based Emissions Trading and the Private Sector	40
2.4. Economic Versus Political Hierarchy in Market-Based Climate Policy?	43
2.4.1. The Theoretical Superiority of Permit Trading in Economics	44
2.4.2. The Problematic Acceptability of Permit Trading in Politics	48

vi   Contents

    2.5. Some Drawbacks of the Existing Literature   50
    2.6. Conclusion   52

Chapter 3. Path Dependence and Lock-In of Market-Based Climate Policy   55
    3.1. Introduction   55
    3.2. Definitions of Institutional Path Dependence and Lock-In   56
    3.3. Conditions for an Institutional Lock-In   58
        3.3.1. The Superior Alternative, Imperfect Markets and Incomplete Information   59
        3.3.2. Self-Reinforcement, Positive Feedbacks and Political Transaction Costs   60
        3.3.3. Probability, Inevitability and Remediableness   67
    3.4. Conditions for an Institutional Breakout   70
        3.4.1. Information, Perceptions and Experiments   70
        3.4.2. Problem-Solving, Crises and Learning   72
        3.4.3. Switching Costs, Legal Compatibilities and Societal Change   72
    3.5. The Superiority of the Superior Alternative Contested   77
    3.6. Novelties of an Institutional Path Dependence Approach   78
    3.7. A Path-Dependent Climate Policy?   80
    3.8. Conclusion   81

Part II.   New Institutional Economics

Chapter 4. Environmental Effectiveness of Market-Based Climate Policy   85
    4.1. Introduction   85
    4.2. Definitions of Environmental Effectiveness and Emission Baseline   86
    4.3. Environmental Effectiveness of Tradeable Emission Rights   87
        4.3.1. Macro-Baseline, Hot Air Trading and Uncertainty   88
        4.3.2. Dynamic Versus Static Perspectives on Hot Air Trading   90
        4.3.3. Options to Limit Hot Air Trading   92
        4.3.4. Non-Compliance and Liability   95
    4.4. Environmental Effectiveness of Project-Based Emissions Trading   97
        4.4.1. Micro-Baseline, Free-Riding and Gaming   98
        4.4.2. Ex Post Corrections of the Micro-Baseline   101
        4.4.3. Standardization of the Micro-Baseline   103
    4.5. Conclusion   107

Chapter 5. Transaction Costs of Market-Based Climate Policy   111
    5.1. Introduction   111
    5.2. Definition of Transaction Costs   112
    5.3. Model Versus Muddle?   114
    5.4. Transaction Costs of Tradeable Emission Rights   116
        5.4.1. Incremental Design, Set-Up Costs and Thin Markets   116

Contents   vii

      5.4.2. Empirical Evidence of Transaction Costs in Permit
            Trading Markets   122
  5.5. Transaction Costs of Project-Based Emissions Trading   124
      5.5.1. Baseline Standardization, Capacity Building and
            Multilateral Funds   124
      5.5.2. Empirical Evidence of Transaction Costs in AIJ Projects   129
  5.6. Methodological Problems of Comparing Transaction Costs   131
      5.6.1. Comparing AIJ Transaction Costs with Permit Trading
            Transaction Costs   131
      5.6.2. Comparing Market Transaction Costs with Political
            Transaction Costs   133
  5.7. Conclusion   136

Part III. Institutional Law and Economics

Chapter 6. WTO Subsidization Law and Distortions of Market-Based Climate Policy   141
  6.1. Introduction   141
  6.2. Definition of Competitive Distortions   142
  6.3. Economic Analysis of Permit Allocation and Competitive
      Distortions   143
      6.3.1. Perfect Competition, Efficiency and Opportunity Costs   144
      6.3.2. Imperfect Competition, Inefficiency and
            Financial Positions   145
      6.3.3. Fair Competition, Equity and Level Playing Field   148
  6.4. Legal Analysis of Permit Allocation and WTO Subsidies Law   151
      6.4.1. Permit Allocation and Actionable Subsidies   152
      6.4.2. Permit Allocation and Non-Actionable Subsidies   155
  6.5. Political Analysis of Perceptions on Subsidization   157
      6.5.1. Perceptions in Political Negotiations on Permit
            Allocation   158
      6.5.2. International Harmonization of Permit Allocation Rules   161
  6.6. Conclusion   163

Chapter 7. EC State Aid Law and Distortions of Market-Based Climate Policy   167
  7.1. Introduction   167
  7.2. Economic Analysis of Permit Allocation and Competitive
      Distortions   168
      7.2.1. Competitive Distortions, Efficiency and
            Opportunity Costs   169
      7.2.2. Competitive Distortions, Equity and Level Playing Field   170
  7.3. Legal Analysis of Permit Allocation and EC State Aid Law   171

7.3.1. Permit Allocation and State Aid Criteria — 173
7.3.2. Permit Allocation and State Aid Exemptions — 175
7.4. Political Analysis of Perceptions on State Aid — 178
  7.4.1. Perceptions in Political Negotiations on Permit Allocation — 179
  7.4.2. The Political Precedent of Emissions Trading in Denmark and the UK — 184
  7.4.3. The Political Outcome of Permit Trading in the EU — 188
7.5. Possible Extensions of the Analysis to the Polluter Pays Principle — 190
7.6. Conclusion — 193

Part IV. Neo-Institutional Economics

Chapter 8. Theoretical Aspects of Restricting Market-Based Climate Policy — 199
8.1. Introduction — 199
8.2. Definition of Supplementarity — 200
8.3. Economic Analyses of the EU Proposal on Supplementarity — 203
  8.3.1. Overall Economic Effects of the EU Proposal on Supplementarity — 203
  8.3.2. Gainers and Losers of the EU Proposal on Supplementarity — 204
8.4. Theoretical Explanations of the EU Proposal on Supplementarity — 207
  8.4.1. Hypotheses on Restricting the Use of the Kyoto Mechanisms — 208
  8.4.2. Alternative Hypotheses and Limitations of the Theoretical Analysis — 222
  8.4.3. Ex Post Clustering of the Hypotheses — 223
8.5. Conclusion — 226

Chapter 9. Empirical Aspects of Restricting Market-Based Climate Policy — 229
9.1. Introduction — 229
9.2. Representativity and Limitations of the Empirical Analysis — 230
9.3. Empirical Analysis of the EU Proposal on Supplementarity — 235
  9.3.1. Content Analysis of EU Documents — 235
  9.3.2. Hypothesis Testing Among Key EU Officials — 239
  9.3.3. Analysis of Questions on Supplementarity Among Key EU Officials — 250
  9.3.4. Bargaining Behavior of the EU at CoP6 — 253
9.4. The Institutional Breakout of EU Climate Policy — 254
  9.4.1. A Path-Dependent History of Market-Based Climate Policy in the EU — 255
  9.4.2. A Path-Dependent Future of Market-Based Climate Policy in the EU? — 259
9.5. Conclusion — 261

Part V. Conclusion

Chapter 10. Conclusion 267
    10.1. Developments in Environmental Economics 267
    10.2. The Institutional Economics of Market-Based Climate Policy 268
    10.3. The New Institutional Economics of Market-Based
         Climate Policy 270
    10.4. The Institutional Law and Economics of Market-Based
         Climate Policy 272
    10.5. The Neo-Institutional Economics of Market-Based Climate
         Policy 275
    10.6. Some Policy Implications 277

Appendix (Questionnaire) 281
    The EU Proposal on Supplementarity: a Questionnaire 282

References 291

Subject Index 315

Foreword

This book is an improved, updated and shortened version of my dissertation. There are many people and organizations that have contributed to my research. I would like to thank the Netherlands Organization for Scientific Research (NWO), and its National Research Programme on Global Air Pollution and Climate Change (NRP), for financial support. I also want to thank the University of Groningen, as well as the University of Twente, for providing me with the facilities and assistance to carry out my research.

I am much indebted to Andries Nentjes who supervised my research through all these years. A true specialist in market-based climate policy — and a wonderful personality. Also Bert Steenge, as well as Dick Ruiter, made some important contributions to this book, for instance by pointing at the relevance of Douglass North's work. I also owe some other colleagues, both in Groningen and Enschede, for helping me to solve some research problems, in particular Oscar Couwenberg, Frans de Vries, Mirjam Koster, Roelof de Jong and Wytze van der Gaast.

I also appreciate the advice, or comments, I received from Jan-Tjeerd Boom, Bouwe Dijkstra, Zhong-Xiang Zhang, Catrinus Jepma, Ger Klaassen, Johan Albrecht, Erik Haites, Axel Michaelowa and René Kemp, among others. Jos Delbeke, Peter Zapfel and especially Peter Vis from the European Commission helped me to complete (and nuance) my empirical analysis on the EU supplementarity proposal. Finally, I want to thank my former teachers Ad van Deemen, Jan van Deth and Jan Verschoor.

But if I can dedicate this book to anyone, it must be to my family and friends. In particular, this book would not have seen the light without the support and understanding of my girlfriend Jacqueline and our daughter Sophie.

Edwin Woerdman

Groningen / Enschede, the Netherlands
February 2004

About the Author

Dr. Edwin Woerdman (1970) was born in Utrecht, the Netherlands, and graduated with honours in political science at Radboud University Nijmegen, where he specialized in economic theory. After writing an introduction to political science in Dutch for Wolters-Noordhoff, he finalized his dissertation at the University of Groningen, where he used the path dependence approach to study the institutional evolution of economic instruments for environmental regulation. In particular, he analyzed the political barriers (including institutional, legal and cultural ones) to implementing the Kyoto Mechanisms in international climate policy. After a joint appointment as a postdoctoral research fellow at the University of Twente and at the University of Groningen, he became associate professor of law and economics at the latter university in 2004.

Woerdman publishes regularly on market-based instruments for environmental regulation both in national and international journals and books, like *Ecological Economics*, *Energy Policy*, *Rationality and Society* and the *Elgar Companion to Law and Economics*. As a private consultant he adviced the Dutch government on greenhouse gas emissions trading. Next to teaching law and economics at the University of Groningen, he has also been a teacher in a postgraduate course on public management at the Academy of Management in Groningen and in a master (as well as a postgraduate) course on institutional law and economics at the Institute for Governance Studies in Enschede.

Woerdman is mainly interested in the interdisciplinary (economic, legal and political) study of governance, institutional change, emerging markets, property rights, transaction costs, path dependencies and lock-in situations. He is also interested in the role of values and equity perceptions in politics and law. Currently, institutional (law and) economics and market-based climate policy, including the Kyoto Mechanisms, are an important focus in his work.

Chapter 1

Introduction

1.1. Introduction

This book studies the institutional economics of market-based climate policy. This type of policy is becoming increasingly popular. Once perceived as politically unacceptable by various governments and non-governmental organizations, tradeable pollution schemes to combat climate change are now in the planning or implementation process in dozens of countries.

The largest institution in the realm of climate policy, both in terms of geographical scope and potential market size, is the Kyoto Protocol of 1997. This legal protocol to the United Nations Framework Convention on Climate Change has been ratified by more than one hundred countries. It imposes absolute emission ceilings on industrialized countries and establishes three market-based instruments, the so-called Kyoto Mechanisms, to meet the emission targets in an economically efficient way.

Market-based climate policy is more than the Kyoto Protocol, however. To enter into force, the number of countries that have ratified the Protocol should account for at least 55% of total CO_2 emissions of industrialized countries in 1990. This condition is not (or is not yet) met at the time of writing. The Russians, for instance, are reluctant to ratify, and the Americans already withdrew from the Protocol in 2001. The United States claimed that the absolute targets agreed upon would harm their economy and argued that large emitters like China should not continue to be exempted from emission ceilings.

But even without the Protocol, the Americans still intend to use market-based instruments, for instance under a greenhouse gas intensity target with the possibility of transferring registered emission reductions between firms. Furthermore, some federal states have expressed their interest in imposing absolute caps, for instance on power plants, and allow for emissions to be traded.[1] In addition, with or without the Kyoto Protocol, the governments of the European

---

[1] Some Russian regions could, at least in theory, do the same thing and develop their own emissions trading schemes if the Russian Federation would decide not to ratify (Grubb, 2003).

Union have decided to implement a cap-and-trade scheme, to start in 2005, where CO_2 emissions can be traded among power generators, steelmakers as well as cement, paper and glass manufacturers.

According to some economists, Kyoto does "too little, too fast" (e.g. Aldy et al., 2003). Other climate policy architectures are thinkable that could provide larger participation, higher effectiveness and lower costs, although there is usually some trade-off between those criteria. However, in spite of its shortcomings, most developed (and developing) countries, including the European Union, Japan as well as several Nordic and Eastern European countries, still support the Kyoto Protocol as an important first step that took years of negotiations. "The Kyoto Protocol of 1997 is and will stay a milestone in the process of ensuring that climate change remains on the political agenda and promoting internationally coordinated action" (Faure et al., 2003: 4). In that setting, the Kyoto Mechanisms "(…) have the potential to become the most important cornerstones of the emerging climate regime (…)" (Oberthür & Ott, 1999: 275). For these reasons, although the theory and concepts used in this book concern market-based climate policy in general, we will frequently (but not only) present applications and examples in the context of the Kyoto Mechanisms.

"The 1997 Kyoto Protocol establishes an international institutional framework for domestic responses to climate change that links emission targets for developed countries to international market mechanisms" (Bernstein, 2002: 203). As such, market-based institutions have moved "center stage" in environmental policy, as Stavins (2002: 15) puts it, but can the same be said about the use of institutional economics to study them? In the past decades, environmental economists have mainly calculated the potential efficiency gains of such instruments and, partly based on experience with real-life (emission) markets, provided design prescriptions that would ensure their efficient and effective functioning (e.g. Tietenberg et al., 1999; Zhang & Nentjes, 1999). The importance and influence of these studies should not be underestimated. Moreover, this literature not only pays attention to institutional considerations, ranging from permit definition to enforcement, but also contains elements of institutional analysis, for instance by taking transaction costs into account.

However, the institutional economics used in these studies is rather limited in scope. The neoclassical approach dominates. This approach certainly has explanatory power, as our book will confirm once again, but it also overlooks at least three crucial aspects of market-based climate policy. First, although some of the literature considers the transaction costs in the market, a traditional institutional economics topic, there are hardly any systematic analyses of the political transaction costs to set up this market. Second, although several authors take the dynamics of the market into account, they hardly ever study the dynamics of the institutions that support them. In particular, they do not recognize that (not just technologies, but also) institutions exhibit patterns of path dependence in an evolutionary process, which might under certain historical circumstances lead to a lock-in of (inefficient) environmental policy instruments. Third, if economists study institutions in market-based climate policy at all, they focus on formal institutions, usually without considering, let alone analyzing, the impact of informal institutions, like political culture. Moreover, when considering formal institutions, surprisingly little use is made of law and economics perspectives.

Obviously, this book tries to fill these gaps. We do not claim that we hereby complete the story on economic instruments and climate institutions. Rather the opposite: much remains to be researched and perspectives other than institutional-economic ones, like international relations theory (just to mention a different field), could lead to relevant new insights. But we do believe that we cover the institutional economics of market-based climate policy in a broader and more systematic way than has been done before. In doing so, we do not reject the neoclassical approach. Instead, we use it and show where and why it is fruitful to employ an institutional approach. The result is that equity is considered next to efficiency, that the evolution and path dependence of both formal and informal climate institutions is studied, and that attention is paid to the politics and law of economic instruments for climate policy, including some new empirical analyses.

We can now formulate the objective and approach of this book. Put briefly, the objective is to analyze the formal and informal institutional barriers that prevent or delay the implementation of market-based climate policy, as well as to provide opportunities to overcome them. The approach is that of institutional economics, with special emphasis on (political) transaction costs and path dependence.

This chapter is organized as follows. Section 1.2 indicates how and when climate change entered the political agenda, describes the Kyoto Protocol including its flexible instruments and sketches the long-term opportunities for market-oriented environmental regulation both with and without the Kyoto Protocol. Section 1.3 traces the intellectual and conceptual origins of market-based climate policy, explains the public good character of reducing greenhouse gas emissions and tries to find out whether tradeable emission entitlements, also those under the Kyoto Protocol, are property rights. Section 1.4 identifies the competitive advantages of the Kyoto Mechanisms based on their negotiated institutional features. Section 1.5 provides a picture of the emerging international greenhouse gas market. Section 1.6 specifies the objective and approach of the book. Finally, Section 1.7 presents an overview of the book.

1.2. Climate Change, the Kyoto Protocol and Beyond

While the first scientific conjecture of an enhanced greenhouse effect resulting from human activities was already formulated at the end of the 19th century, it was not until the late 20th century that climate change moved onto the international political agenda (e.g. Bolin, 1993; Jäger & O'Riordan, 1996). Alarmed by evidence of global warming provided by scientists since the 1960s, governments called for additional research in the beginning of the 1980s, which eventually lead to the establishment of the Intergovernmental Panel on Climate Change (IPCC) in the context of the United Nations (UN) in 1988.

When IPCC scholars reconfirmed the threat of human-induced climate change, for instance caused by the burning of fossil fuels in the industry and transport sector, governments started negotiations to build an international climate change agreement in the beginning of the 1990s. This resulted in the adoption of the UN Framework Convention on Climate Change (FCCC) in 1992 with the objective for industrialized countries (as elaborated in subsequent negotiations) to achieve a stabilization of their greenhouse gas (GHG) emissions-such as carbon dioxide (CO_2), methane (CH_4) and nitrous oxide (N_2O) — at 1990 levels by the year 2000. The developing countries were exempted from emission targets, recognizing that the largest share of historical and current global GHG emissions has originated in the developed countries and that the developing countries need to achieve sustained economic growth and eradicate poverty.

When IPCC reports indicated that the stabilization goal would not be sufficient to prevent a dangerous anthropogenic interference with the climate system, the Parties (governments) to the FCCC decided to formulate emission reduction commitments for the developed countries in the form of a legal protocol, despite the problems they already had to stabilize their emissions (e.g. Oberthür & Ott, 1999). Such a protocol to the FCCC was agreed upon in 1997 in Kyoto (Japan), which has, therefore, been termed the Kyoto Protocol. If this Protocol will be ratified, the industrialized countries shall individually or jointly reduce their overall GHG emission level by at least 5% below 1990 levels in the commitment period 2008–2012 (Article 3.1).

To reach this level, these so-called Annex B Parties (or: Annex I Parties under the FCCC) have adopted differentiated Quantified Emission Limitation or Reduction Commitments (QELRCs), such as an 8% reduction for the European Union (EU), a 6% reduction for Canada and Japan and stabilization for the Russian Federation. The United States (US), which is the largest emitter of CO_2 in the world (IEA, 1999), committed themselves to a 7% reduction target, but in March 2001 the Americans withdrew from the Protocol. The US not only criticized the fact that developing countries are still exempted from the emission ceiling, including China as the second-largest CO_2 emitter in the world (IEA, 1999), but they also claimed that the Kyoto target would harm the American economy (Bush, 2001). Opponents of this stance, both within and outside America, argued that there were and still are sound justice reasons to (temporarily) exempt developing countries from emission ceilings, mainly based on the arguments of historical responsibility and poverty eradication, and that the Kyoto target would cost the US no more than, say, 0.1–2% of its GDP growth (e.g. Banuri et al., 2001: 57).

The Kyoto Protocol allows Annex B Parties to meet their commitments partly by achieving emission reductions abroad. This enables developed countries to improve the cost-effectiveness of emission reduction, because reducing GHG emissions at an emission source in another country may be cheaper than doing so domestically (e.g. Zhang & Nentjes, 1999). Indeed, several authors found that the marginal costs of GHG emission reduction vary greatly among the FCCC Parties (e.g. Hourcade et al., 1996; Kram & Hill, 1996). Moreover, since global warming is caused by the total accumulation of GHGs in the atmosphere, it does not matter where these uniformly mixed pollutants are produced or reduced. If all Parties could make optimal use of these marginal cost differences, without any institutional impediments, the overall costs of combating climate change would be reduced by almost 80% compared with domestic action only (e.g. Richels et al., 1996). To enhance efficiency by means of cross-border emission reduction, Annex B Parties are allowed to purchase emission reduction entitlements from a foreign country by implementing one or more of the so-called Kyoto Mechanisms:

- Joint Implementation (JI) under Article 6;
- Clean Development Mechanism (CDM) under Article 12;
- International Emissions Trading (IET) under Article 17.

An industrialized country can purchase Assigned Amount Units (AAUs) on the basis of IET and/or Emission Reduction Units (ERUs) on the basis of JI from another Annex B country, for instance in Central or Eastern Europe where marginal abatement costs are relatively low. It can also acquire Certified Emission Reductions (CERs) from developing countries based on CDM projects. The Kyoto Protocol (Articles 6.1(d), 12.3(b) and 17) requires that the use of these flexible instruments is "supplemental" to domestic action: each Annex B Party must provide information on how its domestic action is a significant element of the efforts to meet its emission targets.

There are several institutional differences between the Kyoto Mechanisms. IET uses a top-down approach by calculating the emission reductions on the basis of national commitments. The legal text of Article 17 indicates that Annex B governments could trade parts of their assigned amounts. A sovereign government could decide to split up its assigned amounts by allocating permits to private entities (such as firms or sectors) enabling them to trade emissions domestically.

However, it still has to be decided under what conditions firms are allowed to trade directly with each other internationally. JI and the CDM differ from IET, because they are project-based flexible instruments in which an investor receives credits for the achieved emission reductions at the host. In principle, the emission reductions in such projects are not measured top-down from the national commitment, but bottom-up from a baseline which estimates future emissions at the project location if the project had not taken place.

Although both are project based, JI and the CDM also differ from each other. A JI host country has an emission target in contrast with a CDM host country. Furthermore, credits which accrue from CDM projects between 2000 and 2008 can be banked in order to use them for the commitment period (Article 12.10), which is not possible under JI. However, forest management projects (resulting in removal units (RMUs)) which aim at protecting existing forests instead of actually (re)planting trees can be applied to a limited extent under JI Article 6, but these are not eligible as CDM projects. In addition, afforestation and reforestation projects may be fully used for compliance under JI, but only to a limited extent under the CDM. Moreover, the institutional requirements under the CDM in terms of supporting sustainable development in the host countries (and the requirement of a supervising Executive Board) are stronger than under JI.

Next to the Kyoto Mechanisms, the Kyoto Protocol also contains some additional flexibility provisions, notably the establishment of a multi-year commitment period for six GHGs (Article 3.1), the possibility of banking (Article 3.13) and the bubble option (Article 4).

First, instead of a commitment year, the Kyoto Protocol establishes a flexible commitment period in which the target of an Annex B Party must be achieved by calculating its average emissions over 5 years from 2008 to 2012 (Article 3.1). The Kyoto Protocol uses a "basket" of six GHGs (listed in Annex A), which not only includes CO_2 as the major GHG, but also allows reductions in other GHGs, such as CH_4, which are all translated into CO_2-equivalents to produce a single figure.

Second, industrialized countries have the possibility to bank unused parts of their assigned amounts (Article 3.13). If an Annex B Party has lower emissions than its assigned amount in the first commitment period (2008–2012), the difference can be added ("banked") to the allowance for subsequent commitment periods. While such banking is unrestricted for AAUs, the carry-over of ERUs and CERs is restricted to 2.5% of the assigned amount and not allowed for RMUs (CP, 2001b).

Third, Annex B Parties are allowed to form subgroups and reallocate their targets as long as this does not change the total emission ceiling of their original assigned amounts and provided that the FCCC Secretariat is notified of such an agreement (Article 4). The EU has used this "bubble" provision to reallocate its assigned amount among its Member States, which has resulted, for instance, in commitments of 21% reduction for Germany, stabilization for France and 27% allowable emission growth for Portugal. Although this internal burden sharing arrangement could serve to lower compliance costs for the EU, it is not fully efficient because it does not equalize marginal costs among its Member States (Eyckmans & Cornillie, 2000).

Whereas national governments hold the legitimate monopoly of force within a certain territory (Weber, 1976), there is no "world government" in the international political system of sovereign states to bring about and enforce co-operation between governments (Waltz, 1979). After several years of inter-governmental bargaining, co-operation was nevertheless achieved to combat climate change, largely because governments created the Kyoto Mechanisms under the Protocol which would lower their costs of reducing pollution (e.g. Bohm, 1999; Oberthür & Ott, 1999). Although the position of the EU and the developing countries was, at least initially, characterized by market skepsis and moral resistance against trading in the environmental sphere, they accepted the Kyoto Mechanisms, because the latter were a precondition for several other countries, such as the US, to accept an emission reduction target in the first place (e.g. Ringius, 1999). A few years after this compromise was made, the European Commission openly recognized that the Kyoto Protocol put emissions trading on the political agenda of the EU (COM, 2000a: 7). Several historical developments, including internal pressures and external "shocks" (as we will explain later on in this book), eventually lead the EU to adopt an emissions trading scheme of their own, to start in 2005.

The international adoption of the Kyoto Mechanisms in 1997 moved the political process to the implementation stage. In this stage, the details of their design have to be worked out and decided upon to make these flexible instruments operational. However, various institutional barriers hinder the implementation of the Kyoto Mechanisms, including legal ambiguities and cultural objections. Examples of such issues, just to name a few, are the acceptable levels of using sinks and banking, the desirability and methodology of standardizing project baselines, the compatibility of domestic permit allocation with international and European law on state subsidization, the potential and complexities of incorporating households in the trading system, the effect of the international transferability of emissions on the environment and fairness, as well as the corresponding question of whether and how use of the Kyoto Mechanisms should be restricted. It will become clear that a few of these barriers have been negotiated and others not (yet) or only partly, while governments sometimes create additional barriers by posing new demands and by trying to reopen or reinterpret previous international political agreements (e.g. Boyd et al., 2001). The IPCC considers an analysis of institutional barriers to implementing market-based climate policy as a priority area for research (Banuri et al., 2001: 71).

As has been explained in the introduction, however, it is not sure that the Kyoto Protocol will enter into force, given that the number of countries that have ratified do not (yet) account for at least 55% of total CO_2 emissions of industrialized countries in 1990. At the moment of writing, ratification by the Russians, which is still uncertain, would bring total CO_2 emissions over this required threshold. But even without a go-ahead for the Kyoto Protocol, the US still intends to use market-based instruments in climate policy, for instance by transferring registered emission reductions between firms under a greenhouse gas intensity target, while some federal states have expressed their interest in forming a coalition within the US by establishing permit trading schemes and subsequently connect them, for instance for the electricity sector. Moreover, with or without the Kyoto Protocol, the EU will start with a cap-and-trade scheme in 2005, where CO_2 emissions can be traded among power generators, steelmakers as well as cement, paper and glass manufacturers.

If the Kyoto Protocol would enter into force, though, the world's largest market-oriented institution in the realm of climate policy will become reality, both in terms of geographical scope and potential market size. Emissions can then be traded under the Kyoto Mechanisms within developed countries and with developing countries in the first commitment period 2008–2012, and possibly also thereafter as the Parties are required to initiate the consideration of a second commitment period with emission targets for developed countries already in 2005 (Article 3.9), resulting in a potential market value of several billions of US dollars (e.g. Haites, 1998).

Nevertheless, even if the Kyoto Protocol becomes the dominant institution in international climate policy, Parties are free to leave. According to Article 27, at any time after 3 years from the date of entry into force for a Party, that Party may withdraw from the Protocol by giving written notification. In the end, each sovereign state can always choose to construct its own climate policy (or refrain from it all together) and decide to trade emissions with other nations if it perceives this to be beneficial. As many countries have already chosen to construct tradeable pollution schemes, we would then still witness an emerging carbon trading market, albeit a more fragmented one.

1.3. Market-Based Climate Policy, Public Goods and Property Rights

Market-based climate policy, including the Kyoto Mechanisms, allows polluters to reduce the costs of achieving emission targets and has its roots in the tradeable emission rights concept. Dales (1968) is usually seen as the founding father of this concept, Montgomery (1972) as the one who provided formal proof of its efficiency, and Tietenberg (1980) as the one who firmly advocated and established it in environmental economics. Emissions trading can be traced back to the property rights school in economics, according to which externalities should be internalized (e.g. Demsetz, 1967). This means that negative external costs which are not reflected in the market price, like environmental pollution, should be included in this price by allocating property rights.[2]

Starting point of the analysis is the theory of externalities and public goods (e.g. Baumol & Oates, 1988). An externality is a positive or negative external cost which is not reflected in the market price. The emission of GHGs is a negative externality due to the detrimental impact of climate change. The reduction of these GHG emissions has a public good character. Public goods are non-excludable, in contrast with private goods, meaning that nobody can be excluded from consuming it (Olson, 1965).[3] This gives rise to the free-rider problem: an individual (or nation) can enjoy the benefits of the (international) public good, the emission reductions, without having to contribute to the costs of its production. The consequence is a "tragedy of the commons" (Hardin, 1982): the provision of the public good will be sub-optimal and emissions will be too high (e.g. McNutt, 1996). Therefore, property rights theorists advise to transform these public goods into private goods by making them excludable. Polluters will then take the negative external cost of GHG emissions into account. Moreover, in the absence of transaction costs, the allocation of resources is independent of the distribution of these rights (Coase, 1960).

"Since economics is based on property rights, economic solutions to pollution problems also involve property rights solutions" (Dales, 1968: 76). Dales proposed that the government makes these pollution rights transferable by allocating them top-down to polluters, such as firms, so that a market (price) will develop which "(…) ensures that the required reduction in waste discharge will be achieved at the smallest possible cost to society" (Dales, 1968: 107). Several scientists (in particular economists) have advocated the real-life application of the tradeable pollution rights concept in the context of climate change, first mainly North Americans (e.g. Tietenberg, 1980), but later also Europeans (e.g. Koutstaal & Nentjes, 1995) and Asians (e.g. Zhang, 2000a).

In 1975 the US Environment Protection Agency (EPA) began experimenting with emissions trading to control air pollution. Since then the concept and variants

---

[2] Some authors argue that more or less similar ideas can be traced back to as far as John Stuart Mill's work from 1848, who wrote about the possibility of giving air a market price, or Aristotle's work from more than 2000 years ago, who wrote that which is common to the greatest number has the least care bestowed upon it (see Cole, 1999: 105; Yandle, 1999: 17).

[3] Non-excludability is the distinguishing feature between public goods and private goods (Hardin, 1982). Pure public goods are not only non-excludable, but also non-rivalrous in consumption, meaning that the amount of the good is not limited if others consume it as well (McNutt, 1996).

thereof have been used in various other US programs, for instance to reduce ozone-depleting substances under the Montreal Protocol (since 1988) and to reduce SO_2 emissions under the 1990 Clean Air Act Amendments (CAAA) (where such emissions are in fact traded since 1993). Outside the US, some experience was gained mainly with tradeable quota systems, like the tradeable ammonia quota in the Netherlands (since 1994), but the definitive breakthrough of emissions trading outside the US is expected to occur in the context of market-based climate policy, for instance under the Kyoto Mechanisms or in the European emissions trading scheme. In addition, various countries intend to build national tradeable emission rights systems, like Switzerland, Norway, Japan and Canada, which could eventually be linked to each other, and to the European scheme, provided that they mutually recognize their transferable units.

Most economists see tradeable emission rights as property rights, because of their exclusive use against all, market value and incentive effects. In the trading scheme for SO_2 emissions in the US, however, a legal provision was adopted that an emission right, called an "allowance", does not constitute a property right (in section 403(f) of the CAAA). The legislator chose this formulation to avoid that the government would have to compensate polluters for "taking" allowances when the authorities lower the annual emission caps. Both in this scheme and in the European CO_2 emissions trading system, an emission right is basically defined, in legal terms, as an allowance that authorizes a legal entity to emit a certain amount of pollution during a specified period. This is not so much a permanent, private property right, but rather an authorization that can be terminated or limited by the government.

Although some then conclude that emission rights are, and should be, temporary "rights of use" (e.g. Convery et al., 2003), the law and economics literature prefers to characterize allowances as mixed, hybrid or regulatory property rights (e.g. Rose, 1999; Yandle, 1999). Emission rights contain elements of both public and private property rights: instead of common law private rights and liability rules that form over time when conflicts over resource use arise, allowances are non-permanent, government-mandated rights that combine state control over the emission quotas with private freedom for polluters of how to comply (which could be referred to as "command-without-control"). Moreover, although allowances in the American SO_2 emissions trading scheme are not property rights *themselves*, property rights *in* allowances are in fact recognized as emitters can receive, hold and transfer them, while excluding all others, besides the government, from interfering with their possession, use and disposition (Cole, 1999: 113–114).

The Kyoto Mechanisms would create an international market for GHG emissions, but some of these mechanisms are only weak variants of the original property rights concept. The theoretical possibility of international firm-to-firm trading under an emission cap in the context of IET Article 17 comes closest to the original property rights blueprint for environmental pollution as sketched above. Assigned amounts have been allocated (top-down) to governments, which can trade under IET Article 17. Although this part of the Protocol of 1997 left the role of the private sector in international GHG emissions trading initially undefined, the annex on emissions trading in the Marrakesh Accords of 2001 explained that governments may in fact authorize legal entities to transfer and/or acquire emissions under Article 17.

The assigned amounts under the Kyoto Protocol have been defined and allocated for the limited period of 2008–2012, whereas the Parties agreed that the Kyoto Protocol has not created or bestowed any right, title or entitlement to emissions of any kind on Annex B Parties (CP, 2001a: 7). In addition, JI and the CDM do not explicitly assign property rights, but instead provide the legal basis to develop concrete projects in a bottom-up fashion with the aim to reduce GHG emissions in countries where this is relatively cost effective. Nevertheless, it can be argued, again, that although these entitlements are not property rights themselves, property rights in such entitlements are in fact recognized as emitters can receive, hold and transfer them, while excluding all others.

1.4. The Kyoto Mechanisms, Institutional Features and Competitive Advantages

In spite of possible different investment risks, the Kyoto Mechanisms can be defined in monetary units (e.g. dollars) per ton of CO_2-equivalent, which means that they will compete on an international carbon trading market. Buyers are interested in low-cost emission reductions. A Kyoto Mechanism has a competitive advantage if its costs of reducing emissions per ton of CO_2-equivalent are relatively low compared to the other Kyoto Mechanisms. This is the case if a Kyoto Mechanism has the largest emission reduction potential at any given price per ton of CO_2-equivalent. The competitive advantages of the Kyoto Mechanisms depend on their specific formal institutional features negotiated in Kyoto and beyond (Woerdman, 2001a).

These mutual and relative competitive advantages are sketched for each (set of) flexible instrument(s), largely based on existing literature, as summarized in Table 1.1. The qualitative results obtained below are based on the text of the Kyoto Protocol of 1997 including the additions and alterations made by governments thereafter up to and including the seventh Conference of the Parties (CoP) in Marrakesh (Morocco) in 2001. The competitive advantages of the Kyoto Mechanisms could change, therefore, if the CoP decides to alter or elaborate the institutional provisions of the Kyoto Mechanisms in the future.

Table 1.1: Competitive advantages of the Kyoto Mechanisms.

Competitive advantage	Emission ceilings	Pre-budget banking	Sinks option	Transaction costs	Adaptation tax	Export stimulus	Sustainable development	Additionality period
High + ↓	CDM	CDM	JI/IET	JI/IET	JI/IET	JI/CDM	JI	CDM
Low –	JI/IET	JI/IET	CDM	CDM	CDM	IET	CDM	JI

Key: JI, Joint Implementation (Article 6); CDM, Clean Development Mechanism (Article 12); IET, International Emissions Trading (Article 17).

First, the CDM has a competitive advantage relative to JI and IET because of the absence of national emission reduction commitments in developing countries. In contrast with CDM host countries, the Kyoto Protocol does specify such commitments for Annex B Parties which are allowed to act as JI host countries and/or trade parts of their assigned amounts on the basis of IET Article 17. This restricts the supply of both assigned amounts and JI projects and raises their (marginal) costs, since it becomes increasingly difficult for a host country to achieve its own target as it sells more parts of its assigned amount and/or more JI credits (ERUs).

Second, the CDM has a competitive advantage relative to JI and IET, because CDM credits (CERs) can be banked between 2000 and 2008 in order to use them for the commitment period 2008–2012. Neither early emission reductions from IET, nor ERUs which accrue from JI projects between 2000 and 2008 can be banked. A comparable project which starts in 2000 would thus produce credits for 5 years (2008–2012) in a JI host country and for 13 years (2000–2012) in a CDM host country. This means that the same kind of project or emission transfer could yield nearly three times as many emission reduction entitlements under the CDM as under JI or IET.

Third, JI and IET have a competitive advantage relative to the CDM, because sinks (such as forestry projects) are less restricted under these mechanisms than under the CDM. Article 12 does not explicitly mention sinks as eligible CDM projects. This initially triggered a discussion whether sinks are implicitly excluded in the CDM, or whether they are still indirectly eligible under the CDM, because sinks are included in Article 3.3 as a means for Annex B Parties to fulfill their commitments. The answer was provided in 2001, at CoP6 Part II in Bonn (Germany), where the Parties decided to allow for sinks under the CDM, but only to a limited extent (CP, 2001a). Use of sinks was limited, because the perception dominated that the technical methodologies are not sufficiently developed to make sure that the carbon sequestration of maintaining existing forests is calculated and monitored adequately. Since CDM host countries do not have a national emission target, it was feared that allowing sinks, in particular CDM projects aimed at protecting existing forests, could inflate the overall emission ceiling of Annex B Parties. Therefore, forest management is not eligible under the CDM. Afforestation and reforestation are allowed as CDM projects because their additionality is less controversial, but the total of subtractions from and additions to the assigned amount of a Party shall not exceed 1% of its base year emissions times five. IET sellers and JI host countries (such as Eastern European countries hosting JI forestry projects) do have such a national emission ceiling (the assigned amount). This gives them an incentive not to exaggerate the emission reduction in a forestry project, because this would imply a lowering of the national commitment. Therefore, projects aiming at (re)planting trees are not restricted under JI and IET, but the use of forest management is restricted by quota set for each individual country. At CoP7 in 2001 in Marrakesh (Morocco) it was decided to label credits resulting from such forest management or agriculture projects as removal units (RMUs) (CP, 2001b).

Fourth, JI has a competitive advantage over the CDM, because transaction costs in developing countries will be higher than those in Central and Eastern Europe due to both the required technology transfer and the relatively severe informational, institutional and infrastructural constraints (Sokona & Nanasta, 2000). Furthermore, the presence of a national emission target could make baseline determination for JI projects easier than for CDM projects where such targets are absent. A Central or Eastern European government has to define environmental policy targets for its domestic emitters. If it has done so, the JI baseline could be deducted from the defined environmental policy for the host firm or sector involved. Transaction costs typically consist of search costs, negotiation costs, approval costs, monitoring costs, enforcement costs and insurance costs (Dudek & Wiener, 1996). Most economists assume that IET will have relatively low transaction costs if private entities are allowed to trade internationally without too many trading rules (e.g. Tietenberg et al., 1999; Vrolijk & Grubb, 2000).

Fifth, JI and IET have a competitive advantage relative to the CDM, because investors only have to pay an "adaptation tax" for CDM projects (Michaelowa, 1999). Article 12.8 requires that a share of the proceeds from the CDM is used to cover administrative expenses and assist non-Annex B Parties to meet the costs of adaptation against the adverse effects of climate change. Proposals for this fee (as well as estimates of the administration costs) initially ranged from about 1–15% and at CoP6 Part II in 2001 the Parties decided to set the share of proceeds at 2% of the CERs issued for a CDM project. (Projects in least developed countries are exempted from this fee.) The revenues will be used, among other things, to finance concrete adaptation projects and programs in developing countries under the Kyoto Protocol Adaptation Fund. There will also be an additional fee to cover the administrative expenses of the CDM, but its level still has to be determined by the CoP (CP, 2001b).

Sixth, JI and the CDM may have a competitive advantage because, contrary to IET, these project-based approaches have the positive side effect for Annex B Parties of creating the opportunity to export extra capital goods to the host country. In particular through JI and CDM projects, investors may enter possible new export and investment markets in the guest countries (Michaelowa, 1995). Jones (1993) expects that investors will increase their export potential of advanced pollution control technologies.

Seventh, within a subset of the Kyoto Mechanisms, JI may have a competitive advantage over the CDM, because CDM Article 12 as well as Article 10(c) place a relatively strong emphasis, compared to JI Article 6, on sustainable development in and benefits (such as technology transfer) for the developing host country induced by the project. These requirements could make a CDM project in a developing region, ceteris paribus, more expensive than a JI project in a country with an economy in transition.

Eighth, within a subset of the Kyoto Mechanisms, the CDM has a competitive advantage relative to JI, since CDM projects will generate credits for a longer time than JI projects. Often JI and CDM projects only speed up investments that would have been carried out by the host countries themselves in the mid-term (Jepma et al., 1998). In the relatively poor developing countries, this future investment point lies further away in time. Therefore, in determining the baseline, the additionality period for a CDM project is usually longer than for a JI project. Assuming comparable projects (with equal emission reductions per year), this implies that a CDM project reduces more emissions per ton of CO_2-equivalent than a JI project.

Table 1.1 summarizes the rudimentary analysis above by giving qualitative scores to each Kyoto Mechanism on the basis of distinctive negotiated design characteristics. A flexible instrument scores high on a given design feature if this feature, in isolation, would imply relatively low costs of reducing emissions per ton of CO_2-equivalent compared to the other instruments. The table makes clear, among other things, that the CDM has a high competitive advantage compared with JI and IET with respect to its absent emission ceiling and its relatively large banking possibilities, export options and additionality period. However, IET and JI have a high competitive advantage compared to the CDM concerning their broader sink options, lower transaction costs and absent adaptation tax. JI projects have the additional competitive advantage of being subject to relatively moderate sustainability requirements. It depends on institutional design and performance whether IET will beat JI with respect to transaction costs, but the former has the advantage of avoiding the costs of establishing a project baseline.

If the various competitive advantages of the Kyoto Mechanisms would imply that ultimately one of these flexible instruments will have the largest emission reduction potential at any given price per ton of CO_2-equivalent, this may lead to the crowding out of some other emission reduction entitlements on the international market (Woerdman, 1999). This will depend to a large extent on the provisions that decision makers will create for the Kyoto Mechanisms. In relative terms, the CDM may be crowded out, for instance, if the CoP introduces (a variant of) pre-budget banking for JI and IET. Or JI may be crowded out, for example, if the CoP allows unrestricted firm-to-firm trading for IET or if the option of forest management is opened up for CDM projects.

At the end of the last century, just after the Kyoto Protocol had been negotiated, there seemed to be no obvious bias in favor of one of the flexible instruments (Michaelowa, 1999). Each competitive advantage for a specific

Kyoto Mechanism seemed to be offset by a particular competitive disadvantage, while the competitive advantage of one Kyoto Mechanism seemed to neutralize the competitive advantage of another. After conducting experiments with pilot phase projects, so-called Activities Implemented Jointly (AIJ), industrialized countries have begun to prepare or implement JI and CDM projects. However, in part because transaction costs of these projects turned out to be substantial (e.g. Fichtner et al., 2003), at the beginning of the 21st century an increasing number of countries has begun to set up tradeable emission right schemes for private entities under absolute emission ceilings, which offer the prospect of lower transaction costs. Moreover, we have seen that the EU decided to develop a cap-and-trade scheme based on pre-established emission rights and we will see in the next chapters that the international climate negotiation rounds after 2000 have done away with many of the proposed restrictions for private trading under IET Article 17.

Of course it remains to be seen whether this will actually lead to a (partial) crowding out of the project-based instruments in the course of the first commitment period (apart from the question whether that would be desirable), but it is clear that emissions trading faces less formal and informal institutional barriers today than a few years ago. Obviously, the future of the Kyoto Mechanisms, for instance in terms of mutual competitive advantages, potential crowding out effects and transaction costs, will depend on the further elaboration and implementation of their design.

As noted before, it is important to keep in mind that Table 1.1 is largely based on existing literature. Some of this literature writes about (potential) transaction costs in the carbon trading market, but there are hardly any systematic analyses of the initial costs to set up this market. In addition, the table is static and does not portray the evolution and path dependence of the institutions under consideration. In the next chapters, we basically try to extend (and criticize) this table by focusing on political transaction costs and institutional dynamics.

1.5. The Emerging International Greenhouse Gas Market

Because the Kyoto Mechanisms can be defined in monetary units (e.g. dollars) per ton of CO_2-equivalent and allow for transactions across national borders, an international greenhouse gas market will arise and is already emerging, although the market is still fragmented into various sub-markets that are not (yet) connected to each other (Cogen et al., 2003).

In 1998, the sum of investments in AIJ pilot phase projects was already about $425 million for the total of private projects and $133 million for government-financed projects (Woerdman & van der Gaast, 2001: 125). The official number of AIJ projects reached 140 in 2000, involving about a quarter of the Parties, and is still increasing (SB, 2000). Several investments have been planned for the near future. For instance, the multilateral Prototype Carbon Fund of the World Bank will invest $145 million in JI and CDM projects on behalf of 6 countries and 17 companies (JIQ, 2001b).

Another example is the Netherlands which, on the basis of its so-called Emission Reduction Unit Procurement Tender (ERUPT), purchased $32 million worth of emission reductions in 2001 from JI projects to be carried out by legal entities (GGET, 2001). The government of the Netherlands has developed this international procurement procedure, which officially started in May 2000, to support JI investment initiatives by identifying a number of legal entities via an open tender and request them to start JI projects in Central and Eastern Europe. These legal entities (which could be from the host country, from the investor or from a third country) could then initiate a JI project, seek cooperation with a JI host country (which should report the project as JI to the UNFCCC Secretariat) and agree with the host country that they (the entities) will be compensated by that country for the GHG emission reductions. This compensation could either be in the form of money, because the host country wants to use the generated ERUs itself, or in the form of a transfer of ERUs by the host country to the Netherlands. In the latter case the host country basically channels through the Netherlands' payment for the received ERUs to the investing firm which carried out the abatement activity (after the possible retention of some fee). Because banking is not allowed under JI, the project must generate credits during the commitment period 2008–2012. Nevertheless, ERUPT has a prepayment arrangement with the advantage for firms that the government of the Netherlands already pays for so-called "Claims on ERUs" from the date of contracting. The Dutch ERUPT program was extended in 2001 (as Carboncredits.nl) to include tenders for CDM projects as well under the name of CERUPT.

Next to the international project-based mechanisms, several domestic emissions trading schemes are being developed (e.g. Rosenzweig et al., 2002). These are not only cap-and-trade systems, such as the CO_2 permit trading market for electricity producers in Denmark, but also involve combinations of cap-and-trade and baseline-and-credit trading on a national scale, such as the scheme in the United Kingdom (UK) or the scheme that is in preparation in a group of seven US Midwestern states, called the Chicago Climate Exchange (VROM-Raad, 1998; Cooper and Nicholls, 2000; GGET, 2001). Following the Danish and British initiatives, which will be elaborated upon in some of the next chapters, the EU as a whole will start a carbon trading market in 2005 where CO_2 emissions are capped and tradeable for large industrial sectors, including electricity, metal, cement, paper and glass producers. The EU could then reduce its own total abatement costs by about one-third compared to no trading (e.g. Svendsen & Vesterdal, 2003).

The emission reduction entitlements under JI, the CDM and IET are interchangeable (or: "fungible") if these entitlements are defined in tons of carbon equivalent emissions or if some commonly defined conversion measure is applied. To provide a simple, stylized example of the latter, suppose that transactions in assigned amounts would take place in dollars per ton of carbon ($/tC) and that transfers in CDM credits would occur in dollars per ton of CO_2 ($/tCO_2$). If some Annex B country (Party 1) has raised its assigned amount by purchasing a certain amount of CERs from a developing country (Party 2) and wants to sell these (and other) credits to another Annex B country (Party 3) in the form of an intergovernmental transfer of assigned amounts, the first country then has to convert its purchased CERs into dollars per ton of carbon instead of CO_2 according to the equation that 1 $/tCO_2$ = 3.67 $/tC. The fungibility of emission reduction entitlements is also supported by companies, such as the commercial Emissions Market Development Group (EMDG). This Group, which consists of Arthur Andersen, Credit Lyonnais, Natsource and Swiss Re, aims to stimulate efficient international trading by creating a common tradeable carbon unit (GGET, 2001). At CoP7 in 2001, governments actually facilitated the fungibility of AAUs, ERUs, CERs and RMUs by defining them all as units "equal to one metric tonne of carbon dioxide equivalent" (e.g. CP, 2001b Add. 2:57).

It is not surprising, which is also demonstrated by the latter example, that the emerging international carbon trading market already attracts several commercial entrepreneurs, such as brokers, despite the fact that the institutional details of the Kyoto Mechanisms are still under construction by means of intergovernmental negotiations. The potential value of the international carbon trading market is estimated, for instance, at about $5–$30 billion according to Haites (1998) or $9–$17 billion according to Hamwey & Baranzini (1999), depending on the price and quantity of the emissions traded. According to the World Coal Institute, the withdrawal of the US from the Kyoto Protocol in March 2001 means that the market value is now estimated to be half those figures or less (WCI, 2002).

Which (legal entities in the) Annex B countries will trade depends on their marginal abatement costs: those with relatively high marginal abatement costs will buy and those with relatively low marginal abatement costs will sell emission reduction entitlements. Different models assume different marginal abatement cost levels (not only between countries, but also) for each country or set of countries. For instance, the marginal abatement costs of the US vary between 76 and 410 $/tC among several models and those of the EU vary between 20 and 966 $/tC (Banuri et al., 2001: 56). Nevertheless, the general picture which seems to arise from the literature is that the buyers will be the industrialized countries (such as the US, the EU and Japan) and that the sellers will be both the developing countries (such as China and India) and the countries with economies in transition (such as the Russian Federation and Ukraine) (e.g. Zhang, 2000b; Rose & Stevens, 2001).

This does not mean that each and every industrialized country will be a buyer. For instance within the EU, the more detailed picture is that the buyers are likely to be the Netherlands, Belgium and Italy, among others, whereas Germany, France and Spain are expected to be among the sellers (e.g. Ybema et al., 1999).

1.6. Objective and Approach of the Book

An international greenhouse gas market can work well provided that market-based instruments, like the Kyoto Mechanisms, are designed adequately: it demands the participation of private entities, clear trading and enforcement rules as well as information and trade facilities (such as a clearinghouse), for instance to avoid market power, to strengthen compliance and to keep transaction costs low (e.g. Michaelowa & Dutschke, 1998; Tietenberg, 1999). Although market-based climate policy holds the promise of lowering overall compliance costs, history has shown that several institutional barriers hinder its implementation (e.g. Bressers & Huitema, 1999; Dijkstra, 1999).

Therefore, various authoritative scientific organizations such as the IPCC (Banuri et al., 2001: 71), the Energy Research Centre of the Netherlands ECN (Sijm et al., 2000: 45) and the Dutch National Research Programme on Global Air Pollution and Climate Change NRP (Kok & Verweij, 1999: 10), see an analysis of institutional barriers to implementing market-based climate policy as a priority area for research. Next to the necessity of developing a theoretical framework, some have also emphasized the importance of testing hypotheses regarding such barriers empirically (e.g. Wiener, 2000a: 41). In this book, we take up these challenges and perform both theoretical and empirical analyses to study the institutional economics of market-based climate policy. As we have said a few words on this type of environmental policy in the previous sections, we should now indicate which institutions we intend to study with what type(s) of economics.

Nelson & Sampat (2001: 33) argue that there is no "right" definition of institutions. They believe that the concept of institutions extends from laws and organizations to belief systems and political processes, which not only makes a coherent analysis difficult, but also contains the danger of a definition that covers too much conceptual ground (Nelson & Sampat, 2001: 39). A workable, and in fact influential, definition that includes the aforementioned phenomena within confined conceptual borders is provided by North (1990, 1991). He starts by defining institutions as the humanly devised constraints that structure political, economic and social interaction. He continues by making a distinction between formal constraints, including laws and property rights, and informal constraints, including culture and customs. These "legal" and "cultural" constraints usually evolve incrementally throughout history, he argues, and determine the costs of transacting. When North (1990: 51–52) speaks about policy in general, he accentuates that zero transaction cost conditions are scarce enough in the economic world and even scarcer in the political world. If political transaction costs are low, then efficient property rights will result, but the high transaction costs of political markets and subjective perceptions of the actors more often have resulted in inefficient property rights, he writes.

North's distinction between formal and informal constraints as well as his finding that inefficient property rights might result have guided the objective of our book on market-based climate policy, which is:

- to identify and explain the formal and informal institutional barriers that prevent or delay the implementation of market-based climate policy, as well as
- to analyze under what conditions these barriers are (in)effective.

The approach is that of institutional economics, with special emphasis on (political) transaction costs and path dependence. The literature distinguishes new institutional economics from neo-institutional economics (e.g. Groenewegen & Vromen, 1997; Nooteboom, 2000). New institutional economics is an addition to and neo-institutional economics a reaction against neoclassical economics, which focuses on the efficiency of outcomes in which the fittest will survive (or the fitter, for instance due to incomplete information), assuming rational and cost-minimizing actors. Both types of institutional economics are used in this book.

New institutional economics, which is usually associated with transaction cost economics (TCE), as initiated by Williamson (1975), builds upon neoclassical economics by assuming cost minimization, but it also focuses on the efficiency of processes in the context of institutions and recognizes that costs may occur when property rights are transferred. Empirical analysis receives more attention than in the neoclassical approach (Klein, 2000).

Neo-institutional economics extends this framework by leaving the neoclassical optimality assumptions and demonstrates that inefficient outcomes may come about when actors behave in a satisficing manner (bounded rationality) and when selection processes are path dependent (evolutionary analysis). This branch of economics emphasizes the importance of history and learning as well as of perceptions and culture. Transaction costs are not only thought to occur when property rights are transferred, but also when they are established or protected (Allen, 2000).

Transaction costs in the greenhouse gas market itself can be analyzed by using new institutional economics. The political transaction costs to set up such a market, which might depend and build incrementally on the path of earlier choices and events in environmental policy, can be analyzed by using neo-institutional economics. Informal constraints in the form of cultural barriers to the implementation of market-based climate policy, for instance regarding equity, can also be studied with the latter approach, although some insights from political science can be of help (e.g. van Deth & Scarbrough, 1995a, b). However, a serious and detailed study of formal constraints posing legal barriers to this type of policy must be conducted by using institutional law and economics (Medema et al., 2000). In general, law and economics is the economic analysis of the law and mirrors the aforementioned neoclassical and institutional schools and premises in economics (Mackaay, 2000). This means, for instance, that neoclassical law and economics prescribes optimal solutions to legal problems, whereas new institutional law and economics investigates the transaction costs of legal arrangements and neo-institutional law and economics analyzes the historical and value-driven process in which legal-economic structures are worked out. In this way, we use but also go beyond neoclassical economics, because it is widely acknowledged that institutions in market-based climate policy are not only about effectiveness and efficiency, but also or even primarily about equity, distribution, culture, perceptions and law (e.g. Hurrell & Kingsbury, 1992; Barde, 1995; O'Riordan & Jäger, 1996; van der Wurff, 1997; Russell & Powell, 1999; Wiener, 2000a).

Various authors agree that economists and other scientists should pay more attention to the institutional aspects of market-based climate policy (e.g. Bovenberg & Cnossen, 1995; Bressers & Huitema, 1999). One of the reasons for this desire is that the problem of institutional obstacles to carbon trading, according to authors like Ellerman (1998) and Endres (1999), is mainly an equity issue associated with the allocation of emission rights. Looking at equity is important, but not sufficient to explain the institutional barriers to implementing market-based instruments. To illustrate, in the international climate change negotiations at the end of the 1990s, the governments placed equity third on the international political agenda concerning the Kyoto Mechanisms, followed by effectiveness and efficiency in 5th and 14th place, respectively (BAPA, 1998: 23).

To be able to analyze political transaction costs and path dependence in this context, a distinction will be made in the next chapter(s) between several types of market-based climate policy instruments. Permit trading, in which private entities have absolute emission ceilings and are allowed to trade emission rights, is one of them. This instrument is the superior alternative according to neoclassical economics, for instance because its transaction costs, as will be explained (and nuanced), are thought to be relatively low (e.g. Tietenberg et al., 1999). However, on various (but certainly not all) occasions, and during certain periods of time, permit trading has proven to be less politically acceptable than any other (flexible) instrument for climate policy (e.g. Bressers & Huitema, 1999). In other words: the "economic hierarchy" is not necessarily the "political hierarchy" of the market-based instruments under consideration. This is what we try to explain.

The usual explanation for the aforementioned phenomenon is the resistance by interest groups, such as the industry and environmental organizations (e.g. Dijkstra, 1999). However, according to the IPCC, by focusing only or mostly on interest group preferences, this public choice literature tends to neglect the preferences and concerns of governments who ultimately decide which instruments will be used (Banuri et al., 2001: 49). Therefore, we will consider the institutional barriers and opportunities of market-based climate policy in general and the Kyoto Mechanisms in particular, with special emphasis on permit trading, by concentrating on governments rather than lobbyists. This will be done both theoretically and empirically from the perspective of new and neo-institutional (law and) economics, which allows us to pay more attention to equity, attitudes, legal issues and allocation problems, as desired by several authors (e.g. Kuik & Gupta, 1996; Ellerman, 1998; Bressers & Huitema, 1999).

In some countries and/or during some periods of time, ministers and officials, who respectively take and prepare political decisions, are inclined to avoid permit trading and incrementally build sub-optimal flexibility provisions into existing environmental policy. Again following North's (1990) work, path dependence might provide an explanation. In the (economic) literature on technological change, initiated by David (1985) and Arthur (1989), this concept is used to show why and when sub-optimal technologies are difficult or impossible to replace ("lock-in") and when this is possible ("breakout") in the presence of a superior alternative. The survival of the sub-optimal QWERTY-keyboard became a well-known (but also criticized) example of this. Self-reinforcing mechanisms like large set-up costs, increasing returns, co-ordination effects and learning contribute to such a technological lock-in.

North suggested to transform this evolutionary theory in such a way that it can be applied to study (not technological but) institutional continuity and change. North (1990: 95) himself is convinced that all of Arthur's self-reinforcing mechanisms equally apply to institutions, although with somewhat different characteristics, and that institutions are subject to "massive" increasing returns, as he writes. Although we take this idea as a starting point for the theoretical framework of our book, we will also question whether all of Arthur's mechanisms do apply to institutions and try to provide analytical extensions and remedies for any incomplete analogies observed.

As far as we know, we are the first in environmental economics to apply an extended Northian-Arthurian theory on institutional path dependence and lock-in to the theoretical and empirical analysis of both formal and informal institutional barriers which prevent or delay the implementation of market-based climate policy in general and the Kyoto Mechanisms in particular. It is important to note that, unlike most literature on emissions trading, Haddad & Palmisano (2001) also took a much-needed evolutionary perspective by emphasizing the process of establishing greenhouse gas trading mechanisms. However, although they mention Arthur's work, it should be emphasized that they do not apply (let alone elaborate) his particular evolutionary theory in the context of economic instruments for environmental regulation. Moreover, they restrict their analysis to issues of design and lobbying, without considering the impact of specific cultural and legal problems that contribute to the resistance against permit trading and other flexible instruments.

1.7. Overview of the Book

The book is divided in five parts and each part contains two chapters (except the final part). Part I presents the general institutional economics framework to analyze market-based climate policy, both in terms of issues and theory. Part II considers the new institutional economics, Part III the institutional law and economics and Part IV the neo-institutional economics of market-based climate policy. Part V contains the conclusion. The contents of these parts and chapters can be sketched briefly as follows.

1.7.1. Part I: Institutional Economics

Chapter 2 discusses the institutional economics issues of market-based climate policy by making a distinction between various types of flexible instruments, some of which are more efficient than others, and by making a distinction between the economic hierarchy and the political hierarchy of such instruments, which do not necessarily coincide. Chapter 3 presents the theoretical institutional economics framework of the book to explain this phenomenon by extending David's (1985) and Arthur's (1989) work on the path dependence and lock-in of technologies to formal and informal institutions and by elaborating upon North's (1990) notion of political transaction costs.

1.7.2. Part II: New Institutional Economics

Chapter 4 studies the impact of the institutional design and operation of economic instruments for climate policy on environmental effectiveness, while largely confirming but also nuancing the traditional view in environmental economics that flexible instruments other than permit trading are bound to be ineffective. Chapter 5 examines the transaction costs of different types of market-based climate policy instruments and provides an assessment of the empirical literature on this traditional new institutional economics topic, while extending the analysis with a political transaction cost comparison.

1.7.3. Part III: Institutional Law and Economics

Chapter 6 specifies the formal constraints to implementing market-based climate policy by formulating the economic and legal conditions, both in terms of efficiency and equity, under which international differences in the domestic allocation of emission rights lead to competitive distortions and actionable subsidies under World Trade Organization (WTO) law. Chapter 7 extends this analysis to the European context by determining the economic and legal conditions under which the aforementioned allocation differences distort competition and violate the state aid prohibitions and the polluter pays principle under EC (European Community) law, while providing an empirical analysis on the basis of the state aid decisions of the European Commission in the Danish and British emissions trading cases.

1.7.4. Part IV: Neo-Institutional Economics

Chapter 8 specifies the informal constraints to implementing market-based climate policy by elaborating and criticizing various theoretical explanations of the EU (so-called "supplementarity") proposal to quantitatively restrict the use of economic climate policy instruments, including equity as a cultural barrier, in the form of 16 hypotheses. Chapter 9 tests these hypotheses empirically by confronting them with the content of relevant EU documents, the opinions of several high-position EU officials (gathered by means of a questionnaire) and the negotiating behavior of the EU at the international climate negotiations of CoP6, while using the path dependence approach to explain the institutional breakout of the EU towards permit trading.

1.7.5. Part V: Conclusion

Chapter 10 presents the conclusion in which the objective of this book on market-based climate policy is reflected upon, by using the insights gathered in the earlier chapters, against the theoretical background of the institutional economics framework on political transaction costs and path dependence provided before, while discussing some of its policy implications.

PART I
INSTITUTIONAL ECONOMICS

Chapter 2

Design and Implementation of Market-Based Climate Policy

2.1. Introduction

This chapter discusses the design and implementation of market-based climate policy by making a distinction between various types of flexible instruments, some of which are more efficient than others, and by making a distinction between the economic hierarchy and the political hierarchy of such instruments, which do not necessarily coincide.

We do not intend to provide a complete overview of all design issues in this chapter for three reasons. First, most design options will be discussed throughout this book, not so much to provide an overview in itself, but rather to analyze their environmental, economic, legal and political consequences. Second, there are many authors who have already provided overviews of design choices in market-based environmental policy, such as Fisher et al. (1996), Crane et al. (1998), Jepma et al. (1998), Oberthür & Ott (1999), Stewart et al. (1999), Tietenberg et al. (1999) and Zhang & Nentjes (1999). Third, the basic design questions, including the allocation, transfer and enforcement of property rights, revolve around the issue of private sector participation. In general, when a market is created, such as that under the Kyoto Mechanisms, governments determine the formal framework within which legal entities are allowed to operate, such as firms or households. Therefore, this chapter centers on the question how and to what extent governments can (and, from an efficiency perspective, should) let the private sector participate in trading emission entitlements (e.g. Woerdman et al., 2003).

This chapter is organized as follows. Section 2.2 discusses the design of different types of tradeable emission rights systems and their consequences, both for the government and the private sector, in terms of effectiveness, efficiency, transaction costs and administrative costs. Section 2.3 does the same thing for different types of project-based emissions trading schemes. Section 2.4 poses the question whether the economic hierarchy of market-based climate policy instruments is different from the political hierarchy by comparing the theoretical superiority of permit trading in economics with its problematic acceptability and implementation in politics. Section 2.5 points at some drawbacks in the existing literature that studies the difference between the two hierarchies, as well as the recent evolution towards their convergence in some countries (for instance, in the EU), followed by a proposal to explain this process by using the path dependence approach. Finally, Section 2.6 presents the conclusion.

2.2. Tradeable Emission Rights and the Private Sector

A distinction can be made between various types of tradeable emission rights systems. Each institutional form has different consequences for the private sector as well as for the government, for instance, in terms of efficiency, transaction costs and effectiveness. The basic distinction in market-based climate policy is between

- permit trading and
- credit trading.

Under permit trading, a government allocates emission ceilings to private parties, allowing them to trade with each other. This is also referred to as private trading, firm-to-firm trading, allowance trading, inter-source trading or cap-and-trade. Under credit trading, however, one private party can sell credits to another by reducing its own emissions below a baseline, laid down in (energy-efficiency) environmental standards and possibly enforced by covenant. Credit trading is sometimes also referred to as the unilateral approach to project-based emissions trading or, more recently, as performance standard rate trading. The distinction between these two basic types of legal instruments is a crucial one, because according to neoclassical economic theory, permit trading is the superior alternative (e.g. Tietenberg et al., 1999: 106).

Permit trading, which incorporates emission ceilings, is efficient and effective. New-coming and growing firms have to buy permits, also referred to as "allowances", from other firms (or from a government reserve) to cover the additional polluting activities. Those who leave the industry keep their allowances, which they can sell. The system is efficient because every emission allowance that is used to cover the emissions has a price: either the purchase price of new allowances or the revenues that the polluter foregoes by not selling the allowances it already possesses (which are opportunity costs as we will explain in more detail later on). Each unit of emissions therefore has a price, since each unit could be sold. Moreover, if the economy grows, the demand for allowances increases, but the supply remains constant as a result of the emission ceiling. This means not just that the emission target will be achieved, but also that the scarcity of environmental space is reflected in a higher price for carbon-intensive products, thus encouraging technological innovation and an efficient restructuring of the economy in the direction of sustainable energy use.

Credit trading, which does not incorporate emission ceilings, is less efficient and its effectiveness is uncertain. A firm can create credits voluntarily by reducing its emissions below the emission level required by the applicable voluntary or regulatory policies and measures. For instance, if the policy is a performance (or relative) standard which requires a certain quantity of CO_2 per unit of output or energy, a firm should multiply this standard with its production volume to obtain its total emission figure. If this firm emits less CO_2 than this baseline (or benchmark) figure by initiating a certain abatement project, it can sell these credits to another firm.

Although companies can achieve cost savings by selling credits, the environmental scarcity under credit trading is not reflected in a price for each unit of emissions. If the economy grows, the supply of credits also increases because companies do not have an emission ceiling but have to observe an energy-efficiency standard. If an energy-intensive company wants to expand production, or if a newcomer enters the industry, it thus has a right to new emissions. These do not have to be purchased from existing polluters, or from a government reserve, within an environmental consumption space like in the permit trading system. Instead, the company receives its emission credits above and beyond the existing quantity. This is a political advantage if it reduces the resistance of the industry, if any, against climate policy. However, it is also an environmental disadvantage, since the emissions will grow if newcomers arrive or if firms expand their production. Moreover, the consequence is that the social costs of the extra emissions are not fully reflected in the costs per unit of product and thus not in the product price. Carbon-intensive products are therefore priced too cheaply, leading to an inefficient restructuring of production.

The transaction costs of credit trading in the market would not differ much from permit trading because both types make use of the information advantages of the private sector and do not require advance approval of every entitlement transfer. Nevertheless, the determination of the allowed emissions for a given year is more difficult under credit trading because these are not given (as under permit trading), but have to be calculated on the basis of existing climate policy, for example, by multiplying the performance standard with the energy use in that year, which can be done accurately only ex post. Project-based credit trading, such as Joint Implementation (JI) and the Clean Development Mechanisms (CDM) as defined in the Kyoto Protocol, is a different story. In that case, an investor receives credits for achieved emission reductions at a (usually foreign) host. These emission reductions are measured from a baseline that estimates future emissions at the project location if the project had not taken place. These baselines have to approved before the transaction is allowed, which increases such transaction costs.

International permit trading would, obviously, take place among private entities across national borders (e.g. Tietenberg, 1992; Hahn & Stavins, 1999; Zhang, 2000a). To make this possible, it is necessary to develop domestic permit trading schemes first and then connect them under certain conditions, for instance, on monitoring and enforcement, to create an international market (Zhang & Nentjes, 1999). An example of international permit trading is the transfer of ozone depleting substances (ODS) among firms in different countries under the Montreal Protocol of 1987 (e.g. Mullins & Baron, 1997).

Linking domestic permit trading systems internationally requires, among other things, that permits are defined uniformly (for example, in tons of carbon equivalent emissions) or comparably (by using a commonly defined conversion measure). Furthermore, the national supervising agencies should, on a bilateral basis, administer the international permit trades and register the corresponding alterations of their national emission budgets, such as resulting changes in the assigned amounts, in the case of the Kyoto Protocol, which have to be reported to the FCCC Secretariat. It is possible, but not necessary, to establish an international clearinghouse to perform those tasks. The Parties would also have to meet eligibility criteria which specify an adequate national compliance structure, such as binding national emission targets and timetables, reliable national registration and accounting of source-related emissions, accurate emissions monitoring and effective legal enforcement mechanisms. Under Articles 5 and 7 of the Kyoto Protocol, Annex B Parties are required to create national inventory systems for the estimation and registration of GHG emissions and to provide information for compliance purposes. At CoP7 in 2001 it was decided that if a Party does not meet these requirements, it is ineligible to participate in the Kyoto Mechanisms (CP, 2001b).

Some have suggested to shape the international trading of emission rights by means of government trading, in particular during the late 1990s when Article 17 on international emissions trading under the Kyoto Protocol was not yet elaborated. This type of trading, which is also referred to as intergovernmental trading, government-to-government trading or quota trading, would then involve the trading among Annex B Parties of parts of their assigned amounts (e.g. Fisher et al., 1996; Mullins & Baron, 1997; Bohm, 1999). Governments would remain in direct control of transactions with assigned amounts, which have been defined at their level, and it would not be necessary to allocate tradeable permits to firms before international trading can take place, it was argued. This might seem attractive at first sight, but it is important to realize that a government would still have to translate its environmental commitment into domestic policy, for instance, by means of standards or taxes.

There is no real-life example of government trading yet, but an empirical trading simulation by Bohm (1997) between four government teams showed that efficiency gains can indeed be reached among regulators without direct international transactions of firms. However, this result was achieved in an experimental setting under the assumptions, among other things, of complete information and the absence of market power. Both assumptions are problematic, however. First, Gusbin et al. (1999) have calculated that the Russian Federation has the potential to supply 30% of the tradeable units in the case of worldwide trading and even 70% when trade is limited to the developed countries. Second, governments have, in fact, incomplete information on the marginal abatement costs of domestic emitters (e.g. Tietenberg, 1992; Zhang & Nentjes, 1999). The higher the information deficit is, the higher the risk will be that the enacted emissions trading deals are not as cost-effective as would have been possible.

Like permit trading, it also possible to apply credit trading on an international scale if a government wants to involve private entities in emissions trading without capping the emissions of certain sectors, such as the energy-intensive industry (e.g. Palmisano, 1996; Haites, 1997; Crane et al., 1998; Boom, 2000a). There are both similarities and differences between credit trading and JI. First, credit trading mirrors the so-called unilateral approach of the project-based mechanisms (Janssen, 2000), where the firm which funds the project is also the firm where the project is realized (self-financing of emission reductions), while JI could incorporate unilateral, bilateral and/or multilateral investments, as will be explained later on, and has predominantly been used in its bilateral form in pilot projects. Second, credit trading does not necessarily require a pre-approval of transactions if the credits are created and traded after the reductions have taken place (ex post), whereas JI transactions could require pre-approval if the credits are created and traded before the reductions occur (ex ante), in particular, in a context where baselines are not standardized.[1]

In practice, credit trading has not been applied at the international level yet, but it has been implemented on a domestic scale, for instance, in the earliest domestic trading schemes in the US where the emission standards would become the baseline for the reduction of emissions and the generation of credits (Tietenberg, 1999). Although this particular scheme was characterized by pre-approval and trade restrictions (Tietenberg, 1992), it is possible to design a more flexible credit trading scheme as demonstrated by the domestic scheme for tradeable NO_x emission credits in the Netherlands (Nentjes, 2001). This scheme is neither efficient nor likely to be effective because emissions are not capped and the scarcity of licenses to emit is not signaled in a price for every unit of emission, but trade is free and compliance is checked at the end of the year.

---

[1] Nevertheless, credit trading is sometimes called a form of JI, and vice versa, because both have a project-related character.

International permit trading is possible under the Kyoto Protocol. Whereas Article 17 originally did not specify other types of emissions trading than that between developed countries, the Marrakesh Accords of 2001 indicate: "Transfers and acquisitions between national registries shall be made under the responsibility of the Parties concerned in accordance with the provisions (…) for the accounting of assigned amounts (…). A Party that authorizes legal entities to transfer and/or acquire under Article 17 shall remain responsible (…) [and] shall maintain an up-to-date list of such entities and make it available to the secretariat (…)" (CP, 2001b, Add. 2: 53–54). The secretariat, in its turn, as can be read in (point 38 and 42 of) the annex on modalities for the accounting of assigned amounts, "(…) shall establish and maintain an independent transaction log to verify the validity of transactions (…) of ERUs, CERs, AAUs and RMUs (…) [and to check] the eligibility of Parties involved in the transaction (…)" (CP, 2001b, Add. 2: 65–66).

National emission budgets, such as assigned amounts, do not have to be divided into emission caps prior to government trading or credit trading, while the credits accrue bottom-up in the case of credit trading (as well as in the case of JI and the CDM). Permit trading, however, entails a clear and visible top-down (re)allocation of property rights in the form of emission ceilings before trading can begin. The government not only has to decide whether it will sell the permits (auctioning) or give them away for free (grandfathering), but it also has to decide who will receive the permits, for instance, the end-users of fossil fuels or the fossil fuel producers and importers themselves. Although permit trading is the most effective and efficient type of market-based climate policy, the allocation and its consequences are explicit under permit trading, which makes the allocation problem more manifest and pressing under permit trading than under government trading and credit trading (e.g. Shogren & Toman, 2000; Wiener, 2000a). This also implies a longer preparation time and higher set-up costs compared to the other design options.

2.2.1. Domestic Permit Trading Design

The aforementioned allocation problem of auctioning versus grandfathering will receive ample attention later on in this book (in the chapters on WTO and EC rules), so that we will now focus on the question who could be the permit holders. There are different design possibilities (e.g. Jepma et al., 1998; Zhang, 1998c). We make a distinction between

- downstream trading system;
- upstream trading system;
- hybrid trading system;
- mixed trading system.

The regulated entities in a downstream trading system are all energy end-users. The government allocates permits to small emitters (such as households and motorists) as well as large emitters (such as utilities and industrial sources). If all emission sources are included, a large scope will exist for competition, thereby increasing the likelihood of achieving cost-effectiveness and decreasing the possibility of market power. However, administrative costs could be high, not only with regard to allocating the permits to small sources and the transport sector, but also especially with regard to monitoring their emissions and trading patterns. To lower administrative costs, Nentjes (1998) proposes to allocate an amount of permits to each category of small emitters, such as households and motorists, proportional to its historical share in total emissions in a reference year. The individual users within each category of small emitters, e.g. households, receive an amount of permits proportional to its CO_2 emissions resulting from average fuel use per adult in a reference year. People living in small, well-isolated apartments and people without a car, for instance, will end up with a permit surplus at the end of the year, which can either be sold or banked (to cover emissions next year or later). When purchasing fuel or energy, emitters have to hand over their permits to the producers and importers that sell fuel and energy. This means that monitoring can concentrate on the level of producers and importers (instead of small sources), which lowers administrative costs when the system is functioning (e.g. Woerdman et al., 2002). How this alternative works is explained in the next subsection.

In an upstream trading system, permits are allocated to fossil fuel producers and importers. They will pass on their permit costs in a mark-up on the fuel price for both small emitters (such as households and motorists) and large emitters (such as utilities and industrial sources). For consumers, this will look much like some sort of "carbon tax". Administration is facilitated by the relative small number of permit holders, while existing institutions for levying excises on fossil fuels can be used to enforce the scheme. Theoretically, a smaller amount of permit holders reduces market liquidity, tightens the scope for efficiency gains and could also increase the risk of market power relative to the downstream approach. In practice, however, the number of permit holders is still likely to be sufficient to avoid market power: in a small country such as the Netherlands, for instance, the number of producers and importers of fuel is about 40–50 (Koutstaal, 1997).

In a hybrid trading system, a part of total permits is allocated to fossil fuel producers and importers, as in an upstream trading system, who put a mark-up on the fuel price for small emitters (such as households and motorists) equal to their permit costs, while large emitters (such as utilities and industrial sources) receive permits directly, as in a downstream trading system. This means that a hybrid trading system has a moderate performance relative to downstream and upstream trading systems in terms of administrative costs (lower than in a downstream system, higher than in an upstream system). The trading scheme for large emitters must be connected with the scheme for fossil fuel producers and importers to avoid inefficiencies. A complexity of a hybrid system that increases administrative costs is the necessity to avoid double counting (Hargrave, 2000). Fuels consumed by the large sources included in the trading program must be exempt from the indirect fuel tax that is put on the fuel price by the producers through the upstream system.

In a mixed trading system, permit holders are large emitters (such as utilities and industrial sources) as in a downstream system, but small emitters (such as households and motorists) are regulated with other instruments, such as taxes, standards or voluntary agreements. Although command-and-control and voluntary agreements can be effective, depending on the quality of information and enforcement, they are not efficient. Furthermore, taxation is efficient, in principle, and it induces certainty with respect to the price of emission reduction in the form of the tax rate. However, taxation is not necessarily effective due to the trial-and-error tax adjustments induced by imperfect knowledge of the marginal costs of emission reduction of the individual emission sources. Standards or taxes for small emitters could put the government at risk of high costs or non-compliance with its emission commitments, respectively. Moreover, combining tradeable permits for large emitters with taxation for small emitters creates the inefficiency of two prices (tax level and permit price).

A mixed trading scheme will in fact emerge in the EU, starting in 2005. The EU starts with permit trading among large (instead of small) emitters first, such as electricity producers. A reason not to start with a comprehensive scheme is to deal with uncertainties and complexities and to facilitate learning by following "a prudent step-by-step approach" (COM, 2000a: 10). The EU also proposes to maintain standards for the household and transportation sectors, at least initially, and to study the possibility of credit trading and taxation further. If the latter sectors are not incorporated in the permit scheme, the cost-effectiveness potential decreases relative to a full-sector coverage. The characteristics and (dis)advantages of the domestic emissions trading options presented above are summarized in Table 2.1.

The interaction between the trading system design options and the possible allocation methods raises an additional distributive issue with equity considerations for decision makers because the choice between (a combination of) grandfathering and auctioning may depend on the choice between the different design possibilities (downstream, upstream, hybrid or mixed) for a domestic permit trading system. Grandfathering permits (free allocation) imply that emitters only have to pay for the additional costs of emission reduction and not for their emissions as in the case of auctioning. Consequently, as will be explained in detail later on in this book, auctioning permits would increase expenditures for emitters by affecting their financial position relative to grandfathering (see also Woerdman, 2000a).

Table 2.1: Design options for domestic permit trading.

	Downstream	Upstream	Hybrid	Mixed
Permit receivers	Large emitters	Fossil fuel producers and importers (*mark-up price for large and small emitters*)	Large emitters	Large emitters
	Small emitters		Fossil fuel producers and importers (*mark-up price for small emitters*)	Taxes/standards for small emitters
Advantages	Large number of traders increases scope for efficiency gains and reduces risk of market power	Small number of sources reduces administrative costs	Moderate number of sources reduces administrative costs (albeit higher than in upstream system)	Effective and efficient policy for large emitters and certainty of price of emission for small emitters
Disadvantages	Large number of sources may increase administrative costs, but upstream monitoring alleviates this potential problem	Mark-up price looks like "carbon tax" and small number of traders reduces scope for efficiency gains	Large emitters must be exempted from mark-up price ("carbon tax") to avoid double counting, which increases administrative costs	High risk of ineffective (taxes) and inefficient (standards) policy for small emitters as well as the inefficiency of two prices (tax level and permit price)

Suppose that permits are grandfathered to small emitters, as in a downstream trading system, and/or to large emitters, as in a downstream, hybrid and mixed trading system. In those cases it could be maintained that gratis permits are a compensation for the costs end-users have to make to reduce emissions. Some argue that the owners of existing plants are then compensated for the "stranded

Table 2.2: An equity interaction between design and allocation choices.

Choice 2: allocation	Choice 1: design			
	Downstream	**Upstream**	**Hybrid**	**Mixed**
Grandfathering permits	Large emitters		Large emitters	Large emitters
	Small emitters			
Auctioning permits		Fossil fuel producers and importers	Fossil fuel producers and importers	
Tax/standard				Small emitters

costs" they bear as a result of the new requirement to reduce emissions with the introduction of climate policy (e.g. Harrison & Radov, 2002).

However, grandfathering permits to fossil fuel producers and importers (although this gives the same efficiency results as auctioning) may not be desirable from the perspective of the end-users according to the compensation principle of equity (cf. Rose & Stevens, 1993). With auctioning, fossil fuel producers and importers have to pay for the potential emissions contained in the fossil fuels they sell. This additional cost is transferred to consumers as a mark-up on fuel equal to the permit price. With grandfathering, consumers pay a higher fuel price because an emission ceiling creates scarcity and fossil fuel producers and importers can make a profit because they have received the permits for free without having to make the costs of reducing emissions.

The aforementioned profit is undesirable on the basis of the end-user compensation principle of equity. It implies that if upstream or hybrid trading systems are developed (design choice), they should incorporate auctioning (allocation choice) for fossil fuel producers and importers. This view is defended, for instance, by Hargrave (1998). The interaction between the choice of trading design and the subsequent choice of permit allocation from an end-user compensation perspective is summarized in Table 2.2.

2.2.2. Downstream Permit Trading with Upstream Monitoring

Although a downstream permit trading scheme is efficient, it could involve high costs to set up the system, which will be discussed in the next chapters, as well as high administrative costs to monitor many sources when the system is functioning, which is the topic of this subsection. One alternative to bring down these administrative costs is sketched by Woerdman et al. (2002), which builds upon Koutstaal (1997) and Nentjes et al. (2002). Contrary to the common view (e.g. Hamilton, 1998; Hargrave, 1998, 1999; Anderson et al., 1999; Butzengeiger et al., 2001), a downstream system which directly incorporates firms as well as households and car drivers can well be administratively feasible by concentrating the monitoring activities as much as possible on the level of fossil fuel producers and importers (upstream) and by using a generic allocation criterion and chipcard technology for households and car drivers (downstream). The outline of such a "downstream trading and upstream monitoring" approach that focuses on restricting fuel use is sketched below.

The allocation of an amount of permits to each category of emitters, such as households or firms, could be proportional to its historical share in total emissions in a reference year. The individual users within each category of small emitters, e.g. households, receive an amount of permits proportional to its CO_2 emissions resulting from average fuel use per adult person in a reference year. For every ton of fossil fuel a firm or household purchases from distributors, it has to hand over an equivalent number of carbon permits. Distributors in turn can only obtain fuels from their suppliers in exchange for carbon permits. In this way all permits will end up in the hands of producers and importers of fuel, including the permits purchased by distributors to cover their fuel supply to consumers and other small users. Producers and importers of fuel are placed under the obligation to turn over to the environmental authorities carbon permits for the carbon contained in the fossil fuels they have sold on the market.

Permit allocation occurs downstream, but monitoring of emissions (fuel sales) and checking whether they match with permits concentrates upstream on producers and importers of fuel whose number is usually limited (in the Netherlands about 40–50 (Koutstaal, 1997) as we already indicated). The bookkeeping of these fuel producers and importers is checked at the end of the emission year. It is determined how many permits are actually present and how many they should have by calculating the number of required permits on the basis of the administration of fuel sales. In the case of a determined shortage of permits, the fuel producer or importer gets one month to obtain (and thus buy) the necessary permits. If it is not able or willing to do so, the company receives a fine which is a multiple of the highest expected market price, while it remains obligatory for the company to hand over the lacking permits to the authorities.

From an institutional law and economics perspective, it is important to realize that the system is to a large extent self-enforcing. In this design, fuel producers and importers (as well as distributors) have an interest to receive the correct number of permits alongside their fuel sales: the supplier does not want to deliver fuel without the transfer of permits by the buyer. It is not necessary that the national agency monitors the millions of fuel users, which brings down the costs of monitoring and enforcement. The monitoring scheme fits in with existing institutions for levying excises on fossil fuels, present in most Western countries. For instance, in the Netherlands, traders and suppliers of mineral oils are obliged to have a license and to report each month the quantity they have supplied to the market, while they have to turn over the excise tax to the authorities. This administrative system of self-reporting is supplemented by occasional physical checks (Koutstaal, 1997).

Permits can not only be grandfathered to (big and small) firms, but also to households. The large number of households incorporated in the trading scheme makes market power unlikely. In the beginning of the year, these end-users receive the permits for the coming year for stationary and mobile sources together on their permit account. The national permit agency, where all participants are registered, also sends a chipcard. (Instead of sending a separate chipcard it might be possible to combine it with existing chipcards from banks.) Households can uprate the chipcard at the expense of their permit account. In principle, it is also possible to grandfather permits to the distributors who will pass on the permit costs in a mark-up on the fuel price, thereby avoiding allocation to households. However, as indicated in the previous subsection, grandfathering is then not likely to be politically acceptable because it would create a profit for the distributors, while the consumers pay for the emission reductions. Households are better off if they (instead of the distributors) get permits for free, not only because they receive a wealth transfer, but also because it enables consumers to make a profit by selling permits if they succeed in using less energy and fuel.

When purchasing fuels, the end-user has to transfer an amount of permits (which corresponds with the carbon content of the acquired fuel) to the permit account which the distributor holds at the national permit agency. For the mobile sources the transfer occurs by using the chipcard which households can fill by lowering their permit account. An alternative is a permit pincard which enables permits to be transferred directly from one's own account to the fuel supplier. Upgrading or writing off from the account is only possible in the case of a positive permit balance. A car driver who buys fuel can choose to transfer part of his own permits himself or buy the fuel with a mark-up price which reflects the price of permits that the distributor has bought as a kind of service for customers (for instance, for those clients that have forgotten to take their chipcard or pincard with them). For stationary sources the transfer of permits is enacted by connecting the permit transfer to the mailing of the yearly gas- and electricity bill of the distribution company. If a household does not have a sufficient number of permits, the distributor has the right to buy the required permits and to recover the costs from the client.

When a car driver goes to the cash desk, he or she not only pays the money for filling up the tank with fuel, but also transfers an amount of permits (which corresponds with the carbon content of the acquired fuel) to the permit account which the distributor holds at the national permit agency. At filling stations and at other strategic places, machines are installed where one can electronically upgrade the permit chipcard (buy) or write off from the permit chipcard (sell) at the current market price. The automated machines are exploited by companies who trade professionally in carbon permits. The current market price arises from the transactions of and between the permit trading companies. At the end of the year the national agency establishes the balance of the permit account for every user unit. This is equal to: grandfathered permits (via chipcard or account) plus the purchased permits minus the permits sold minus the permits used and transferred. This balance can be positive, but not negative. The positive balance is added to the permit account for the next year. These can be sold by the account holder or they can be kept as an investment.

The introduction of the permit chipcard requires investments in automated machines and a telecommunication network. The investment costs are comparable to the costs of installing a pincard or chipcard system of a bank with millions of account holders. Possibly, these costs can be shared between the permit registry and the banks when the permit chipper is combined, if desirable, with other existing chipcards from banks. The large-scale character and the intensive use of the machines will result in low costs per transaction. Next to the aforementioned costs of the chipcard technology (depreciation, interest and exploitation), the time costs of the extra permit action have to be taken into account when paying the fuel bill at the filling station.

The domestic implementation costs consist of the registration of the participants as well as the yearly allocation of permits and mailing of chipcards. For European countries, we roughly estimate this to be several million euros, which implies a few euros per chipcard. The monitoring focuses on the limited number of car fuel importers and producers who already have a detailed administration of their fuel sales for commercial and fiscal reasons. The monitoring costs will therefore be limited to no more than several million euros. The political process will, however, induce set-up costs, as various examples later on in this book will underline, but these initial costs are unavoidable and necessary to reap the environmental and economic benefits of permit trading.

To summarize, administration costs can be kept low (a) by distributing the permits to large and small end-users, the latter via a generic allocation criterion, (b) by using chipcard technology for households and car drivers (downstream) and (c) by concentrating monitoring and enforcement on the level of fossil fuel producers and importers where all permits end up (upstream). For car drivers the permit transfer occurs by using a chipcard which they can fill by lowering their permit account. For households the permit transfer is connected to the mailing of the yearly gas- and electricity bill of the distribution company. The high number of traders makes market power unlikely. This particular design, which combines downstream and upstream elements, is technically feasible and economically desirable. Whether it is also politically acceptable, causing an institutional "breakout", depends on several factors, in particular the willingness of the industry to accept emission limits and the willingness of households and politicians to accept market instruments for environmental policy. The formal and informal institutional barriers that partly underlie this (un)willingness will be discussed thoroughly in later chapters.

2.3. Project-Based Emissions Trading and the Private Sector

Project-based emissions trading, such as JI and CDM projects under the Kyoto Protocol, is a variant of credit trading (which is less efficient and effective than permit trading, as discussed above). Both credit trading and emission reduction projects allow for the transfer of credits, but projects usually require pre-approval to check the environmental integrity of the project baseline, thereby raising transaction costs, which is not necessary under credit trading where the baseline is existing environmental policy (like energy-efficiency standards), so that compliance can be checked at the end of the year. Moreover, the firm which funds the reductions under credit trading is also the firm where the reductions are realized, but in the case of project-based emissions trading, three design options are available:

- multilateral approach;
- bilateral approach;
- unilateral approach.

In the multilateral approach, an international fund would be created in which Annex B private and/or public entities are required to pool their investments (Dutschke & Michaelowa, 1999; Stewart et al., 1999). The institution that administrates this multilateral fund selects and invests in emission reduction projects and the investors receive credits proportional to their share of the portfolio. Before the credits are given to the investors, the administrative body of the fund could take a portion of the credits (and/or other revenues earned) as a fee. A multilateral fund has the advantage that it can spread project risks, achieve scale economies, reduce the transaction costs for small investors and strengthen the bargaining position of the (relatively small) host countries compared to bilateral negotiations with a big investor. However, the disadvantage is that a large fund can become bureaucratic and administration costs can be high. A real-life example of a mutual fund in the context of climate change is the so-called Prototype Carbon Fund (PCF) of the World Bank in which investors pool capital to be invested in

GHG emission reduction projects in cooperation with potential host countries. Although the CDM has some elements of the multilateral approach, like the imposition of an adaptation tax, the supervisory Executive Board of the CDM does not actively take part in project investments.

The bilateral model places more emphasis on private investment and market forces as project selection and implementation are left to the participants (Dutschke & Michaelowa, 1999; Stewart et al., 1999). A project can be negotiated freely on a case-by-case basis between private entities, for instance, in two different countries where the governments have to approve of the deal. An international institution could then function (not as a multilateral fund but) as a clearinghouse or project exchange to match potential investors with partners in host countries. The bilateral approach has the advantage of keeping administration costs low, reducing transaction costs for big firms (if they invest in several projects) and selecting cost-effective projects. The disadvantage is that small investors face relatively high transaction costs, small or high-risk projects and countries with relatively underdeveloped markets and capacities have less chance to be selected and small host countries may fear a weak bargaining position when facing a big investor. JI seems to lean towards this approach, in particular because legal entities are explicitly allowed to participate subject to the approval of the Parties involved. In practice, bilateral investments dominated in projects under the pilot phase of Activities Implemented Jointly (AIJ) without crediting which started in 1995 to experiment with the project-based approach.

In the unilateral model, a (legal entity within the) host government generates the credits on its own without foreign direct investment (Dutschke & Michaelowa, 1999: 52; Oberthür & Ott, 1999: 177; Stewart et al., 1999: 10). The host Party selects, develops and invests in a private or public project on its own territory after which it can bank the credits or sell them to foreign entities (for instance, by means of an auction). The advantage of this self-financing approach is that it promotes government autonomy and oversight, but it has the disadvantage that it requires substantial host country project development and financing capacities. This makes the option, on average, less suitable for developing countries under the CDM, for instance, where such institutional capacities are relatively low (Karani, 1997), than for JI host countries in Central and Eastern Europe. Nevertheless, developing countries with sufficient institutional capacity may still be interested in this approach. This has actually been the case during the AIJ pilot phase for Costa Rica, which tried to sell so-called Certified Tradable Offsets (CTOs) generated in forestry and energy-efficiency projects that were financed by means of a fuel tax.

The models are not mutually exclusive, but they rather represent different levels of supervision and control over the market (Oberthür & Ott, 1999). For instance, a clearinghouse need not only be established in the bilateral system, but can also be created in the multilateral or unilateral model. Moreover, unilateral and/or bilateral projects are necessary, since multilateral funds are unlikely to manage the complete (e.g. CDM) market (Denne, 2000). Certification of the emission reductions will be performed on a periodic basis when the project is implemented by operational entities accredited by the CoP. Both private and government investments are possible in every model, but the emphasis lies on the former. Freedom of choice for private investors with respect to selecting projects is the highest in the bilateral model and lowest in the multilateral approach.

There is a difference between the definition of JI and the CDM concerning private sector participation. JI Article 6 defines, among other things, the role of the national governments of Annex B Parties and the potential role of legal entities in GHG abatement project co-operation between Annex B Parties. The role of the private sector in the CDM is defined less strictly. There is no passage in Article 12 saying that credits can only be transferred to or acquired from Parties.

JI Article 6.3 of the Kyoto Protocol makes clear that an Annex B Party "(…) may authorize legal entities to participate, under its responsibility, in actions leading to the generation, transfer or acquisition (…)" of ERUs. In (point 29 of) the annex on the guidelines for the implementation of Article 6 in the Marrakesh Accords of 2001 it can be read: "A Party that authorizes legal entities to participate in Article 6 projects shall remain responsible for the fulfillment of its obligations under the Kyoto Protocol (…)" (CP, 2001b, Add. 2: 13). Although the business sector is to play a key role in the generation of ERUs, the actual transfer and acquisition can be made by and to State Parties only.

Article 12.9 allows for the participation of "private and/or public entities" in the CDM, subject to the guidance of the Executive Board, which means that the transfer and acquisition of credits is not limited to states as under JI Article 6. An underlying difference is the absence of an assigned amount for a CDM host country in contrast with a JI host country. At CoP7 in 2001, it was decided that the Executive Board shall supervise the CDM, under the authority and guidance of the CoP, for instance, by accrediting operational entities (which validate, verify and certify emission reductions), by maintaining a CDM registry (to monitor the creation and transfer of CERs), by approving new baseline methodologies and by making recommendations to the CoP on further modalities and procedures (CP, 2001b). Public entities could be involved, for instance, in the implementation of CDM projects in countries where the private sector is relatively underdeveloped.

The allocation problem of credit sharing resurfaces under each (model) of both project-based mechanisms. This may not only include the sharing of the credits themselves, but also of project revenues (and risks). JI host countries may use the credits for compliance now and CDM host countries could bank the credits and use them once, and if, they accept an emission (growth) target in the future. Hosts can also sell them to third parties on the market (if the CoP does not prevent this possibility). In the traditional view, the investor is supposed to get all the credits from a project (and the host the revenues), so that there is no credit sharing (Pearce, 1995).

The introduction of credit sharing reduces the cost-effectiveness of the project for the investor and could be seen as a tax: when a project is undertaken, some percentage of the credits generated is retained by the host country (Denne, 2000). Reasons to share the project value are to compensate for the supposedly strong bargaining power of (private) investors from industrialized countries and to reduce the potential effect that the investors pick all "low-hanging fruits" and leave the host country only with expensive mitigation options in the future (Rose et al., 1999).

Under the AIJ pilot phase, different credit sharing arrangements were negotiated bilaterally between investor and host, such as 50%–50% (e.g. Netherlands–Honduras), 80%–20% (e.g. Netherlands–Russian Federation) or 65%–35% (e.g. Netherlands–Romania) in energy-efficiency projects (Gosseries, 1999; JIQ, 2000a). An Annex B host country has an incentive to increase its share of the credits if it has a relatively stringent emission target or to lower its share to compete with relatively cheap CDM projects (Jepma & van der Gaast, 1999).

2.4. Economic Versus Political Hierarchy in Market-Based Climate Policy?

When it comes to design, several economists have suggested, either explicitly or implicitly, that there is a hierarchy among the economic instruments for environmental policy, including the Kyoto Mechanisms. As explained before, permit trading is at the top of this economic hierarchy (e.g. Baumol & Oates, 1988; Koutstaal & Nentjes, 1995; UNCTAD, 1995; Holtsmark & Alfsen, 1998; Anderson et al., 1999; Tietenberg et al., 1999; Haites, 2000). For instance, Tietenberg et al. (1999: 106) literally call permit trading "superior" to credit trading in terms of economic and environmental results. These authors basically argue that permit trading is more effective and efficient, and has lower transaction costs, than all other design options.

However, when it comes to implementation, politicians do not always or immediately opt for permit trading, but may choose to set up sub-optimal arrangements such as credit trading. The usual political economy explanation for this is that credit trading has advantages for certain interest groups, such as the industry which does not have to purchase extra emission rights if companies seek to expand their production (e.g. Dijkstra, 1999). There are, however, also advantages of credit trading for the politicians themselves. Permit trading sets emission ceilings by explicitly (re)distributing property rights, while credit trading uses existing environmental policy to calculate the tradable emission reductions.

An institutional economics explanation for the political attraction of credit trading in developed countries is that the start-up "capital" or political transaction costs of permit trading are relatively high since it comes to replace existing environmental policy, while credit trading builds incrementally on extant (possibly ineffective and inefficient) arrangements. Another explanation for the political attraction of credit trading is that under permit trading, a choice must be made between auctioning emission allowances or give them away free (e.g. "grandfathering" based on historical emissions). Under credit trading, emissions are always given away free, thus lowering the political visibility of the (re)distribution issue.

Apparently, the economic hierarchy and the political hierarchy in market-based climate policy do not necessarily coincide. Both elements will be worked out in more detail in the next two subsections.

2.4.1. The Theoretical Superiority of Permit Trading in Economics

According to economic studies (e.g. Tietenberg et al., 1999), permit trading is superior to other flexible instruments in terms of

- effectiveness;
- efficiency;
- transaction costs.

First, permit trading is considered to be more effective than government trading and (project-based) credit trading including JI and the CDM. Assuming that monitoring and enforcement are adequately organized (which is required under every instrument for environmental policy), effectiveness is achieved because the total of tradeable permits form an emission ceiling. However, when governments trade, they still have to translate their emission ceiling into domestic environmental policy for private entities, which is sometimes done by means of taxes and usually by means of energy-efficiency standards or covenants (voluntary agreements).

In theory, taxation is as effective as permit trading assuming a perfect world where the regulator knows exactly how high to set the tax rate to reach the emission goals. In practice, however, information is incomplete, so that governments become involved in a trial-and-error process of adjusting and readjusting the tax rate in an attempt to reach the emission target. Moreover, newcomers have to pay the tax, but emissions will grow, whereas newcomers in a permit market have to buy their permits from the existing firms or from a government reserve. Furthermore, in the case of inflation, the real value of a tax decreases and sources will increase their emissions, contrary to a permit market where only the price of allowances will increase. When domestic policy is an energy-efficiency standard, either or not in the form of a covenant, emissions will grow if firms expand their production and if newcomers arrive. Only a non-tradeable emission cap for firms can be as effective as a tradeable emission cap, but these are usually not the cornerstone of environmental policy, whereas making these caps tradeable is rather thought to facilitate compliance because it lowers abatement costs.

Credit trading and JI use the existing environmental policy as the baseline from which to calculate the (tradeable) emission reductions. It follows from the remarks above that credit trading and JI will not be fully effective if this policy is an energy-efficiency standard or a covenant. Moreover, there may be several plausible ways to calculate the baseline for emission reduction projects, in particular in JI host countries where domestic climate change policy is just being developed and in CDM host countries where such policy is (largely) absent. The problem is that the baseline emissions which would have occurred without the project will never be known because the project is implemented. Effectiveness can be undermined if future emissions are overestimated by inflating the baseline to claim more credits. This incentive is strongest for the investor and host under the CDM (both for the Parties and/or legal entities involved) — irrespective of whether the investment is unilateral, bilateral or multilateral. This incentive also exists on a micro-level for legal entities involved in a JI project, but not for the JI host Party government which has an assigned amount and runs the risk of being in non-compliance by transferring too many credits.

An additional environmental disadvantage for the CDM relative to the other forms of trading is the absence of an emission ceiling in developing countries under the Kyoto Protocol. Permits are traded under an emission ceiling (which is assumed to be lower than actual and/or future business-as-usual emissions) for each participating legal entity, whereas assigned amounts, credits, ERUs and RMUs are traded under a national emission ceiling. However, even if CERs are generated on the basis of genuine emission reductions achieved at the project location, emissions may still increase in the CDM host country outside this location. An example of such "carbon leakage" (Jepma & Munasinghe, 1998: 313) is that emission reduction policies in Annex B countries lower world demand for fossil fuels, leading to lower energy prices and higher fossil fuel use in developing countries. For competitiveness reasons, energy-intensive industries then also have an incentive to relocate to developing countries.

Second, permit trading is considered to be more efficient than government trading and (project-based) credit trading including JI and the CDM. Permit trading is efficient because marginal abatement costs are equalized and every unit of emission will have a price, since each unit has the opportunity of being sold. When the economy grows, the demand for emission permits rises, but their supply remains constant as a result of the emission ceiling. This means that the scarcity of environmental space will be reflected in a higher price for carbon-intensive products, which also stimulates technological innovation and an efficient restructuring of the economy towards sustainable energy use. Government trading and taxation are likely to be less efficient because governments have incomplete information on the marginal abatement costs of domestic emitters. Moreover, costs are not minimized if a government formulates its domestic climate policy by means of non-economic instruments, such as (energy-efficiency) standards and covenants, because they do not reduce emissions where it is cheapest to do so.

Although credit trading as well as JI and CDM projects could generate such cost savings by making use of the marginal abatement cost differences among emitters, the environmental scarcity would not be signaled in a price for every unit of emission. When the economy grows, the supply of emission credits also rises (contrary to permit trading), since firms do not have an emission ceiling. If an energy-intensive firm expands its production or enters the industry as a newcomer, it is licensed to new emissions (as defined by the environmental — e.g. energy-efficiency — norms) and it does not have to buy (a part of) its emission rights from existing polluters out of the emission space, but receives new emissions for free (apart from the cost of an environmental license). The consequence is that the social costs of additional emissions from economic growth are not fully reflected in the costs per unit of product, and thus also not in the product price, leading to an inefficient restructuring of production as carbon-intensive products are priced too low. A related difference between tradeable permits and transferable credits concerns the incentive for carbon-intensive firms with low profitability to close down: in a permit trading scheme, a firm that closes down can sell its permits, but in a credit trading scheme, there are no credits to sell simply because its baseline emissions have become zero.

In the unilateral variant of JI and the CDM (as in credit trading), the host is also the investor, so that disaggregated information is available to execute the project efficiently. However, when legal entities in JI and CDM host countries to some extent lack the money and knowledge to invest themselves, a foreign investor first has to identify and execute an emission reduction project in the host country, usually on a bilateral or multilateral basis. From the perspective of the investor, this means that the efforts to obtain information and the costs to find and negotiate with a project partner are likely to be higher than in the case of permit (and credit) trading where it is not necessary to identify a project before trading can occur.

Third, the latter point brings us to the assertion in the economic literature that permit trading is expected to have lower transaction costs than government trading and (project-based) credit trading including JI and the CDM. Although in theory each type of flexible instrument could equalize marginal abatement costs across countries, some will be more apt to use the international cost saving potential in practice because of their institutional features. In particular, JI and the CDM are considered to have higher transaction costs (such as search costs, negotiation costs and approval costs) than permit trading. It is argued that such projects usually require a pre-approval and independent verification of every single transaction in contrast with transfers in a permit trading system which are automatically registered and can be checked at the end of the year.

Although the project's transaction costs could be somewhat reduced if the seller finances the abatement itself (unilateral approach) or if the buyers pool their investments (multilateral approach), the hierarchy is well illustrated by the figures and comparisons found in the emissions trading literature. With a view to international carbon trading, reference is not only made to the high transaction costs of the international AIJ pilot phase projects as well as of the domestic (and restrictive) early credit trading arrangements in the US, but also to the low transaction costs of domestic SO_2 permit trading in the US. The latter transaction costs are estimated to be around 5% of the transaction value (e.g. Klaassen & Nentjes, 1997), whereas the transaction costs are roughly assumed to be 15% for JI projects and 25% for CDM projects in the reference case of a model by Haites (2000). The transaction costs of AIJ projects range from 1 to as high as 89% (e.g. Fichtner et al., 1999, 2003). In general, CDM projects are thought to have the highest transaction costs due to the relatively weak institutional capacities of developing countries. Less attention has been paid in the literature to the transaction costs of government trading and credit trading. We will discuss them briefly below.

Government trading is believed to have higher transaction costs than permit trading. On the one hand, government transactions are expected to involve larger emission quantities than private trades, which reduces the transaction costs per unit of emissions traded for government trading compared to permit trading (Boom & Nentjes, 2000). On the other hand, the bargaining process could be more time-consuming when governments negotiate instead of private entities. Governments are smaller in number and are expected to trade less frequently than firms, so that price uncertainty is likely to be higher under government trading than under permit trading. This price uncertainty complicates the bargaining process and raises negotiation costs. Moreover, businessmen would have more and better information about their emissions and marginal abatement costs than government officials, which adds to uncertainty and increases the costs of gathering reliable information.

The transactions costs of credit trading would not differ much from permit trading because both types make use of the information advantage of the private sector and do not require advance approval of every entitlement transfer. Nevertheless, the determination of the allowed emissions for a given year is more difficult under credit trading, because these are not given (as under permit trading), but have to be calculated on the basis of the existing climate policy, for example, by multiplying the performance standard with the energy use in that year, which can be done accurately only ex post.

In contrast with the political hierarchy, which is an empirical one, some authors have challenged the validity of the economic hierarchy, which is a theoretical one. Haddad & Palmisano (2001: 427), for instance, claim that credit trading is "superior" with regard to adaptability and fairness. However, we want to keep economics and politics conceptually separated, and take neoclassical economics seriously, by using the criteria of efficiency and effectiveness as the starting point of defining superiority. Nevertheless, based on institutional considerations, we will nuance the efficiency and effectiveness properties that make permit trading the superior alternative, investigate the formal and informal barriers that prevent or slow down its implementation and, rather than presenting a normative equity argument, analyze equity from a positive-theoretical point of view.

2.4.2. The Problematic Acceptability of Permit Trading in Politics

The general picture that emerges from the aforementioned economic literature on emissions trading is clear: permit trading is the superior alternative and ranks first in the economic hierarchy of market-based climate policy instruments. One would then expect that (economically rational) politicians accept permit trading as the leading instrument for climate policy. However, it is a well-known phenomenon that this has not been the case (e.g. Heller, 1998; Bressers & Huitema, 1999; Dijkstra, 1999; Russell & Powell, 1999).

On the international level, permit trading was mainly developed by American (and European) scientists and, in the context of the Kyoto Protocol, advocated by US negotiators and other JUSCANZ countries such as Canada, Australia and Japan.[2] Permit trading and in fact any flexible instrument was highly disputed in international politics, inspiring developing countries in the early 1990s to accuse the industrialized countries of "carbon colonialism" (Kuik & Gupta, 1996) and leading the EU, around the turn of the century, supported by countries like China and India, to propose a quantitative restriction on the use of the Kyoto Mechanisms (SBSTA/SBI, 2000).

---

[2] The so-called JUSCANZ group is a more or less occasional coalition formed in the context of the international climate change negotiations, which incorporates Japan, the United States, Canada, Australia and New Zealand. The US also discusses the possible implementation of an emissions trading scheme with the so-called "Umbrella" group incorporating the JUSCANZ countries (of which most are likely to be potential buyers) as well as Norway (an early European advocate of emissions trading) and the Russian Federation and Ukraine (which are potential sellers).

In 1995, the FCCC Parties accepted a weak version of project-based flexibility in the form of the AIJ pilot phase where crediting is absent. Prior to CoP3 in 1997 in Kyoto (Japan), the US advocated permit trading for the developed countries and project-based flexibility mechanisms for the developing countries, but Western and Eastern European governments (such as the EU and the Russian Federation) found it too early to start with a permit trading scheme and favored JI for the whole industrialized world. The developed countries, organized in the G77, rejected the use of any flexible instrument (Kuik & Gupta, 1996; AGBM, 1997).

During the negotiations in Kyoto, the Parties accepted JI for industrialized countries, as well as the CDM (as a more sustainable and equitable version of the JI concept) for developing countries. They were willing to accept a text on international emissions trading because that was seen as a precondition for the JUSCANZ coalition to sign a legal protocol on climate policy in the first place, but the Parties refused to include a text which explicitly allowed permit trading, as favored by this coalition, because it would require more elaborate rules than government trading, for instance, on permit allocation and compliance (Oberthür & Ott, 1999: 196). Furthermore, it was decided at CoP4 in 1998 in Buenos Aires (Argentina), as desired by the developing countries, to give priority to elaborating the CDM, not only because the developing countries are most vulnerable to climate change, but also because CDM projects could already be credited from 2000 onwards (BAPA, 1998). At CoP6 in The Hague in 2000, the EU made an attempt to quantitatively restrict the use of emissions trading (EU Council, 1999). International political pressure from various industrialized countries nevertheless forced them to reject this proposal in 2001.

Also on a national level, permit trading has not been the dominant instrument. Developed countries usually implemented domestic climate policy, if developed at all, by means of taxes or relative (energy-efficiency) standards, sometimes in the framework of covenants (e.g. COM, 2000e; Vermeulen & Kok, 2002). To introduce flexibility, various governments mainly used (or planned to use) credit trading or some combination of credit trading and permit trading, like in the Netherlands, Belgium, the UK and, to some extent, Germany (e.g. VROM-Raad, 1998; Cooper & Nicholls, 2000; GGET, 2001; Cools, 2003). Even the US, an early advocate of trading CO_2 permits (similar to their SO_2 allowance trading scheme), withdrew from the Protocol in 2001 and proposed to use voluntary climate measures under a GHG intensity target with the possibility of transferring registered GHG reductions between firms (Bush, 2002; see Pew Center (2002) for an analysis). Several economists have argued that grandfathering would increase the political acceptability of permit trading for firms (e.g. Koutstaal & Nentjes, 1995; Baumol & Oates, 1988; Tietenberg et al., 1999), but it appears that the energy-intensive industries do not want to see their emissions capped in the first place, because expanding production would not allow them to obtain the additional emission rights for free.

Still, this is only half of the story. History has shown that the political hierarchy of market-based climate policy instruments is not necessarily static, but can evolve over time. First, on a domestic level, political opposition can be overcome as proven, for instance, by the domestic permit trading scheme for SO_2 emissions in the US since 1995, where electricity producers, backed by some Congressmen, successfully lobbied to raise the emission ceilings (Conrad & Kohn, 1996; Klaassen & Nentjes, 1997; Schmalensee et al., 1998). Second, the international community has moved from the experimental AIJ scheme established in 1995, where crediting was still prohibited, to the inclusion of the flexible mechanisms in the Kyoto Protocol in 1997, which allows the trading of emissions. Third, when it came to elaborating the design of these instruments, some actors, notably the Member States of the EU, have moved from strong resistance to formal acceptance of permit trading, which will be implemented in Europe in 2005.

Three questions come to mind when reading these rather amazing pieces of history that exhibit both impediments and incentives for evolution to efficiency. The first question is backward-looking: why did various governments oppose the use of permit trading for so long and why did some of them, like the EU Member States, suddenly make a U-turn in climate policy by accepting the instrument? The second question is forward-looking: will those governments that have accepted emissions trading implement the permit trading blueprint as developed by economists for all sectors included in this market, not only before, but also after 2012? The third question is an overlapping one: can we answer the previous questions by using an overarching theoretical framework? That is, can we avoid, or at least embed, the summing up of (more or less relevant) ad hoc explanations and/or projections, as some authors have done, for instance, with regard to the EU case (like Christiansen & Wettestad, 2003; or Convery et al., 2003)?

2.5. Some Drawbacks of the Existing Literature

Why is it possible that permit trading, which ranks highest in the economic hierarchy, may rank low in the political hierarchy of market-based climate policy instruments, and when do both hierarchies converge?

The usual explanation is the resistance or support by interest groups, such as the industry and environmental organizations (e.g. Svendsen, 1998; Dijkstra, 1999), but this public choice literature, as we have seen in the previous chapter, tends to neglect the preferences and concerns of governments who ultimately decide which instruments will be used (Banuri et al., 2001: 49). Moreover, a few authors who do consider the preferences of bureaucrats and politicians with regard to environmental policy instruments, like Nentjes & Dijkstra (1994), do not take into account the legal or cultural aspects of their preferences, for instance, related to equity, and focus on domestic rather than international political economy settings.

Another problem of the existing literature is that most analyses are static. Those who do take a more evolutionary perspective usually do not consider the possible impact of path dependence. Tietenberg & Victor (1994), the UNCTAD (1995) and Ellerman (1998), for instance, believe that (project-based) credit trading schemes for CO_2 emissions can evolve, either or not via government trading, into international permit trading schemes. They do not take into account the risk that starting with credit trading could result in an institutional lock-in, reinforcing a path from which it may be difficult to escape. Some authors do mention Arthur's work on path dependence, like Haddad & Palmisano (2001), but they do not apply, let alone elaborate his approach in the context of market-based climate policy. Moreover, they restrict their analysis to issues of design and lobbying, without considering the impact of formal and informal institutional barriers that contribute to the (possibly temporary) resistance by governments against permit trading and other flexible instruments.[3]

Another example is the article by Damro & Méndez (2003), who explain the (un)acceptability of emissions trading as a (slow) process of "policy transfer", arguing that the instrument had to be transferred from the US to other countries, including those in Europe. However, this is more a description than an explanation of the adoption of emissions trading by those countries. Furthermore, although these authors acknowledge the role of sunk costs and learning, for instance, they fail to analyze political transaction costs, scale advantages, drivers of cultural change and possible institutional lock-in effects. They even forget to make a distinction between permit trading and credit trading, which is crucial to understand the subtleties in the history and implementation of emissions trading in America, as many authors underline (e.g. Tietenberg et al., 1999), and more recently also of that in Europe (e.g. Woerdman, 2004b).

In an attempt to avoid these drawbacks, we will try to find out whether the path dependence approach is able to take us an analytical step further. Path dependence means that policy outcomes are dependent on the (sometimes coincidental) starting point and specific course of an historical decision-making process. As the development proceeds, the costs of reversing or altering previous decisions may increase, narrowing the decision-making scope. By focusing on set-up costs, formal and informal constraints, scale effects and learning, for instance, this theoretical perspective also explains why the choice for a sub-optimal design, like

---

[3] Despite these and other shortcomings, it should be acknowledged that Palmisano (1996) was one of the first to analyze the implementation differences between permit trading and credit trading.

credit trading, may "lock-in" institutionally rather than evolve into an optimal one, and why an institutional "breakout" might occur, in the direction of permit trading.

As far as we know, there are only a handful of authors in the climate change literature who have picked up the ideas of path dependence and lock-in. In principle, it is possible to distinguish three strains of literature. The first strain only or mainly considers the possibilities of institutional path dependence and "carbon" lock-in, as it is called, to the extent that they strengthen a particular technological lock-in. Unruh (2000) falls in this group. The second strain is more or primarily interested in the institutional lock-in itself, like we are, but does not come further than either mentioning the possibility of such a lock-in or mentioning Arthur's self-reinforcing mechanisms without substantially extending, let alone questioning and altering, the arguments provided. Dietz & Vollebergh (1999), Haddad & Palmisano (2001) and Foxon (2002) fall within this group. The third strain not only mentions the arguments behind institutional path dependence and lock-in, but also questions whether all of Arthur's technological self-reinforcing mechanisms equally apply to institutions and provides analytical extensions and remedies for the incomplete analogies observed. Here is where we place our book.

2.6. Conclusion

When designing market-based climate policy, a choice must be made between various types of flexible instruments. Permit trading is superior according to neoclassical economics. Credit-based instruments, including performance standard rate trading and the project-based Kyoto Mechanisms, are inefficient and their effectiveness is uncertain. The environmental scarcity is not reflected in a price for each unit of emissions: when the economy grows, the supply of credits increases as well because polluters do not have an emission ceiling. Moreover, the project-based instruments have relatively high transaction costs because transactions require pre-approval.

Under permit trading, also called allowance trading or cap-and-trade, polluters do have an emission ceiling. This design option is both efficient and effective: when the economy grows, the demand for emission rights increases, but the supply of such rights remains constant because of the emission ceiling. Without pre-approval, transaction costs in the market are relatively low. The administrative costs of permit trading for large and small emitters (downstream) can also be reduced by concentrating monitoring and enforcement on the level of fossil fuel producers and importers (upstream).

However, when implementing market-based climate policy, politicians are tempted to make existing environmental policy more flexible by adding credit trading to it. An example is the EU where several Member States developed such plans. Also the international community initially started with (experimental) emission reduction projects and avoided the trading of emission allowances. Apparently, the economic hierarchy and the political hierarchy of market-based climate policy instruments do not necessarily coincide. Nevertheless, history has shown that they may converge. The international community now accepts permit trading under the Kyoto Protocol, next to the project-based mechanisms, and the EU will start its own permit trading scheme in 2005.

Three questions emerge from this. The first one is backward-looking: why did various governments oppose the use of permit trading for so long and why did some of them, notably the EU, make a U-turn in climate policy by accepting the instrument? The second question is forward-looking: will those governments that have accepted emissions trading implement the permit trading blueprint as favored by economists for all sectors included in this market, not only before, but also after 2012? The third question is an overlapping one: can we answer the previous questions by using an overarching theoretical framework that considers both economic and institutional aspects?

Here is where most literature falls short. Some authors sum up more or less relevant ad hoc explanations or projections, but do not embed them in a theoretical framework at all. Others use public choice and focus on the resistance or support by interest groups, but tend to neglect the preferences of governments and the legal and cultural (equity) barriers that they face. Moreover, although most analyses are static, the evolutionary writings on emissions trading usually do not consider path dependence or the risk that starting with credit trading could result in an institutional lock-in. Those authors who do consider the path dependence and lock-in of climate institutions usually mention the possibility, but do not substantially extend, let alone question and alter, the arguments provided by authors like Arthur and North.

In an attempt to avoid these drawbacks, the next chapter tries to work out the rudiments of an economic theory of institutional path dependence and lock-in. The path dependence approach basically explains why decision makers often change policy incrementally by building upon existing regulation, ineffective and inefficient as it may be. This should also explain why the choice for a sub-optimal design like credit trading may "lock-in" institutionally and when an institutional "breakout" in the direction of permit trading might occur.

Chapter 3

Path Dependence and Lock-In of Market-Based Climate Policy

3.1. Introduction

This chapter presents the overarching theoretical framework of the book, not only by elaborating upon North's (1990) notion of political transaction costs, but also by following his suggestion to extend David's (1985) and Arthur's (1989) work on the path dependence and lock-in of technologies to that of formal and informal institutions.

The path dependence approach builds upon David's well-known (but also criticized) story of the QWERTY-keyboard, a sub-optimal technology, that proved to be difficult to replace. Late 19th century, this keyboard was invented for typewriters as a remedy for the problem that typebars often clashed and jammed if struck in rapid succession. Since then, various technological improvements and ergonomically superior designs have been developed, such as the sequence DHIATENSOR that would facilitate faster typing, but none were implemented. Even late 20th century when the typewriter was replaced by the computer, which obviously does not operate with (potentially clashing) typebars, the old-fashioned QWERTY arrangement of keys remained dominant. And it still is dominant on our keyboards today. Self-reinforcing mechanisms like large set-up costs, increasing returns, co-ordination effects and learning are thought to explain this.

North suggested to use this approach in an institutional context. This suggestion has been welcomed, not only in various branches of economics, like institutional economics (e.g. Magnusson & Ottosson, 1997) and law and economics (e.g. Field, 2000), but also in fields like political science (e.g. Pierson, 2000) and sociology (e.g. Mahoney, 2000). North (1990: 95) himself is convinced that all of the aforementioned self-reinforcing mechanisms equally apply to institutions, although with somewhat different characteristics, and that institutions are subject to "massive" increasing returns, as he writes. In this chapter, we make the first steps to build such an economic theory of institutional path dependence. We want to see whether it can explain why policy makers find it difficult to switch to permit trading and rather add some form of credit trading, a sub-optimal design, to the existing environmental policy framework.

This chapter is organized as follows. Section 3.2 presents the definitions of institutional path dependence and lock-in. Section 3.3 defines the conditions for an institutional lock-in by paying attention to incomplete information, positive feedbacks, self-reinforcement, political transaction costs and probabilities, for instance. Section 3.4 defines the conditions for an institutional breakout by considering such explanatory factors as (decreasing) switching costs, crises, societal change, experiments and learning. Section 3.5 discusses what happens when the superiority of the superior alternative is contested. Section 3.6 highlights the novelties of an institutional path dependence approach. Section 3.7 applies the theoretical framework to our case and wonders whether there is such thing as a path-dependent climate policy. Finally, Section 3.8 presents the conclusion.

3.2. Definitions of Institutional Path Dependence and Lock-In

According to some versions of evolutionary theory, history always moves in the direction of a superior alternative (e.g. Fisher, 1958). According to the lock-in concept from the path dependence approach, however, this evolutionary process can get stuck — either or not temporarily. The analysis of path dependence is becoming increasingly popular in the literature on institutional evolution (Nelson & Sampat, 2001: 37), but its interpretation is actively disputed (Williamson, 2000: 611). As far as we know, Mahoney (2000) and Pierson (2000) are the only ones who have tried to build a general theory of some form of institutional path dependence of their own.

To start with the latter, Mahoney (2000: 514) basically defines path dependence as every (contingent) outcome that, on the basis of prior events or conditions, cannot be predicted by a particular theory (such as neoclassical economics). This definition, however, is too broad: it is almost always possible to find some theory that cannot explain a particular outcome. After making a legitimate case against broad definitions and concept stretching, Pierson (2000: 252) then claims to use a narrow definition of path dependence by defining it as increasing returns. In our view, however, he actually uses a broad definition of the increasing returns concept itself, as we will explain later on. Next to this, we will not only look at institutional rigidity ("lock-in"), as Pierson confines himself to, but we will also try to find the conditions for institutional change ("breakout"). Finally, unlike these authors, we will make explicit the nature of the analogies made between technological and institutional lock-in situations.

Path dependence generally refers to situations in which decision-making processes (partly) depend on earlier choices and events, but it is more than just a recognition that "history matters". That element is only part of the story. The path dependence approach not only recognizes the impact of history, but also shows that a decision-making process can exhibit self-reinforcing dynamics, so that an evolution over time to the most efficient alternative not necessarily occurs. In addition, it not only matters where you come from, but also where you have started.

This idea can also be applied to institutions. Institutional path dependence then recognizes that a choice, say between a number of policy instruments, is not made in some historical and institutional void just by looking at the characteristics and expected effects of the alternatives, but also by taking into account how much each alternative deviates from current institutional arrangements that have developed over time. The path dependence approach puts forward that this historical path of choices has the character of a branching process with self-enforcing properties that cause the costs of reversing previous decisions to increase, and the scope for reversing them to narrow sequentially, as the development proceeds.

The possibility of an institutional lock-in is central to this approach. A lock-in, in general, can be defined as the dominance of a sub-optimal situation in the presence of a superior alternative. Optimality is defined in terms of efficiency. Efficiency, in a neoclassical sense, refers to minimizing the costs (and, if measurable, maximizing the benefits) of running a process and/or of performing transactions in a market. An institution can be defined as the humanly devised constraints that shape human interaction or, less formally, the rules of the game in society (North, 1990). Institutions can be informal, such as cultural values, or formal, such as legal rules. Cultural constraints emerge, whereas legal constraints are designed, imposed and enforced by the government that holds the legitimate monopoly of force within a certain territory (e.g. Weber, 1976). We will focus our analysis on formal institutions and consider informal institutions in so far they have an impact on formal institutional rigidity and change.

Institutional arrangements refer to legally embedded regulations in the form of policy instruments to achieve certain policy targets, such as an environmental policy instrument to achieve an environmental target by constraining polluting behavior.[1] An institutional lock-in then refers to the dominance of a sub-optimal institutional arrangement, such as a (set of) inefficient policy instrument(s), in the presence of a superior institutional arrangement. An institutional arrangement is thought to be dominant when it is (formally adopted and) effectively implemented,

---

[1] This definition of institutional arrangements is different from the definition offered by North & Thomas (1973) who describe them as organizations and social practices, while they use the term institutional environment to refer to laws and regulations, among other things. The term institutional environment to refer to legal constraints makes sense if the firm is the unit of analysis, as for instance in Williamson (1975), but not if the focal point is the government that arranges and enforces these constraints itself (see also Groenewegen 1996: 9; Nooteboom 2000: 93).

while its alternative is not. By doing so, we avoid that any form of institutional persistency or absence of political change is called an institutional lock-in (like Alexander (2001) seems to do, for instance), which would make the concept too broad and imprecise. An institutional breakout then means that the superior institutional arrangement is adopted and implemented. Now that we have given these definitions, we can try to find the conditions for institutional lock-in and breakout situations to occur.

3.3. Conditions for an Institutional Lock-In

The path dependence approach originally stems from the (economic) literature on technological change. From Arthur (1994) it can be inferred that there are three general and necessary conditions for a lock-in to occur, namely the presence of (a) a superior alternative, (b) an imperfect market and (c) self-reinforcing mechanisms. The latter point is basically what distinguishes the path dependence approach from other economic perspectives.

According to Arthur (1994), self-reinforcing mechanisms regarding technologies or products are usually variants of, or derive from, large set-up costs, increasing returns, coordination effects, learning effects and adaptive expectations. Authors like Kemp (1995) and Windrum (1999) implicitly extend this list by considering, albeit not always systematically or thoroughly, the potential self-reinforcing impact of legal rules, cultural values, vested interests, perceptions and problem-solving capacities. We will try to find out whether these factors also bear relevance regarding institutional arrangements and how they relate to each other.

Liebowitz & Margolis (2000: 983) implicitly make a distinction between complete and incomplete analogies. We find it useful, which will become clear hereafter, to make this distinction explicit, because contrary to what some authors expect (e.g. Pierson, 2000), not all concepts from Arthur's theory can be copied to institutions without modifications. We consider an analogy to be complete if a certain condition is present both in the technological context and in the institutional context, also if its functioning or effects differ. An analogy is incomplete if this condition is only present in one context and not in the other, also if its functioning or effects are similar. This means that when two conditions are analogous, while the settings in which they take place are not, we still talk of a complete analogy. We will argue below not only that complete analogies can be made between the factors that contribute to techno-economic and institutional lock-in situations, but also that there are two important exceptions: there is an incomplete (albeit not absent) analogy both with imperfect markets and increasing returns when comparing technologies, or products, with institutions.

3.3.1. The Superior Alternative, Imperfect Markets and Incomplete Information

The first condition for the lock-in of technologies or products is the existence of a superior alternative where a sub-optimal situation dominates. The analogy with institutions is complete: an institutional lock-in only occurs when there is a dominant and sub-optimal institutional arrangement, while a superior alternative exists that is not being adopted and implemented. The superior alternative is more efficient, in terms of running and/or transaction costs, than the dominant one. A superior alternative is said to "exist" in two cases. First, the alternative may be present in theory. In that case, (a subgroup of) the scientific community has developed a superior alternative which is, however, not (yet) adopted and implemented. Second, this innovation may already have been adopted and implemented in a particular institutional setting, but not in the setting under consideration. In that case, the alternative is used in another policy area and/or in another country.

The second condition for a technological or economic lock-in is the existence of an imperfect market. In a perfect market, market forces ensure that the most efficient technology will be adopted. However, a lock-in can occur if the market is not perfect, mainly due to (a) the existence of monopolies or (b) high degrees of knowledge gaps and uncertainty. When this condition is translated to an institutional context, an incomplete analogy emerges between competing technologies and competing institutional arrangements. We will first consider monopolies and then incomplete information.

What makes the analogy with monopolies incomplete is that institutional arrangements, such as policy instruments, (unlike products) are not bought and sold on a market against a price. The government is not a buyer of such arrangements (and the seller is unidentifiable). Rather, formal institutions are decided upon and imposed by the government. The market analogy is not entirely absent, though. The government can be characterized as a monopolist if society is seen as a "buyer", or "consumer", of rules from the government, which is (and should be) the only one that imposes and enforces public law. Moreover, new policy instruments, for instance, can be said to compete with existing ones to become selected by the government. The fact that the government is a monopolist, in these respects, means that it does not have strong incentives to make a switch to different, including more efficient, institutional arrangements.[2] Monopolies make producers lazy.

---
[2] International competition among governments could provide some incentive, for instance if firms move abroad to avoid unfavorable national laws, but with respect to environmental regulation, firms appear to "vote with their feet" only to a limited extent (e.g. van den Bergh & van Beers, 1997).

The analogy with limited knowledge and uncertainty causing a lock-in is complete. They are present both in technological and institutional contexts. The superior alternative must not only exist (either in theory or in some other concrete setting), but it must also be known by those who choose. It is possible to distinguish three cases: (1) the superior alternative does not exist, (2) it exists but it is not or hardly known among those who choose or (3) it exists and it is fully or largely known by those who choose. In a similar way, Liebowitz & Margolis (2000) make a distinction between three degrees of path dependence.

In the case of first degree path dependence there is no error: the outcome is optimal. There is no lock-in, because the best alternative is chosen and a superior alternative does not exist. In the case of second degree path dependence there is an error: actors think they choose the optimal path, but the outcome turns out to be sub-optimal. This happens when the superior alternative exists but is not or hardly known by those who choose at the moment the decision was made. This does not preclude, as explained above, that (some) scientists or, for instance, decision makers in other countries are already familiar with the innovative option. The latter are in a position to refer to the sub-optimal situation as locked-in. In the case of third degree path dependence there is a remediable error: the outcome is sub-optimal, while those who made the choice had sufficient information about the existence of a superior alternative.[3] Because developing, gathering and absorbing information is a continuous process, a lock-in may (but need not) gradually shift from second degree to third degree path dependence.

Third degree path dependence occurs when most decision makers do not choose a superior alternative, while they know of its existence, largely understand its characteristics and perceive its uncertainty to be more or less acceptable. In that case, which attracts most scientific attention because it challenges neoclassical economic analysis, incomplete information alone cannot explain a lock-in. Such an explanation is offered by the functioning and impact of self-reinforcing mechanisms, which will be discussed in the next subsection.

3.3.2. Self-Reinforcement, Positive Feedbacks and Political Transaction Costs

Following Arthur's (1994: 27) conclusion that increasing returns are a necessary condition for the path dependence and lock-in of technologies or products, North (1990: 95) is convinced that institutions are subject to increasing returns as well.

---

[3] It is not appropriate to call this error "avoidable" as Liebowitz & Margolis (2000: 986) do, precisely because the erroneous outcome turned out to be unavoidable on the path chosen. Rather, in the words of Williamson (1993), the outcome is "remediable", because the outcome could be changed for the better based on the known existence of a superior alternative.

Likewise, when writing about politics, Pierson (2000: 252) claims to use a narrow definition of path dependence by defining it in terms of increasing returns. In our view, however, he actually uses a broad definition of the increasing returns concept itself. Pierson (2000: 251, 252, 262, 263) repeatedly speaks of "increasing returns arguments", "increasing returns tendencies" and "increasing returns or path-dependent processes", which he describes as "self-reinforcing or positive feedback processes". Moreover, institutions are not typewriters, so to say, which makes QWERTY different from politics. In fact, contrary to what North and Pierson believe, we argue that there is an incomplete (but not absent) analogy with increasing returns to scale in an institutional setting.

In an economic context, increasing returns imply a decline in unit production costs as fixed costs are spread over an increasing production volume. In other words: increasing returns in economics ("scale economies") is about production quantities. The firm then has an advantage, put simply, if it produces more of the same. In an institutional context, however, increasing returns is not about production quantities. The government can be seen as to "produce" regulation or policy and the "production costs" are its administrative costs. The advantage for the government of building upon existing policy arrangements does not originate from producing larger quantities of (similar) rules, as a complete analogy would require. What matters, though, is that the differential administrative costs (the extra costs of adding another collection of units) decline as the institutional scale increases. This can be done, as in Fig. 3.1, by expanding an existing policy instrument (horizontally) to cover extra target groups, such as more segments of industry, or the government itself can expand the instrument (vertically) by incrementally adding another element to it, for instance by allowing the target groups a more flexible application of the instrument. These are the coordination advantages or positive network externalities that Arthur speaks about. Just as these externalities give rise to increasing returns in an economic setting (e.g. Katz & Shapiro, 1985: 425), we argue that these externalities give rise to decreasing differential administration costs in an institutional setting.

Self-reinforcing mechanisms are "highly correlated" and, therefore, subject to mutual reinforcement (Kemp, 1995: 268), which causes the interactions among these mechanisms to strengthen the institutional barrier to change. An example is the role of expectations, which have a strengthening (or weakening) effect on positive network externalities (David, 1985; Katz & Shapiro, 1985). In a technological or economic context, the expectation of consumers that a product will have a large share of the market induces producers to put large quantities of this particular product on the market, which fulfills the original expectations. The analogy is complete in an institutional context, where the pervasive use of a particular institutional arrangement enhances the expectation of both regulator and regulated entities that it will be applied in the future without fundamental changes.

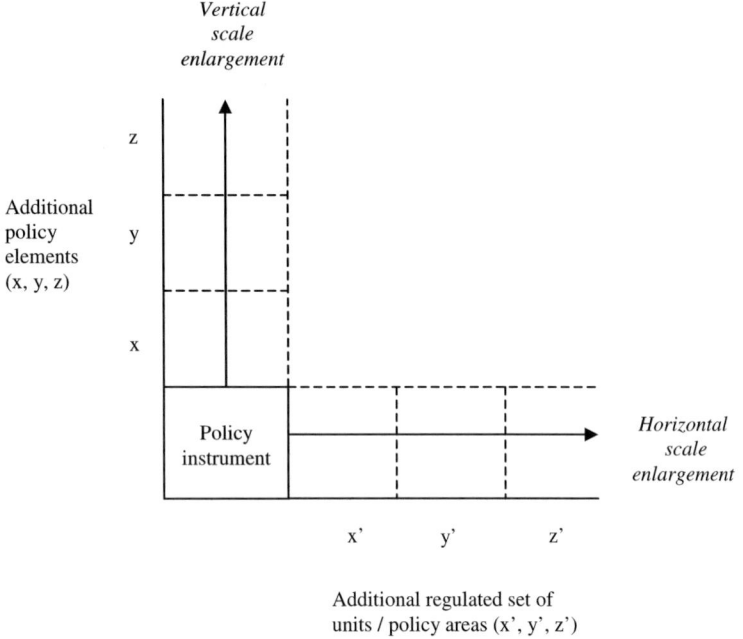

Figure 3.1: Institutional scale enlargement.

Expectations of a regulation's permanence can lead the regulator to impose a policy instrument on more entities and induce the regulated entities to lobby for incremental (instead of fundamental) improvements. The institutional arrangement has become a "self-fulfilling prophecy" (Wortman, 1995: 1391).

Despite the analogy, positive network externalities work differently in an institutional context. Coordination effects are advantages to cooperation with other economic agents taking "similar action" (Arthur, 1994: 112), but in an institutional context the perspective shifts from consumers and producers to the hierarchical relation between regulator and regulated entities. Furthermore, rather than emerging unintentionally, institutional coordination advantages can be created intentionally from the side of the government, as we have seen, by imposing a policy instrument on more entities or by improving its design. The government creates an additional coordination advantage by making sure that the same and transparent type or set of policy instruments is imposed on the regulated entities. If the property rights subsumed under these instruments are tradeable (as has been proposed in climate policy, for instance), this results in lower transaction costs in the market for the users, such as lower search and communication costs.

When talking about administrative costs, a distinction must be made between the set-up costs of establishing an institutional arrangement and the running costs of continuing it. The average administration costs of running the system become lower as knowledge and experience increase through learning. Set-up costs are the costs involved in establishing or changing an institutional arrangement. They rise as complexity increases. In institutional economics, set-up costs are also referred to as political transaction costs (e.g. North, 1990; Furubotn & Richter, 1997). Whereas transaction costs originate when property rights are transferred between parties in a market, political transaction costs arise when such rights are created (or protected) through political, administrative or judicial decisions (Allen, 2000). This reflects the idea that "(…) property rights themselves are costly (sometimes too costly) to impose (…)" (Cole, 2000: 306) and that "(…) abandoning a previously chosen path is not likely to occur without costs" (Magnusson & Ottosson, 1997: 2–3).

The political transaction costs to set-up an institutional arrangement are subdivided into sunk costs (of the existing arrangement) and switching costs (of a new arrangement). Sunk costs are not relevant for the decision whether or not to continue and extend the existing arrangement because they were made in the past ("bygones are forever bygones", according to economic theory), but switching costs are relevant when establishing a new one because they still have to be made. Examples of set-up costs that the government incurs are the costs of gathering and processing information, the costs of developing the required legal framework, the costs of (re)allocating property rights and the costs of dealing with lobbying efforts and cultural resistance.

Pierson (2000: 259) is wrong to argue that the "(…) sunk costs (…) terminology is unfortunate [because the] whole point of path dependence (…) is that these previous choices often are relevant to current action". Although we agree with this general description of path dependence, we disagree that this description would undermine the use of the sunk costs terminology. The point of sunk costs is namely that, from the perspective of set-up costs, continuation of the status quo is for free. Sunk costs are not included in the costs to be taken into account when planning and deciding on the next move. Choosing for a new design and introducing it, however, is not costless (as Pierson also acknowledges). The perceived costs of switching to the superior alternative arise, among other things, from legal problems and cultural resistance. Of course, costs are not the nature of legal requirements or cultural values themselves, but they do perform the role of switching costs when (and to the extent that) their content is unfavorable to change. Such switching costs play a more important role in institutional change than in technological change, because institutions, in North's framework, are in essence made up of formal (legal) and informal (cultural) constraints.

Analogous to the argument of technological compatibility (e.g. Mariñoso, 2001), the introduction of a new institutional arrangement must be compatible with existing laws and when it is not, it requires a change in design of the new arrangement or a change of the law itself. When information is complete, the (in)compatibility can be ascertained with certainty, but when knowledge is imperfect, this is a matter of approach and perspective (as will be demonstrated in some of the next chapters). In general, replacing the dominant institutional arrangement is likely to be less compatible with (and to require larger changes of) existing laws than extending it with an additional element. Both checking compatibility and changing the law add to the switching costs for the administration.

Culture can be self-reinforcing if the values in a society somehow favor the dominant institution and somehow reject the superior institution as unethical, just as in the case of dominant versus superior technologies. A lock-in is strengthened when "pulling values" exist that block change, for instance if a majority in society perceives an efficient alternative as less equitable than the dominant institutional arrangement, whereas a breakout is facilitated when "pushing values" make change possible by lowering switching costs.[4] Society in particular refers to those who prepare and make the decisions, such as the administration, the government and indirectly the voters, but also to lobbying groups trying to influence decision making. The more of these subgroups of society share common values against the superior alternative, the higher the information, bargaining and decision-making costs are and the stronger the lock-in is likely to be.

There is a complete analogy between technologies and institutions regarding the role of vested interests (e.g. Kemp, 1995: 268; Cowan & Hultén, 1996: 65). In both settings, there are actors that are interested in maintaining the sub-optimal status quo, for instance the industry, which has more financial resources and is better organized than the larger and more diffused group of individuals or households (e.g. Olson, 1965). A group with vested interests has a weak incentive to adopt the superior alternative when its economic losses are perceived to outweigh its economic gains compared to the existing situation or compared to other alternatives. By (continuing its) lobbying in defense of the dominant sub-optimal institution, the industry contributes to the switching costs and intensifies the lock-in situation. The government faces the (opportunity) costs of bargaining with interest groups and might provide side-payments to break down their opposition.

Apart from some exceptions (e.g. Mariñoso, 2001), most literature on technological and economic path dependence tends to focus on sunk costs (e.g. Pitelis, 1993: 10; Arthur, 1994: 112; Kemp, 1995: 268). The opposite seems to be the case in the literature on institutional change, where more emphasis is placed on the self-reinforcing effect of switching costs (e.g. Wortman, 1995: 1394; Pierson, 2000: 259). Although some authors do not fully understand the logic behind these cost concepts, they are at least right to focus on switching costs, not only because the level of sunk costs, that were made in the past, is irrelevant for the

---

[4] In an implicit way, also North (1997: 150, 153) acknowledges that there are values which lower transaction costs and values which raise such costs.

decision today whether or not to continue the established policy, but also because sunk costs simply do not have to be made, now or in the future, when switching to a superior institutional arrangement. Of course, continuing the existing institutional arrangement involves running costs. Extending it, for instance with additional policy elements or target groups, is not costless either, but that is (much) less costly than switching to a new institutional arrangement which has to be established from scratch. Moving from less to more complex governance entails incurring added bureaucratic costs, as Williamson (2000: 603) puts it.

Set-up costs (sunk costs and switching costs) are difficult to quantify (and could even be unknown), in particular if there is uncertainty, which makes it more complicated and less straightforward to perform a neoclassical analysis (Magnusson & Ottosson, 1997: 3). Switching costs may be even more difficult to measure than sunk costs, because the latter have already been made contrary to the former. Imperfect knowledge about switching costs also implies that decision makers will, subjectively, act upon their perception of such costs (e.g. Rizzo, 1994; Simon, 1997). Contrary to Liebowitz & Margolis (2000), who define a lock-in for situations where switching costs are not high, we regard upon (sunk and) switching costs as one of the essential factors to explain the lock-in of institutional arrangements. These costs could make it more easy to build policy upon the existing institutional arrangement than to introduce a superior one.

A necessary lock-in condition is that set-up costs are "large", as Arthur (1994: 112) writes. How large such political transaction costs must be for a lock-in to occur can be explained on the basis of Fig. 3.2. If no institutional arrangement would be in place yet, the government would choose for the arrangement with

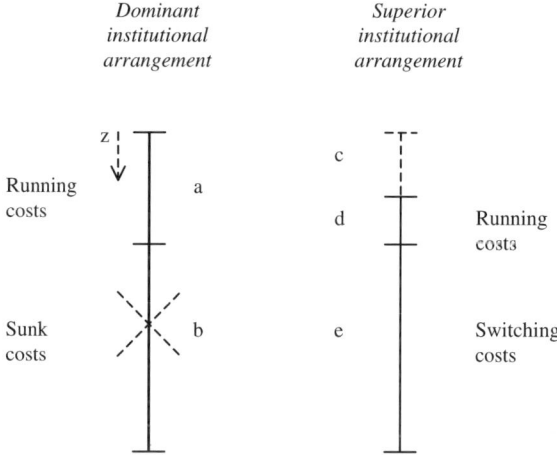

Figure 3.2: Institutional lock-in due to switching costs.

lowest total costs. Neoclassical economic analysis usually does not consider the role of set-up costs and would then only look at the efficiency gains to be made with respect to running costs, whereas an institutional economic analysis would also compare the switching costs of both alternatives. If the political transaction costs of both institutional arrangements are equal, the government would choose the superior one because it has lower running costs.

However, if an institutional arrangement is already in place, as the path dependence approach acknowledges, continuation of the dominant arrangement involves the annual (differential and average) costs of running it (denoted as a), which decline over the years as a result of network externalities and learning (denoted as z), while the level of sunk costs (denoted as b) is irrelevant for the decision today whether or not to continue the arrangement (portrayed as a cross through line b). By definition, a switch to a superior institutional arrangement is called superior because it would imply efficiency gains (denoted as c) in running the arrangement and thus lower running costs (denoted as d) compared to the running costs of the existing arrangement. However, switching to a superior institutional arrangement would also involve switching costs (denoted as e), which are, in fact, relevant for the decision today whether or not to continue the existing arrangement, because the set-up costs for the superior institutional arrangement still have to be made. In sum: an institutional lock-in occurs if switching costs are perceived to be "large", that is if $d + e > a$.

It follows from the analysis above that set-up costs in the form of sunk costs and switching costs contribute to an institutional lock-in. We have also demonstrated that self-reinforcing mechanisms do not arise because of increasing returns, as most authors believe, but because they generate positive feedbacks which lower the running costs (as opposed to the set-up costs) of the dominant arrangement. Next to the advantages of increasing the institutional scale, either horizontally or vertically, learning effects lower the average costs of running the established system. Such advantages could also accrue to the superior arrangement once established, but its establishment is made more difficult precisely because people benefit from learning and experience with the dominant sub-optimal arrangement. In that respect, the superior alternative must not only exist (either in theory or in some other concrete setting), but it must also be fully or largely known by those who choose. The other side of the coin is thus that incomplete information can contribute to an institutional lock-in if a superior alternative exists (for instance when it is used in some policy setting in another country), but is not or hardly known among those who choose. Because knowledge is always imperfect, we have also seen that perceptions work through all elements of self-reinforcement. Where "objective" theories and data are only one input in the formation of "subjective" perceptions, other inputs are, for instance, beliefs and expectations shaped by both personal and collective experiences and culture (e.g. North, 1990: 102).

Windrum (1999) adds that self-reinforcement of a sub-optimal technology is obtained if its problem-solving capacity is perceived to be growing or stable. A complete analogy can be made with institutions: the lock-in of a dominant sub-optimal institutional arrangement is strengthened if its problem-solving capacity (or: effectiveness), for instance in the light of some policy target, is perceived to be growing or stable. "Satisficing" rather than "optimizing" government representatives and officials, in a situation of bounded rationality, will then be less receptive to or even indifferent about any alternative arrangements, including theoretically superior ones. They become more receptive for (and pay more attention to) such alternatives when the perceived problem-solving capacity of the dominant arrangement is decreasing.

3.3.3. Probability, Inevitability and Remediableness

Rather than mentioning a list of factors that may contribute to a lock-in, like most authors do (e.g. Kemp, 1995; Pierson, 2000), we have tried to search for the necessary and sufficient conditions for this situation to occur. Fig. 3.3 sketches the results. Left in the figure is the past and right is a possible better future, namely the efficiency gains of a superior institutional arrangement. To take the path from the dominant to the superior institution, society must overcome the barrier of switching costs caused by legal problems, cultural resistance and vested interests. Going to the superior institutional arrangement in Fig. 3.3 is not only difficult because society has to "climb the hill" of switching costs (visualized by going up), but also because it carries the "luggage" of the past, both in the form of sunk costs and in the form of decreasing (differential and average) administration costs caused by learning effects and positive network externalities (visualized by going down). Expectations magnify network effects and perceptions work through all self-reinforcing mechanisms, in particular regarding future switching costs and benefits.

In general, it can be inferred from Arthur's theory of path dependence that the presence of a superior alternative, incomplete information and self-reinforcing mechanisms are necessary conditions for an institutional lock-in. The former two conditions are circled in Fig. 3.3. Incomplete information is placed in the middle of the figure, not only because information is never fully perfect so that perceptions play a crucial role, but also because a lock-in may already occur when information about the superior alternative is incomplete and insufficient to understand its characteristics and consequences (second degree path dependence). The next question is then which self-reinforcing mechanisms, visualized by means of boxes, are necessary and sufficient to cause an institutional lock-in when decision makers do not choose a superior alternative which they know and largely understand (third degree path dependence).

68   The Institutional Economics of Market-Based Climate Policy

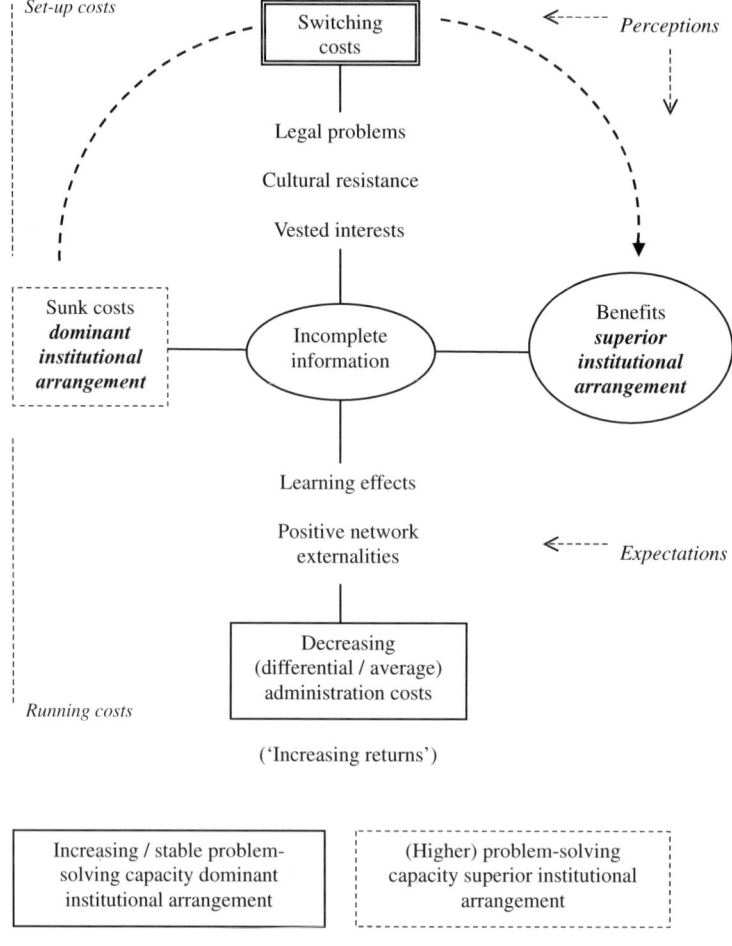

Figure 3.3: Institutional lock-in and self-reinforcing mechanisms.

First, an increasing or stable problem-solving capacity of the dominant arrangement is a necessary self-reinforcing mechanism, as explained before. Second, running costs for the administration may have decreased, but they must still be higher than the running costs of the superior alternative in order for the latter to be called "superior" (in the neoclassical sense). Consequently, despite their self-reinforcing effect, decreasing (differential and average) administration costs are not a necessary condition for an institutional lock-in. Third, because a lock-in requires the aforementioned superiority of the alternative in terms of running costs, it is necessary that perceived switching costs make the alternative more costly than the existing institution to cause a lock-in. Sunk costs made in the past, however, are placed in a box with dotted lines, because these set-up costs are irrelevant for today's choice between continuing the existing arrangement or switching to a superior one. The other self-reinforcing mechanisms are secondary, portrayed in Fig. 3.3 without boxes or circles, in the sense that they determine the magnitude of these set-up and running costs.

Switching costs must be large for an institutional lock-in (third degree path dependence) to occur. Whether it is also sufficient depends on the case. In general, a lock-in becomes more likely when there are more self-reinforcing mechanisms at work, for instance when not only switching costs are perceived to be large, but also when there are substantial network externalities and learning effects that lower the running costs of the dominant institutional arrangement. Nevertheless, the costs of switching to a superior institutional arrangement, for instance arising from legal problems and cultural resistance, are likely to play a crucial role in issues of institutional change, precisely because institutions are made up of formal and informal constraints (e.g. North, 1990: 68).

In sum, the following conditions are necessary to cause an institutional lock-in:

(a) superior alternative;
(b) increasing or stable problem-solving capacity;
(c) incomplete information; and/or
(d) large switching costs.

Conditions (a), (b) and (c) are necessary to cause a lock-in, whereas conditions (a), (b) and (d) are necessary to cause a lock-in based on third degree path dependence. This not only indicates that conditions (a)–(c) can already be sufficient to cause a lock-in, but also that a lock-in becomes more likely when conditions (a)–(d) are met. The "and/or"-operation must be read as follows: when condition (c) is not met in the sense that there is enough information to make an informed choice, it is necessary (and may be sufficient) that condition (d) is fulfilled.

Now that we have defined the necessary conditions for a regulatory lock-in, we want to know when they are sufficient to cause institutional rigidity. This is a difficult (if not impossible) task. The lock-in of technologies and products, for instance, is impossible to predict because there are multiple equilibria, as Arthur (1994: 14) concludes. Likewise, the lock-in of institutions is hard to predict when (perceived) set-up and running costs are difficult to measure and when information, for instance about the benefits of the superior institution, is fragmentary. Nevertheless, it is possible to say something about the probability of a lock-in, which becomes larger if, and to the extent that, more conditions for a lock-in are fulfilled.[5]

---

[5] Interestingly, in a chapter on lock-in, Arthur (1994: 14) starts by pointing at the non-predictability of outcomes in (lock-in) situations of increasing returns, but in the course of his text he weakens his proposition by writing that outcomes are not "entirely" predictable in advance (Arthur, 1994: 25).

Fortunately, however, path dependence does not imply inevitability (North, 1990: 98). The lock-in was unavoidable given the institutional choices made in the past (e.g. Langlois, 1994), but it is also remediable (e.g. Williamson, 1993), because the outcome can be changed for the better based on the (known) existence of a superior alternative. To put it in our terminology: an institutional breakout may occur.

3.4. Conditions for an Institutional Breakout

The conditions for an institutional lock-in are the existence of a superior alternative, incomplete information, a problem-solving capacity of existing policy which is perceived to be increasing or stable, as well as large set-up costs. The conditions for an institutional breakout mirror those of a lock-in as long as they are reversible. This exercise, which is also referred to in the technological literature as unlocking lock-in or as exit or escape from lock-in (e.g. Wortman, 1995; Windrum, 1999; Gerlagh & Hofkes, 2002), puts us in a position to analyze the path-dependent evolution of institutional arrangements. Put briefly, the chances for the superior alternative improve when information quality is enhanced and when set-up costs decrease against the background of a deteriorating problem-solving capacity of extant policy. External shocks can also provide strong pressures for policy change (Licht, 2001: 201).

Perhaps surprisingly, much of the lock-in literature (e.g. Arthur, 1994; Liebowitz & Margolis, 2000; Unruh, 2000) is characterized by a lack of attention to unlocking lock-in situations. Also Pierson (2000) tends to focus on inertia and does not list the conditions for institutional change. In a first attempt to fill this gap in the literature, we shall discuss the conditions for an institutional breakout hereafter. These define the opportunities to overcome institutional barriers.

3.4.1. Information, Perceptions and Experiments

First, the existence of a (known) superior alternative is the reason to unlock the lock-in situation. Analogous to technological examples (e.g. Windrum, 1999: 31), this means that its existence is not only a necessary condition for an institutional lock-in, but also for an institutional breakout.

Second, where incomplete information was one of the elements to cause the government to make a sub-optimal choice (second degree path dependence), its reversal, namely complete information, would lead to the optimal choice. Although perfect knowledge and full certainty are impossible to attain in reality, improving the information is likely to contribute to a breakout towards the superior institutional arrangement. This can be done by actively reducing uncertainty about a potential switch, for instance by providing more and better information about the superior alternative or by making this information known with more and preferably all decision makers. This information could be provided externally by actors without formal decision-making power, such as interest groups, NGOs or research centers, or internally by sub-communities of the decision makers themselves.

However, it may be difficult for those gathering and presenting the information to provide a perfect, complete and objective outline of the superior alternative. Likewise, it may be difficult for those receiving the information to separate its objective elements from its (at least partly) subjective and incomplete presentation and interpretation, and subsequently, to understand the characteristics and advantages of the superior institutional arrangement. Because information is never fully perfect, perceptions are likely to color the information based upon the interests and values of those who disseminate it.

To make sure that the superior alternative is known and that decision makers and other actors understand its functioning, they must learn, either organized or not, on the basis of information from theoretical studies and/or practical experiences with the superior alternative elsewhere (in other geographical or policy areas) or in the form of small-scale experiments (e.g. Bressers & Huitema, 1999: 193). In this respect, the technological lock-in literature refers to scientific results and, in an incomplete analogy, to entrepreneurial activities in niche markets (e.g. Kemp, 1995: 273; Cowan & Hultén, 1996: 65). The perceived balance of theoretical and practical evidence of studies and experiences, respectively, must be positive and, in itself, the incentive for a breakout becomes larger when its perceived benefits increase. These perceived benefits increase, not only when new and reliable studies and experiences are conducted which place the superior alternative in a (more) positive light, but also when international political developments or agreements stimulate its adoption in other countries. If the alternative will also be used in other countries, additional benefits may accrue, for instance, from positive network externalities between those countries. Via the perception of these benefits (what some might even call "perceived efficiency"), science and international politics may contribute to a breakout.

Third, where self-reinforcing mechanisms are necessary for an institutional lock-in (third degree path dependence), breaking those mechanisms or overcoming them by means of additional incentives is required to unlock the lock-in situation. Arthur (1994: 118) learns that exit from a sub-optimal situation depends on the source of the self-reinforcing mechanisms and on the degree to which they are reversible (or transferable).[6] With respect to institutional change, we identified two primary self-reinforcing mechanisms necessary for a lock-in, namely an increasing or stable problem-solving capacity and large switching

---

[6] This also means that Liebowitz & Margolis (1995: 205) in a way caricature the path dependence literature when they say that part of its claim is the irreversibility of lock-in situations.

costs. Are they reversible as well as necessary and sufficient for an institutional breakout? Both issues will be treated separately in the next subsections.

3.4.2. Problem-Solving, Crises and Learning

An increasing or stable problem-solving capacity of the dominant institutional arrangement, which is a condition for a lock-in, is reversible in the sense that it can be decreasing. The problem-solving capacity of existing policy decreases when it becomes less (or ceases to be) effective with a view to a particular policy target. In a technological context, it is necessary (but not sufficient) for a breakout to occur that the problem-solving capacity of a technology begins to slow to a standstill or deteriorates (Windrum, 1999: 13). This is most obvious if there is some crisis associated with the existing technology (Cowan & Hultén, 1996: 65). Also regarding institutional arrangements, a crisis, due to its visibility, is more likely to attract attention by politicians and the public than a gradual deterioration of effectiveness (e.g. Bressers & Huitema, 1999: 180; Nooteboom, 2000: 101).

Next to problem-solving capacities and switching costs, there are two other self-reinforcing mechanisms identified earlier: sunk costs and decreasing (differential and average) administration costs. The latter are not necessary conditions for a lock-in, but they may strengthen a sub-optimal situation. First, sunk costs are irreversible: they are simply there (although it may not be straightforward to quantify them). Second, learning effects, that cause average administration costs to fall, are usually also irreversible (Arthur, 1994: 118). Politicians and civil servants are not likely to forget what they have learned because they work frequently with the dominant design and create institutional structures to support it. Learning by doing lowers running costs for the administration, for instance when monitoring the regulated entities. In principle, a breakout is facilitated if government and the bureaucracy would learn about the potential benefits of the superior alternative (e.g. Aidt & Dutta, 2001). Third, adding new policy units, areas or elements to an existing institutional arrangement, thereby lowering differential administration costs, is reversible, in principle, which is also suggested by Arthur (1994: 118), but it is not in the interest of the politicians and officials to reverse those advantages by removing the dominant institutional arrangement from certain regulated entities or by developing policy which does not build upon the dominant design. They usually will not cease to implement and extend a sub-optimal institution as long as its problem-solving capacity is perceived to be satisfactory.

3.4.3. Switching Costs, Legal Compatibilities and Societal Change

It was demonstrated before that large switching costs are necessary to cause an institutional lock-in (third degree path dependence). Switching costs are reversible in the sense that they can be lowered. They must be "reversed" until they become small, namely to the level where the perceived gains of switching to a superior alternative outweigh the perceived costs of making such a switch. In that case, a breakout results, which facilitates the adoption and implementation of the superior institutional arrangement. This can be explained in some more detail on the basis of Fig. 3.4.

A comparison between Fig. 3.4 and Fig. 3.2 shows that switching costs (denoted as e) are lowered with the amount f. All other elements of the figure have remained equal. The result is that the running costs of continuing the dominant institutional arrangement (denoted as a), despite its learning effects and scale advantages (denoted as z), are now higher than the costs of both running (denoted as d) and switching to another institutional arrangement, which is superior (by definition) because of its efficiency gains (denoted as c). Sunk costs (denoted as b) do not play a role, because they are bygones. Where large switching costs caused the institutional lock-in in Fig. 3.2, switching costs have now been lowered in Fig. 3.4 to a level where they do not hinder an institutional breakout. Because an institutional lock-in occurs if perceived switching costs are large (that is if $d + e > a$), an institutional breakout occurs when switching costs are perceived to be small, that is if $d + e < a$.

It is not straightforward to determine the institutional reversibility of switching costs. Rather, such political transaction costs are "reversible" in the sense that they can be lowered, namely to the level where a switch can be made at negligible costs (Arthur, 1994: 118). Nevertheless, the factors which determine the level of these switching costs, namely legal problems, cultural resistance and defensive lobbying efforts, are reversible in different ways, as we will explain below.

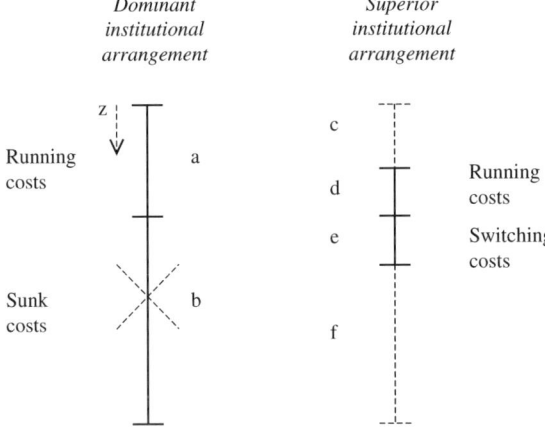

Figure 3.4: Institutional breakout by lowering switching costs.

First, to lower the level of switching costs, legal problems can be solved. There is a legal problem if the introduction of the new and superior institutional arrangement is perceived to be (potentially) incompatible with the existing legal framework. The switching costs result from the efforts required to change the law. These costs can be lowered if the superior alternative is made or perceived to be (more) compatible with existing law. Another "solution" to the problem is a legal decision that the superior institutional arrangement (after some period of analyzing its legal complexities) is declared compatible with existing law, which may happen if judging its compatibility is a matter of approach and perspective. In practice, changing (either the design of the superior alternative or) the law may be a politically difficult or even undesirable exercise, which largely depends on the distribution of values and interests in society. Those points will be treated separately hereafter.

Second, where a society's culture adds to the switching costs if this culture is unfavorable to the superior institutional arrangement, cultural change lowers these perceived costs. In a similar fashion, the technological and economic path dependence literature pays attention to the role of changing "consumer preferences" (e.g. Cowan & Hultén, 1996: 65) and the overcoming of "social opposition" (Kemp, 1995: 273) in escaping from lock-in. In an institutional context, cultural change requires that the balance in society shifts from, what we have called, "pulling" (change-blocking) to "pushing" (change-facilitating) attitudes, which could occur if a majority of those who prepare and make the decisions (but also of those who try to influence decision making) stops seeing the superior alternative as inequitable, for instance. The less cultural opposition there is in society against the superior alternative, the lower the associated switching costs and the weaker the lock-in will be. Cultural change is usually a slow process, which can take several years (or even decades). When the dominant culture already favors the introduction of the superior institution, however, there may be other factors that hinder a breakout. Vested interests could be such a factor.

Third, switching costs are lowered when societal actors representing vested interests reduce or cease their lobbying efforts in defense of the dominant institution. Expectations and perceptions are important here. Interest groups will put less time and money in lobbying if they can be convinced that they will lose the battle and that the adoption and implementation of the new institutional arrangement is inevitable. The extent to which this belief can be fed also depends on the presence, number, stake and lobbying efforts of actors representing new interests that plead for the superior alternative. Although the lobbying of new interests does not reduce, but rather increases the pressure on politicians and civil servants, it does lower switching costs by altering the "balance of interests": the actors representing vested interests now face some competition in the political arena from emerging actors representing different interests. Nevertheless, the vested interests, such as the industry, will generally weigh heavier than the new interests, such as small entrepreneurs, for instance in terms of the (size of actual instead of potential) capital and jobs they represent.

The government can try to change these interests to reduce switching costs, for instance by giving the industry some stake in the institutional change on the basis of incentives or side-payments. These are factors that contribute to a breakout, but which are additional in the sense that they do not mirror a particular lock-in condition (unless one would argue that the absence of incentives is not only a potential characteristic, but even a precondition for lock-in). To stimulate a transition, the techno-economic path dependence literature mentions regulation (Cowan & Hultén, 1996: 65) or legislation (Cowan & Gunby, 1996: 539), for instance in the form of some subsidy (Arthur, 1994: 118) or tax incentive (Wortman, 1995: 1396). A precondition, however, is that the government must be in favor of switching to the superior alternative. This also means that the government is not likely to provide incentives if it prefers the dominant institutional arrangement, for instance because of its sunk costs and learning advantages, and if it does not like the superior alternative, for instance because of its legal problems or ethical consequences. The best situation, here, is obviously when "(…) traditional firms possessing (…) large financial means commit themselves to the development of this [superior] trajectory" (Kemp, 1995: 255).

More or less similar to Williamson's (2000: 597) levels of social analysis, Nooteboom (2000: 101) argues that "(…) markets (…) are more easily changed (…) than laws (…), while basic cultural categories (…) are most difficult to change". Licht (2001: 149) even calls a nation's culture "the mother of all path dependencies". Cultural resistance may well be the largest impediment for a democratic government to provide incentives to stimulate change towards the superior institutional arrangement, because such resistance reduces the motivation to solve legal problems and to overcome the vested interests. Cultural resistance is also stronger in the sense that it is likely to be more widely shared both in the government and throughout civil society than legal problems and vested interests. Legal problems to the introduction of a new institutional arrangement are more a concern for the government and usually do not attract the attention of the general public, for instance because of its detailed and complex nature. Vested interests are more appealing to the general public than legal problems, for instance when there is a potential for job losses under the new institutional arrangement, but the government may still have (partly) different interests (like achieving an environmental policy target). Moreover, when actors have an "objective" interest in, but "subjective" cultural objections against an institutional breakout, they will not plead for it, whereas they are likely do so when they find it morally desirable even though it is not in their interest.

However, culture is not only a potential impediment to change, but it may also facilitate change when attitudes (are or) become less favorable to the status quo. Although cultural change is usually slow, it can change rapidly when there are strong pressures for change (Nelson & Sampat, 2001: 52) or external shocks (Licht, 2001: 201). In our framework, these pressures may be a (sudden) decrease in the effectiveness of the dominant design or international political developments (in favor of a switch) which are difficult if not impossible to influence.

In sum, the following four conditions are necessary to cause an institutional breakout:

(a) superior alternative;
(b) decreasing problem-solving capacity;
(c) improving information; and/or
(d) lowering switching costs.

For a breakout, there must be a known superior alternative. A (sudden) decrease in the problem-solving capacity of existing policy is required to make the dominant institutional arrangement become unattractive. Furthermore, to make the alternative known and understood, decision makers (as well as those who prepare and those who try to influence decision making) must learn and improve information on the basis of scientific studies, small-scale experiments and/or experiences elsewhere. Its perceived benefits can be increased by new and optimistic studies and experiences as well as by international political developments that work in favor of the superior institutional arrangement. Closing information gaps may be sufficient for a breakout (second degree path dependence), but is not likely when perceived set-up costs are substantial (third degree path dependence). Crucial for a breakout, then, is that switching costs are lowered. Such political transaction costs decline when legal problems are solved, when culture changes and when the opposition of vested interests is reduced, for instance by means of incentives.

Although these factors are necessary for a breakout, it is more difficult to judge whether they are also sufficient to bring about the transition to the superior institutional arrangement. In the technological lock-in literature, a breakout is considered to be impossible to predict due to multiple equilibria (Windrum, 1999: 31). In the institutional sphere, however, the probability of a breakout becomes higher if, and to the extent that, more conditions for a breakout are fulfilled. This means that, to estimate the probability of a breakout, one has to keep track of policy and legal developments as well as of the changes in the perceptions, cultural values and power positions of the actors involved.

A related issue is when self-reinforcing mechanisms are capable of actually preventing a breakout and when they can do no more than delaying a breakout. The answer based on our framework is that a breakout is likely to be prevented when the set-up and running costs of the superior institution are higher than the running costs of the dominant institution. However, a breakout is only a matter of time when the problem-solving capacity of the dominant arrangement deteriorates and when perceived switching costs continue to become smaller than they were, for instance because cultural resistance against the superior alternative diminishes or because the perception takes hold that legal incompatibilities can be lifted. A breakout actually comes about when these set-up costs have become small enough to make the superior alternative, including its running costs, less expensive than the costs of continuing and/or extending the dominant institutional arrangement. However, such changes can take several years (or even decades) to materialize.

On a practical level, extrapolating simple rules of thumb from the theoretical analysis above, it can be concluded that the following developments make an institutional breakout more probable in the case of an institutional lock-in:

- deteriorating effectiveness of the dominant institutional arrangement;
- more and better information about the superior institutional arrangement;
- international political developments in favor of the superior institutional arrangement;
- larger legal compatibility of the superior institutional arrangement;
- cultural change in favor of the superior institutional arrangement;
- less lobbying by vested interests and more lobbying by new interests;
- use of incentives to stimulate the superior institutional arrangement.

This makes clear that institutional change may not only arise from an unplanned evolutionary process, as emphasized in the Austrian and German tradition of institutional economics, but also from planned and conscious action, as emphasized by American institutional economists (Rutherford, 1994: 529; Nelson & Sampat, 2001: 36).

3.5. The Superiority of the Superior Alternative Contested

An essential feature of the institutional lock-in concept is the known existence or emergence of a superior (more efficient) alternative that makes a breakout desirable. Part of the explanation for the persistence of a sub-optimal institutional arrangement may be that the superior alternative is not as superior as presumed by the underlying (economic) theory. In fact, its superiority may be subject to societal or theoretical ambiguity.

The theoretical foundations behind the assessment that one alternative is superior to another are not rarely criticized by certain groups of scientists and/or actors in society. For instance, Liebowitz & Margolis (2000) offer some technological arguments which suggest that the Dvorak keyboard was not superior to the QWERTY keyboard as David (1985) claimed. In a similar fashion, the technical superiority of Betamax over the VHS video system, or of GEOS and other computer systems over Windows, as examples of lock-in brought up by Arthur (1994) and Unruh (2000), respectively, is still a topic of discussion. It is telling that Williamson (1999: 316) refers to these superior alternatives as "would-be-rivals". A situation could be perceived sub-optimal by some and optimal by others, so that the latter group will not refer to the situation as locked-in, whereas the former will. This may have at least three causes.

First, the social sciences rarely produce theoretical consensus and many of its scientific concepts are essentially contested. Also the theoretical judgment that one institutional arrangement is superior to another is made within certain theoretical boundaries and assumptions, in particular those of neoclassical economics that focuses on the efficiency of institutions. Theories with other considerations and assumptions may come to (partly) different conclusions or nuances about the superiority of this alternative. An example could be (neo-) institutional economics, which also takes into account the costs of setting-up institutional arrangements (including their equity consequences).

Second, Windrum (1999) points at what he calls the "relative fitness problem". This problem arises when one alternative is only superior to another in some but not in all respects. The perceptions among different actors of its superiority will differ if they attach different weights to these characteristics. These weights are likely to be shaped by the actors' value orientations. This means that not only the theoretical characteristics of the alternatives are relevant, but also the associated "social preference functions" (Windrum, 1999: 23).

Third, when determining the characteristics of the alternatives there may be uncertainty. Arthur, for instance, repeatedly speaks of technologies which are "possibly" or "probably" inferior (e.g. Arthur, 1994: 15, 118). With respect to this, Wortman (1995: 1402) argues that there may be open questions and uncertainties in the pioneering stage when implementing the superior alternative. The theoretical (and/or societal) ambiguity about the superiority, or inferiority, of particular alternatives poses a barrier to change in itself, because it makes the benefits of change uncertain. This particular barrier becomes smaller if either uncertainty decreases or decision makers become less risk-averse (e.g. Kemp, 1995: 254; Unruh, 2000: 825).

3.6. Novelties of an Institutional Path Dependence Approach

Although the explanations offered by the path dependence approach regarding institutional continuity and change to some extent draw from familiar perspectives and concepts, it does contain some new and additional elements and insights.

First, similar to some authors in the field of transaction cost economics (e.g. Dixit, 1996), the institutional path dependence approach pays attention to the incremental role of political transaction costs, in the form of costs to set-up the superior institutional arrangement. Nevertheless, path dependence goes beyond transaction costs not only by approaching political transaction costs from a historical and evolutionary perspective in the form of sunk and switching costs, but also by considering positive network externalities, self-reinforcing mechanisms and learning (e.g. Arthur, 1994). The approach recognizes that historical evolution is more than just a process of transaction cost minimization (Magnusson & Ottosson, 1996: 351). Contrary to the path dependence approach, transaction cost economics lacks the dynamics of learning, lacks institutions in the form of cultural values and does not explain the survival of sub-optimal institutions (Nooteboom, 2000: 105, 112). The role of culture and perceptions (for instance in the form of perceived switching costs which are difficult or even impossible to quantify) cannot be analyzed with neoclassical or new institutional economics (Magnusson & Ottosson, 1997: 3).

Second, the observation that institutions evolve slowly and institutional change is difficult also appears in the form of the incrementalism concept, for instance in institutional economics and political science (e.g. Lindblom, 1959). "There are no rigid laws in politics or policymaking, but there are tendencies. Incrementalism is one of the strongest" (Sharkansky, 2002: 116). Democratic political systems tend toward gradual political change as politicians often reduce complexity and uncertainty by considering alternative policies that differ only marginally from the status quo. However, according to Weiss & Woodhouse (1992), the concept of incrementalism — albeit cited frequently — has not served as a basis for a cumulatively developing line of research. Therefore, they propose to reframe the original concept. One possible way of doing this is by elaborating the evolutionary concepts and conditions of institutional path dependence and lock-in. Because of the positive feedback mechanisms identified above, path dependence is more than the incremental process of institutional evolution (North, 1991: 109). The institutional path dependence approach not only explains why policy making often leads to non-decisions or incremental changes by taking self-reinforcing mechanisms into account, but also formulates the conditions under which a switch to new institutions and instruments might occur. In a similar way, the institutional path dependence approach also clarifies the so-called "parallel institutionalization hypothesis" (e.g. Ruiter, 2000: 26) by indicating why and when new institutional arrangements will exhibit a high degree of parallelism to the old arrangements.

Third, although historical research plays a role in some (classical) traditions of economics, the institutional path dependence approach entails a clear revaluation of the idea that "history matters" in explaining continuity and change by acknowledging that decisions are not made in an institutional void, but in a dynamic setting of existing institutions where evolution to efficiency not necessarily occurs. However, when extending David's and Arthur's work on the path dependence and lock-in of technologies and products to that of institutions in such an evolutionary context, formal and informal constraints start to play a more prominent role (North, 1990: 68). This leads to another revaluation in economics, as desired by several authors (including Williamson, 2000: 610–611), namely that "culture matters" and that the "law matters" in explaining institutional lock-in and breakout situations.

3.7. A Path-Dependent Climate Policy?

Whereas Liebowitz & Margolis (2000: 995) believe that, in general, "(...) lock-ins are rare or nonexistent (...)", Magnusson & Ottosson (1997: 3) contend the opposite, namely that "(...) social scientists dealing with applied matters could most certainly list an endless number of instances where path dependency also in the third sense might be evidenced". Pierson (2000: 256) is also critical of Liebowitz and Margolis' claim. In this book, we want to find out who's right in the case of market-based climate policy. We will also investigate the middle position of (neither an absent, nor a permanent, but) a temporary institutional lock-in.

There are a few other case studies in the field of institutional change based on the path dependence approach. In the (institutional) law and economics literature, for instance, Hathaway (2003) finds that the doctrine of *stare decisis* creates path dependence in common law, not only because each legal decision increases the probability that the next will take a particular form, but also because significant costs may arise out of the reliance on precedent. Another example is provided by Wortman (1995) who not only finds that positive network externalities and learning benefits, in tandem, caused an absence of change from sub-optimal to superior corporation laws, but also acknowledges the potential importance of sunk costs and switching costs to explain this specific legal lock-in situation.

In environmental policy, Bressers & Huitema (1999: 180) observe that "(...) new 'economic' instruments are often based on existing legal instruments (...)". We believe that institutional path dependence and lock-in are an important, but often overlooked part of the explanation. Therefore, we use the institutional path dependence approach in this book to explain why various politicians, in particular in Europe, have long favored economically sub-optimal designs in climate policy, such as credit trading without absolute emission ceilings, whereas permit trading between private parties under an emission ceiling is economically superior. Moreover, the book highlights the risk that introducing (elements of) credit trading now could create an institutional lock-in that will make it more difficult to switch to permit trading in the future. Finally, we will conduct some theoretical and empirical research on the switching costs from traditional to market-based climate policy in relation to the formal constraints posed by WTO and EC competition law and the informal constraints posed by (European) political culture.

3.8. Conclusion

The institutional path dependence approach provides us with an economic underpinning of the conditions that either lead to institutional continuity (lock-in) or change (breakout), in particular by defining the role of set-up costs, network externalities and learning. The approach places the concept of political transaction costs in a historical and evolutionary setting, not only by distinguishing sunk costs from switching costs, but also by considering positive feedbacks, self-reinforcing mechanisms and lock-in effects. This explains why policy making often leads to non-decisions or incremental changes (resulting in an elaboration and reorientation of the concept of incrementalism) and makes clear when a switch to new institutions and instruments might occur. By considering not only set-up costs, but also the role of self-reinforcement, sustained inefficiency, culture and learning, the institutional path dependence approach (builds upon but) goes beyond transaction cost theory.

Although history moves in the direction of a superior alternative according to some versions of evolutionary theory, the path dependence approach demonstrates that this evolutionary process can get stuck, temporarily or not. This approach, mainly building upon North (1990) and Arthur (1989), not only recognizes the impact of history, but also shows that a decision-making process can exhibit self-reinforcing dynamics, so that an evolution over time to the most efficient alternative not necessarily occurs. An institutional lock-in refers to the dominance of a sub-optimal institutional arrangement, such as a (set of) inefficient policy instrument(s), in the presence of a superior institutional arrangement. An institutional breakout then means that the superior institutional arrangement is, in fact, adopted and implemented.

The conditions for an institutional lock-in are the existence of a superior alternative, incomplete information, a problem-solving capacity of existing policy which is perceived to be increasing or stable, as well as large set-up costs. Such a lock-in is strengthened when the superiority of the superior alternative is contested, for instance due to theoretical ambiguity or uncertainty. The conditions for an institutional breakout mirror those of a lock-in as long as they are reversible. This method, which is also referred to in the literature as unlocking lock-in or as exit or escape from lock-in, puts us in a position to analyze the path-dependent evolution of institutional arrangements. The chances for the superior alternative improve when information quality is enhanced and when set-up costs decrease against the background of a deteriorating problem-solving capacity of extant policy. External (political) shocks and additional (economic) incentives can also provide pressures for regulatory change.

Contrary to the popular notion of scientists like Pierson (2000), we demonstrate that there is an incomplete (but not absent) analogy with increasing returns to scale in an institutional setting. Increasing returns imply a decline in unit production costs as fixed costs are spread over an increasing production volume, giving the firm an advantage if it produces more of the same. In an institutional context, however, increasing returns is not about production quantities. The advantage for the government of building upon extant policy arrangements does not originate from producing larger quantities of (similar) rules, but from increasing the institutional scale of an existing policy instrument to cover extra target groups or by incrementally adding another element to it, leading to a decline in differential administrative costs (the extra costs of adding another collection of units).

The political transaction costs that have to be made to create a more efficient institutional arrangement are an important element in explaining an institutional lock-in. Examples of such set-up costs that the government incurs are the costs of gathering and processing information, the costs of developing the required legal framework, the costs of (re)allocating property rights and the costs of dealing with lobbying efforts and cultural resistance. Switching costs that arise from legal problems and cultural resistance play a more important role in institutional change than in technological change, because institutions, in North's framework, are in essence made up of formal (legal) and informal (cultural) constraints. The costs arising from these constraints are therefore extensively analyzed in the next chapters.

In environmental policy, Bressers & Huitema (1999: 180) observe that "(…) new 'economic' instruments are often based on existing legal instruments (…)". We believe that institutional path dependence and lock-in are an important, but often overlooked part of the explanation. Whereas Liebowitz & Margolis (2000: 995) believe that lock-ins are rare or non-existent, Magnusson & Ottosson (1997: 3) contend the opposite by claiming that it is possible to list an endless number of lock-in situations. A middle position could be that some lock-ins rather have a temporary character. The rest of this book should make clear whether the idea of path-dependent institutions, which is still under construction, bears any relevance in explaining the developments in market-based climate policy.

PART II
NEW INSTITUTIONAL ECONOMICS

Chapter 4

Environmental Effectiveness of Market-Based Climate Policy

4.1. Introduction

This chapter studies the impact of the institutional design and operation of market-based climate policy on environmental effectiveness, while largely confirming but also nuancing the traditional view in environmental economics that flexible instruments other than permit trading are bound to be ineffective.

Efficiency and effectiveness are the traditional decision and judgment criteria in neoclassical economics. Where the previous chapters mainly focused on efficiency, we will now study the effectiveness of different market-based climate policy instruments. Neoclassical analysis is clear in this respect. Permit trading is most effective because emission sources have an emission ceiling, whereas the effectiveness is uncertain for instruments without emission ceilings, such as credit trading (e.g. Tietenberg et al., 1999). However, just like Liebowitz & Margolis (2000) clarified the survival of the QWERTY-keyboard by contesting the superiority of its alternatives, we will partly explain the inclination of politicians to add sub-optimal designs to existing environmental policy by nuancing the effectiveness advantages of permit trading and the presumed ineffectiveness of credit-based approaches. To that end, an institutional economics perspective is taken by extending the neoclassical analysis of such instruments with several institutional factors. Perceptions and equity considerations will also be taken into account.

This chapter is organized as follows. Section 4.2 defines the concepts of environmental effectiveness and emission baselines. Section 4.3 nuances the environmental effectiveness of tradeable emission rights systems by considering hot air trading and non-compliance. The formal and informal character of those problems is evaluated, including some options to reduce them, like taxes and liability rules. Section 4.4 highlights the institutional opportunities to enhance the effectiveness of project-based emissions trading, notably the environmental and transaction cost consequences of ex post baseline corrections and baseline standardization. Finally, Section 4.5 presents the conclusion.

4.2. Definitions of Environmental Effectiveness and Emission Baseline

Environmental effectiveness can be described in a "static" sense as the achievement of a pre-defined policy target, such as a certain emission level, but also in a "dynamic" sense as the extent to which this policy target is attained. Therefore, it is possible that environmental effectiveness is improved, achieved or reduced.

In the context of containing and trading GHG emissions, it is possible to distinguish two interpretations of the concept of environmental effectiveness. In a formal interpretation, environmental effectiveness is achieved if the official aggregate emission target is attained, such as the 5% reduction by industrialized countries, as required by Article 3.1 of the Kyoto Protocol, of overall GHG emissions below 1990 levels between 2008 and 2012. In (what might be called) an ethical interpretation, environmental effectiveness is achieved if aggregate emissions are reduced below the official target by refraining from those economically attractive actions that are legally possible but that would result in higher emissions or less emission reductions than without those actions. Although this distinction may seem unconventional at first sight, it will prove to be useful to understand the arguments used in societal or theoretical discussions on the environmental performance of market-based climate policy instruments. If environmental effectiveness is mentioned in this chapter, it should be considered in its (usual) formal interpretation, unless its ethical interpretation is referred to explicitly.

An emission baseline (or simply baseline) is defined as an estimate of future emissions at one or more points in time under "business-as-usual" conditions. To clarify the discussion, we make a distinction between macro-baselines and micro-baselines. A macro-baseline is constructed at the national level and estimates the future emissions of (the total of emission sources in) a country at one or more points in time in the absence of an emission ceiling. A micro-baseline is used in emission reduction projects, like JI or CDM projects under the Kyoto Protocol, and estimates future emissions at the project location at one or more points in time in the absence of the project. A macro-baseline is constructed on the basis of aggregated national projections of economic growth, energy use and technological development, among other things. A micro-baseline also incorporates project-specific data, for instance, the average (projected) emission level of comparable emission sources in the sector in which the project will be implemented.

In 2001 in Marrakesh (Morocco), governments reached an international agreement on the definitions of JI and CDM project baselines. JI projects should have a baseline that is, among other things, "appropriate", "reasonable" and "transparent". The CoP defines the baseline for an Article 6 project as "(…) the scenario that reasonably represents the anthropogenic emissions by sources or anthropogenic removals by sinks of greenhouse gases that would occur in the absence of the proposed project" (CP, 2001b: 18). These emissions and removals should not only be estimated within the project boundary, but also outside the project boundary provided that they are "(…) significant and reasonably attributable to the project during the crediting period" (CP, 2001b: 19). The project-specific JI baseline shall take into account relevant national and/or sectoral policies and circumstances (CP, 2001b: 18).

CDM projects should have a baseline that is, among other things, "conservative", "reliable" and "transparent". Project participants must assess "(…) the environmental impacts of the project activity, including transboundary impacts [if] considered significant by the project participants or the host Party (…)" and [if] "(…) reasonably attributable to the CDM project activity" (CP, 2001b: 34, 37). The CoP defines the baseline for a CDM project activity as "(…) the scenario that reasonably represents the anthropogenic emissions by sources of greenhouse gases that would occur in the absence of the proposed project activity" (CP, 2001b: 36). The project-specific CDM baseline shall take into account relevant national and/or sectoral policies and circumstances (CP, 2001b: 37).

4.3. Environmental Effectiveness of Tradeable Emission Rights

Just like the superiority of alternatives to the QWERTY-keyboard has been contested in some technological writings, which might help to explain its survival (e.g. Liebowitz & Margolis, 2000), the neoclassical superiority of permit trading over other flexible instruments in terms of effectiveness has been subject to societal and theoretical ambiguity on the basis of institutional considerations. This partly explains why politicians hesitated to adopt permit trading right away. The information they received was neither complete nor unidirectional. The same type of arguments applies as in the technological lock-in literature (discussed in the previous chapter), namely the absence of consensus, the relative fitness problem and uncertainty. There was no perfect consensus, neither among scientists nor among political actors, that permit trading would yield optimal environmental integrity and part of the emissions trading literature considered permit trading as superior in some, but not all respects, depending on the perspective taken, leading to uncertainty for policy makers on its environmental benefits.

4.3.1. Macro-Baseline, Hot Air Trading and Uncertainty

The most important effectiveness problem of tradeable emission rights systems was considered to be that of hot air trading. This was thought to be "a serious problem on the practical level", posing an institutional barrier to accepting emissions trading in the international climate negotiations (Oberthür & Ott, 1999: 189). If there is a gap between the official emission ceiling and business-as-usual emissions, referred to as "hot air", permits may be traded and used to cover emissions that might have remained unused without emissions trading. According to the definitions provided above, this might be called a macro-baseline problem. If the hot air is traded under the emission ceiling, effectiveness is still achieved in its formal interpretation, but not in its ethical interpretation. Trading hot air is economically attractive and legally possible, but without such trading it could be that actual emissions of all emission sources together are lower than the overall target.

The effectiveness of permit trading is considered to be guaranteed because emission sources operate under an emission ceiling (e.g. Baumol & Oates, 1988; Tietenberg, 1992). This ceiling is assumed to be lower than business-as-usual emissions (e.g. Anderson et al., 1999: 118). However, this assumption might not hold in the real world. To obtain emission reductions, the emission ceiling must be set lower than the "macro-baseline" of business-as-usual emissions. In real-life (inter)national environmental agreements, emission sources (firms of states) may receive generous emission budgets due to different kinds of political and institutional factors. Examples of such factors are incomplete information (such as incorrect macro-baseline projections), the exertion of negotiating power, or considerations of equity and political acceptability.

The hot air problem emerges in emissions trading systems if a country's macro-baseline emissions (probably) remain below its negotiated emission budget. Under the Kyoto Protocol, for instance, if the assigned amounts of some countries exceed their expected business-as-usual emission figures, they may end up with unused assigned amount units (AAUs) that can either be banked or transferred, without having to take mitigation efforts. Without trading, assuming that the other countries use their assigned amounts completely, total GHG emissions in 2008–2012 would be even lower than agreed upon in the Kyoto Protocol. However, when emissions trading is allowed, the hot air countries will be able to sell this surplus to other countries that will use it to cover emissions that would not have been allowed without the transfer of this surplus. Because hot air trading then occurs under the emission ceiling of the Kyoto Protocol, it is legally allowed, but it might be considered ethically undesirable as it would make overall emissions higher than without such trading. The issue of hot air makes emissions trading problematic because with emissions trading the hot air can be sold.

With respect to the Kyoto case, most analysts expect that hot air will be available, in particular, from the Russian Federation and the Ukraine that have not only managed to negotiate stabilization targets (for the commitment period 2008–2012 based on 1990 levels), but have also faced strong economic decline due to problems of transition and economic restructuring, so that they will probably not reach this emission level in a business-as-usual scenario. This would allow them to sell parts of their assigned amounts to other Annex B Parties without having to reduce emissions.[1] This is not possible in JI projects, as some authors wished to emphasize (e.g. Jepma & van der Gaast, 1999), which involve real reductions based on the assumption (which might also not hold as we will discuss later on) that the micro-baseline used in such a project is correct. Although hot air trading could be seen as unethical (because the result is that emissions can be covered that would otherwise not have been covered), Bashmakov (1999) has the opposite view and considers the tradeable hot air in Eastern Europe as a legitimate compensation for the emission reductions induced by the economic decline which resulted from the disintegration of the centrally planned economic system.

It is always difficult to calculate the precise amount of hot air trading in advance due to the inherent uncertainty of macro-baseline emission estimates. Therefore, the projections of hot air under the Kyoto Protocol differ considerably. Michaelowa & Koch (1999a) rightly emphasize that there is a "range of forecasts" which originates from several studies using diverse assumptions and different data. In a survey of several different models by Zhang (2000b), for instance, the hot air projections under the Kyoto Protocol vary between 92 and 374 MtC-eq, roughly somewhere between 0 and 50% of the required Kyoto reduction efforts, depending on the estimated level of business-as-usual emissions.

To illustrate the level of uncertainty that policy makers faced, Haites (1998) calculated that 165 million tons of carbon could be sold as hot air (compared with an annual reduction from business-as-usual emissions of 1,000 million tons of carbon), while Victor et al. (1998) expected that the carbon "bubble", as they called it, could be as much as 1,000 million tons of carbon (in the central scenario). Still, the possibility was not ruled out that the Russian Federation and the Ukraine would experience higher economic growth rates than anticipated, in which case the hot air might not become available at all. A guesstimate would be that the expected magnitude of hot air under the Kyoto Protocol appears to lie roughly between 10 and 30% of the reduction efforts necessary to meet the aggregate emission target.

---

[1] Since the US (when it had not yet withdrawn from the Kyoto Protocol) was expected to be a large net buyer of (hot air) assigned amounts, one commentator cynically remarked that "American Cadillacs will be fuelled by Russian depression" (in Hamilton, 1998).

The hot air problem probably also applies to those developing countries that might wish to take up commitments somewhere in the future and engage in emissions trading (Michaelowa & Koch, 1999a).[2] On the one hand, emissions trading lowers compliance costs, providing an incentive to accept a more stringent target than without trading. On the other hand, emissions trading provides an incentive to negotiate a generous emission ceiling in order to maximize the economic gains from trading (O'Connor et al., 1997: 28). Furthermore, according to Baumert et al. (1999), it is likely that excess emissions for new participants would be welcomed by some industrialized countries, since it would make compliance less expensive for them. Therefore, it is naïve to believe that "[hot air] is a temporary phenomenon because it is unlikely to happen again in a future budget period" (Metz et al., 2001: 175).

Politicians and officials were harassed with different figures and studies according to which there was likely to be a hot air problem, big or small, if they would choose to implement permit (or government) trading rather than emission reduction projects. Environmental problem or not, this put doubt on the effectiveness advantages of tradeable emission rights systems in general and made permit trading between private entities look suspicious instead of superior. Although hot air trading would not disturb effectiveness in its formal interpretation, but only in an ethical one, it does mean that trading pollution rights was considered relatively "fit" in some, but not all respects, which helps to explain the (temporary) institutional lock-in.

4.3.2. Dynamic Versus Static Perspectives on Hot Air Trading

The view that hot air is problematic, because it can be sold and used to cover emissions elsewhere that might not have been covered without trading, assumes an ex post perspective on the negotiated emission targets by taking those targets as given. However, (trade in) hot air is *not* problematic from an ex ante perspective on the negotiations in which the level of the negotiated emission targets depends on the level of flexibility created. In this dynamic institutional setting, the higher the level of flexibility created, the higher the level of the accepted emission targets. Without the hot air, for instance, under the Kyoto Protocol, the emission targets could have been less stringent, to an extent that might even exceed the volume of hot air.

Some authors consider the allocation of hot air as a side-payment (or "bribe") for the acceptance of the cap-and-trade provisions under the Kyoto Protocol

---

[2] This can be referred to as "tropical air" (by analogy with the trading of "hot air"), although Goldberg et al., (1998) rather uses this term to refer to baseline inflation in projects in developing countries.

(e.g. Shogren & Toman, 2000: 32). Although the US initially signaled only to be willing to accept a stabilization target, it adopted a reduction target in 1997 because emissions trading was included in the Protocol with the prospect of buying cheap hot air from the Russian Federation and the Ukraine (e.g. Oberthür & Ott, 1999). These former Soviet countries were finally persuaded to adopt a target because they had the prospect of being able to sell (some) emission space that they will not use anyway to the industrialized West. It could be argued, therefore, that the allocation of hot air was necessary for some countries to make their emission limits acceptable. Taking away this hot air might have prevented them to accept the specific negotiated emission targets (e.g. Baumert et al., 1999) and may prevent them to ratify the Kyoto Protocol (e.g. Bohm, 1999).

In a simple rational choice model, Boom (2000b) demonstrates that the US has only been willing to accept a relatively stringent cap on its emissions in the expectation that the emission reduction could be implemented in Central and Eastern Europe by way of emissions trading that includes hot air (which of course is also in the interest of the Russian Federation and the Ukraine as potential sellers). The analysis also reveals that if the EU had blocked the allocation of hot air to Central and Eastern Europe, the US would have committed itself to a much less stringent emission ceiling which might even have prevented an agreement in 1997 in Kyoto in the first place. Importantly, from an environmental perspective, when hot air would not have been allocated, this model points at the possibility of a less stringent US emission cap that exceeds the volume of hot air in the Kyoto Protocol.

This ex ante perspective on the negotiations (in which the targets are not seen as given) would seem to suggest that hot air trading is more an opportunity than a problem in establishing emission ceilings. This is true for the decision-making stage, but once trading and hot air are established to ensure that countries accept certain emission targets, some actors may try to block the implementation of hot air trading. This actually happened, not only in the form of green NGO opposition, but also in the form of the EU proposal to limit hot air by placing a quantitative restriction on the use of the Kyoto Mechanisms (SBSTA/SBI, 2000). This shows that the ethical interpretation of the effectiveness of hot air trading might resurface in the implementation stage after the targets have been set. Hot air is an institutional feature that becomes an institutional barrier to get emissions trading functioning once actors start to block its implementation.

Another dynamic aspect of hot air is that banking encloses (or "institutionalizes") the initially negotiated hot air permanently into the trading system, since banking allows for the transfer of unused hot air to future commitment periods. Under the Kyoto Protocol, for instance, the carry-over of hot air is possible on the basis of Protocol Article 3.13 stating that the emissions of an Annex B Party which are less than its assigned amount in a commitment period can be added to its assigned amount for subsequent commitment periods. While the carry-over of ERUs and CERs is restricted to 2.5% of the assigned amount, banking to the next commitment period is unrestricted for any AAUs held by an Annex B Party in its national registry which have not been retired or cancelled (CP, 2001b: 61).

Neoclassical economic analyses either neglect institutional features such as hot air (e.g. Montgomery, 1972), or recognize such features but see hot air as an unproblematic allocation aspect that neither affects efficiency nor formal effectiveness (e.g. Tietenberg et al., 1999). From that perspective, it can be defended that trading, and thus also hot air trading, would generate a price per tonne of CO_2, which provides the Russian Federation and the Ukraine with an incentive to reduce emissions if the associated costs are below the market price. The environmental issue of hot air becomes unquestionable once it is redefined as an allocation issue.

However, some policy makers, also in Europe, associated and confused emissions trading (that makes effectiveness cheaper and easier to achieve) with hot air trading (that undermines effectiveness, not from a formal and dynamic perspective, but from an ethical and static perspective by making emissions higher than without trading). If such a perception takes hold, the credibility of emissions trading is undermined (e.g. Butzengeiger et al., 2001), which may hinder the implementation of such a scheme. Environmental problem or not, depending on the perspective taken, several options have been proposed to reduce hot air trading.

4.3.3. Options to Limit Hot Air Trading

Various options have been proposed in the literature to cope with the hot air problem in emissions trading schemes, mainly in the context of the Kyoto Protocol. We will discuss the advantages and disadvantages of seven of them:

- renegotiating the targets;
- transaction tax;
- pre-budget banking;
- quantitative restriction on trading;
- hot air purchase and retirement;
- excess emission reductions system;
- eligibility requirements.

First, at least in theory, it is possible to renegotiate the assigned amounts of the Russian Federation and the Ukraine. To reduce the hot air problem, their emission targets can be strengthened from stabilization to reduction commitments. However, the Russian Federation and the Ukraine would then not only lose the competitive advantage of being able to sell hot air, which they might also perceive as unfair (e.g. Morlot, 1998; Bashmakov, 1999), but potential buyers (such as the EU or Japan) would also lose some of the options to purchase relatively cheap assigned amounts from abroad. Given these mutual economic interests to preserve the hot air, a renegotiation of the targets for the Russian Federation and the Ukraine is unlikely to be politically acceptable.

Second, a transaction tax could be imposed solely on emissions trading with Central and Eastern European Parties in order to reduce the demand for hot air during the first commitment period (Zhang, 1998a). The tax rate could be imposed on the buyer side only, with zero (or low) rate for transactions between legal entities within the advanced OECD countries, but with a higher rate for transactions between them and legal entities in countries with economies in transition. According to Tietenberg et al. (1999), buyer countries' governments could use the revenues from the tax for several (environmental) purposes, including subsidizing technology transfer to developing countries, stimulating R&D investments in climate-friendly technologies or even retiring hot air allowances from the market. Although a transaction tax raises total costs of meeting the Kyoto commitments, it reduces demand for hot air and it would be less trade-restrictive than imposing a percentage limitation on the use of emissions trading (as was proposed by the EU at that time). However, it does provide a disincentive for cost reductions via trading, yielding a high probability of international political opposition.

Third, it is possible to include pre-budget banking of emission reductions which have been achieved only in industrialized countries (between 2000 and 2008), while explicitly excluding Central and Eastern European countries from this possibility (Zhang, 1998a). Not only are the latter likely to perceive this proposal as unfair because it only favors those countries that are relatively prosperous already, but the proposal would also neither limit the trading of hot air between 2008 and 2012 nor its carry-over to subsequent commitment periods. In addition, the pre-budget banking of (some) Annex B Parties' emission reductions between 2000 and 2008 would inflate the overall emission ceiling of the first commitment period.

Fourth, the EU (supported by countries such as China and India) proposed to limit hot air by placing a quantitative restriction on the use of the Kyoto Mechanisms (SBSTA/SBI, 2000). However, after fierce opposition from other industrialized countries, the EU gave up its proposal and accepted the unspecified requirement that domestic action shall be a "significant element" of Annex B countries' climate policy (CP, 2001a). If accepted, the trade restriction would probably have limited hot air to one-third of its potential magnitude (Baron et al., 1999). In addition, the hot air countries, such as the Russian Federation, still would have had the opportunity to use the hot air in the future by banking it to a subsequent commitment period on the basis of Article 3.13.

Fifth, to avoid these disadvantages, the EU could buy and retire the hot air from the market, while accepting unrestricted trading (Nentjes & Woerdman, 2000). Although the EU would obtain a Protocol without hot air and all (non-)Annex B Parties could gain from unrestricted trading, it would be difficult for the EU to predict the available amount of hot air and to decide how to provide the money for the acquisition and retirement of hot air. An option could be to (partly) finance it from the revenues of a transaction tax on all Kyoto mechanisms (similar to the current adaptation tax defined only for CDM transactions in Article 12.8), which would be less trade-restrictive than a quantitative ceiling on trade (as the EU had proposed earlier).

Sixth, another alternative to deal with the hot air issue was provided by Switzerland who proposed to only make those units of assigned amount eligible for transfer via Protocol Article 17 which are "backed up" by GHG emission reductions beyond business-as-usual emissions.[3] The practical disadvantage of the Swiss proposal is that it remains rather difficult to determine the business-as-usual scenario for a country. In addition, if such an "excess emissions reductions" system would be applied to emissions trading, the amount of excess reductions should have to be known beforehand in order not to frustrate the scope for early transfers under Article 17, which seems to be a difficult option as long as clear internationally agreed rules are absent on how (and by whom) the amount of such excess reductions will be determined.

Seventh, a more indirect policy option to possibly exclude the hot air is to demand that participants must satisfy certain eligibility criteria, for instance, with respect to accurate monitoring and adequate enforcement, before they are allowed to trade. Although this is usually advocated irrespective of the hot air problem in order to obtain a credible trading system (e.g. Tietenberg et al., 1999), it could de facto exclude the hot air countries like the Russian Federation and the Ukraine when they do not meet the requirements in the short term (before and/or during the first commitment period). However, a disadvantage for the potential buyers is that it would probably exclude those Annex B countries from the trading system that have the cheapest abatement options, while depriving some potential sellers of the possibility to reduce the costs of meeting their Kyoto commitments by means of trading.

These policy options to eliminate hot air demonstrate that each alternative contains one or more dilemmas to be solved or trade-offs to be made, for instance, between efficiency and equity or between effectiveness and acceptability.

---

[3] An example may clarify this proposal. Suppose a Party has been assigned with an amount of 100 units of GHG emissions per year during 2008–2012. According to its business-as-usual emissions scenario, the emissions turn out to be only 90 units per year. Under the proposal this Party could only transfer assigned amount units via IET if it reduces its emissions to a level lower than 90 units per year. The 10 units business-as-usual reduction would, in this system, not be eligible for trading.

In addition, none of the proposals are capable of completely eliminating the hot air from the Kyoto Protocol. Nevertheless, the withdrawal of the US from the Kyoto Protocol in March 2001 changed the game and seems to have increased the acceptance of hot air by the EU and the green NGO community as a "necessary evil" to keep Annex B Parties such as Japan and the Russian Federation on board of the Protocol. It also demonstrates that hot air was seen by some actors as an environmental "evil" in the first place, casting doubt, in their (ethical) view, as we have seen, on the superiority of permit trading.

4.3.4. Non-Compliance and Liability

Various environmental economics textbooks, largely based on neoclassical theory, postulated that "(…) under a permit scheme (…) there is, in principle, no problem in achieving the target" (Baumol & Oates, 1988: 178). The idea in this literature that effectiveness is guaranteed because permits are traded under an emission ceiling, giving the instrument superior environmental properties, is based on the often implicit assumption of perfect compliance (procedures). This assumption was questioned, however, by several decision makers in the international climate negotiations (Oberthür & Ott, 1999: 204).

Non-compliance by an Annex B Party to the Kyoto Protocol affects environmental effectiveness (in its formal interpretation). For that Party, the non-compliance procedures of the Marrakesh Accords of 2001 imply, among other things, a suspension of the eligibility to trade under Article 17 and a deduction of 1.3 times the amount of its excess emissions from a second commitment period. On the one hand, it is a remarkable achievement that these rules have been adopted, not only because penalties are rarely agreed upon or used by sovereign states, but also because compliance-incentive measures (such as penalties) are considered to be less politically acceptable than compliance-facilitating measures (like the requirement of publicly available emissions data) (e.g. Morlot, 1998). On the other hand, these non-compliance procedures are probably insufficient because a non-complying Party could try to strategically negotiate a higher emission budget for the second commitment period, or carry-over its excess emissions from one commitment period to another based on Article 3.13, or even withdraw from the Kyoto Protocol by following the exit provision under Article 27.

A central question is to what extent emissions trading helps or hinders to achieve compliance. In the context of the Kyoto Protocol, there are two opposing views. The first is that emissions trading improves compliance, since it lowers emission reduction costs and therefore reduces the benefits of not complying. The second is that emissions trading deteriorates compliance, since it gives an Annex B Party the incentive to oversell assigned amounts beyond its emission budget.

To prevent the overselling of emission rights, policy makers and scientists, including neoclassical economists for that matter, began to consider the impact of liability rules, a classic law and economics theme (e.g. Calabresi & Melamed, 1972).

A basic distinction is made between seller liability and buyer liability. Seller liability means that any allowances acquired by the buyer are valid regardless of whether the seller is in compliance with its emission commitments. Buyer liability implies that the buyer is liable for the non-compliance of the seller. The advantage of adding buyer liability is not that it would solve the compliance problem, but that it strengthens compliance incentives. It discourages buyers (countries or firms) to purchase tons of emission reductions from countries that appear to be heading towards non-compliance. The disadvantage is that it would also erode the commodity nature of permits by allowing them to be retroactively devalued. This raises transaction costs by creating price uncertainty until the moment that compliance is checked. In this discussion, some authors argued that seller liability was also already in place under Article 17 and would suffice to combat non-compliance (e.g. Baron, 1999a; Tietenberg et al., 1999), whereas others argued in favor of some (limited) form of buyer liability (e.g. Zhang, 1998b; Michaelowa & Koch, 1999b).

Tietenberg et al. (1999) argue that the choice for buyer or seller liability should depend on the quality of enforcement. If this quality is high (as would be the ideal case), they prefer seller liability because of the efficiency advantages mentioned above. However, Nentjes & Klaassen (2004) add that the choice for buyer or seller liability should also depend on the governments' willingness to comply.[4] In their model, emissions trading does not change the compliance gap that would emerge without emissions trading, if buyers and sellers have an equally low propensity to comply. The sellers might be willing to oversell at a sufficiently high price, but no buyer is willing to pay that price. However, if the buyers have a higher propensity to comply than the sellers, the former are, in fact, willing to pay that price. Buyers then compensate the sellers for their subjective costs of being branded as unreliable or irresponsible Annex B Parties which do not sufficiently reduce their emissions to cover their permit sales. In that case, buyers will reduce their emissions less than they would have done without emissions trading by purchasing emission rights that are not covered by emission reductions of the seller.

If the enforcement system is weak and if the compliance culture is stronger developed in Annex B countries that are potential buyers, like the EU and Japan, than in Annex B countries that are potential sellers, like the Russian Federation and the Ukraine, which some believe to be the case, it would mean that emissions

---

[4] They define willingness to comply as the maximum marginal control cost a Party is willing to make. For a potential buyer it is the maximum price he would accept and for the potential seller its minimum price.

trading might indeed deteriorate compliance. To deal with this potential threat, it is possible to shift liability from the seller to the buyer, or to construct a hybrid seller/buyer liability arrangement (e.g. Zhang, 1998b, 1999a, 2001), but that turned out to be politically unacceptable so far. Seller liability is still in place. The fact that buyer liability raises transaction costs could explain this, as it would have made those governments in favor of an efficient carbon trading scheme less willing to ratify the Kyoto Protocol (in particular after the withdrawal of the US from the Protocol in March 2001).

The aforementioned formal and informal institutions shed a different light on the superior effectiveness properties of permit trading. They gave rise to some kind of "relative fitness problem" by making the instrument look effective in some respects, for instance, considering its emission ceiling, but not in all respects, for instance, considering the issues of hot air, non-compliance and liability. These issues, which were surrounded by uncertainty and absence of consensus, even contributed to developing a certain "mistrust" against permit trading among some decision makers, for instance, in Europe (Oberthür & Ott, 1999: 189–190). The complex institutional conditions for an effective tradeable emission rights scheme not only eroded its presumed environmental superiority, but also strengthened the perception that "(…) the institutional set-up of such a mechanism (…) will be a tremendous task (…)" (Oberthür & Ott, 1999: 204). Institutional considerations, both formal and informal, regarding its effectiveness made decision makers reluctant to switch to permit trading.

4.4. Environmental Effectiveness of Project-Based Emissions Trading

Under the Kyoto Protocol, for instance, the international cost-savings potential could be tapped, in principle, by means of JI and CDM projects (e.g. Pearce, 1995), but the general textbook prefers an (international) permit trading system to reduce GHG emissions (e.g. Tietenberg et al., 1999). JI and the CDM face the micro-baseline problem of measuring emission reductions by estimating future emissions at the project site as if the project had not taken place (e.g. Woerdman et al., 2003). The emission reduction is calculated as the difference between these micro-baseline emissions and the (lower) emissions measured during the project's lifetime. The micro-baseline of a JI or CDM project is constructed on the basis of reasonable assumptions about specific project-type related features and future local (as well as relevant regional and national) developments with respect to, for instance, economic growth, energy use and available technology.

Traditional analyses in environmental economics are right to conclude that project-based emissions trading does not guarantee effectiveness on an (inter)-national level. For instance, if future emissions are overestimated by setting the baseline too high, emission reductions will be credited that have probably not occurred. However, some neoclassical writings on this subject do tend to neglect or underestimate the institutional opportunities to improve the effectiveness of credit-based approaches. These opportunities, which will be discussed in the following subsections, put the environmental inferiority of JI and CDM projects in a different (less negative) perspective. To some extent, this also helps to explain why politicians are tempted to use credit-based approaches, building upon existing environmental policy, instead of switching to permit trading. We will start by paying some more attention to the baseline issue of project-based emissions trading.

4.4.1. Micro-Baseline, Free-Riding and Gaming

Under certainty, the micro-baseline emissions, also referred to as reference scenario or simply "baseline", correspond to what the emissions would have been at the project site in the absence of the project. However, under uncertainty, which is present in the real world, baseline emissions have to be estimated. Because it is an estimate, the environmental accuracy of the project may not be perfect. Moreover, the baseline itself already describes a situation that will never exist because of the project. Therefore, the baseline can be characterized, for instance, as counterfactual (e.g. Jepma et al., 1998), hypothetical (e.g. Tietenberg et al., 1999), virtual (e.g. Matsuo, 1999), never happening (e.g. Ellis, 1999a) and therefore unobservable (e.g. Chomitz, 1999). This creates an ex ante uncertainty with respect to the environmental effectiveness of JI and CDM projects.

The baseline is constructed on the basis of default values of several key parameters, such as expected (and current) future fuel and electricity prices, pollution charges or regulations, and capital costs or target rates of return (Chomitz, 1999). Different project types require different data sets, such as fuel emission factors and combustion efficiency for energy sector projects, or carbon density of the land and the time lag between planting and sequestration for biotic projects (Ellis, 1999a).

It is imaginable to derive several different ex ante baselines on the basis of seemingly reasonable arguments (e.g. Jepma, 1999b). Ellis (1999a), for instance, showed that it is possible to rationalize various baselines for the Swedish-financed Daugavriva boiler conversion project (from gas/diesel oil to biomass). Different assumptions concerning this AIJ project's additionality period and lifetime as well as future energy demand appeared to yield a range of projected GHG mitigation between 130 and 477 kt CO_2.

The leakage problem of indirect effects, or systems boundary issue, should also be taken into account when constructing a micro-baseline (e.g. Michaelowa, 1998a). In the context of the CDM (CP, 2001b: 37), leakage is officially defined as the net change of anthropogenic emissions by sources of greenhouse gases which occurs outside the project boundary and which is measurable and attributable to the project. The project boundary is defined as all anthropogenic emissions by sources of greenhouse gases under the control of the project participants that are significant and reasonably attributable to the project. An example of such indirect effects are "snapback" price effects. For instance, if carbon-rich fuels are largely substituted by low-carbon fuels, the price of the former falls, which provides an incentive for greater use of carbon-rich fuels causing less emission reduction than initially foreseen.

By definition, different baselines lead to different amounts of credited emission reductions. If future emissions are overestimated by setting the baseline too high (i.e. higher than what actually would have happened in the absence of the project), emission reductions will be credited that have not in fact occurred. This is called the additionality problem. In that case, if actual emissions of the project equal planned emissions, environmental effectiveness is achieved at the project level in its formal interpretation, but not in its ethical interpretation. Overstated baselines also divert rents away from projects with relatively accurate baselines. In addition, a Party that systematically grants too many credits for JI projects could find it difficult, in the end, to realize its Kyoto emission target, so that overcrediting at the project level threatens effectiveness at the national level. A credited *non-additional* GHG emission reduction project is sometimes referred to as free-riding (e.g. Chomitz, 1999; Ellis & Bosi, 1999).[5]

It could be argued that baseline determination is less problematic for JI than for the CDM. A CDM host country does not have an emission ceiling, but the host country of a JI project is an Annex B Party that has committed itself to an assigned amount. First, this implies that a JI host country has a stronger incentive than a CDM host country *not* to overstate individual project baselines, since it becomes increasingly difficult for an Annex B host country to achieve its own target as it sells more ERUs. Second, to ensure compliance with its assigned amount, a JI host country has to define environmental policy targets for its domestic emitters. If it has done so, the JI baseline to calculate the additional emission reductions could be derived from the existing environmental policy for the host firm or sector involved. In that case, this not only could, but also should be done because the CoP requires (among other things) that the project-specific JI baseline shall take into

---

[5] In principle, the reverse also holds: if future emissions are underestimated by setting the baseline too low, too little emission reductions will be credited (e.g. Heller, 1999). If this would occur systematically, it might turn out that the Party's actual emissions over the commitment period have remained substantially below its emission target.

account relevant national and/or sectoral policies (CP, 2001b: 18).[6] In this way, project-based emissions trading can build upon existing (possibly sub-optimal) environmental regulation, if present, without having to switch to the more efficient system of permit trading.

However, there is still a risk of a "micro–macro mismatch" for JI baseline determination. Baseline inflation is not in the interest of both Parties under the double bookkeeping of assigned amounts, but deriving baselines from existing environmental policy does not remove the incentive for project developers at the micro-level to inflate the baseline in order to claim more credits (Jepma, 1999b). The incentive on the project level to exaggerate baseline emissions is called gaming (e.g. Fisher et al., 1996; Jepma, 1999a; Matsuo, 1999). Exaggerating would increase both the amount of credits for the investor and the amount of money received by the host partner. The climate would lose, since an inflated baseline represents, for some part, fake emission reductions, depending on (as well as reflecting) the degree of inflation. If the investors of JI projects would succeed in claiming too many emission reductions, the host countries would have to compensate for this at home later on.

Therefore, it goes too far to say that "(...) additionality largely disappears as a concern in a capped emissions system" (Trexler & Gibbons, 1999: 126) or that "(...) an emissions ceiling provides confidence that emissions reductions corresponding to the AAUs sold will be made somewhere in the host country" (Hargrave et al., 1999a: 98). The government must demand that project developers report their baselines, so that the administration can collect and check those baselines to provide some sort of countervailing power against gaming at the project level. Nevertheless, apart from committing fraud, inflating the baselines is difficult when they are deducted from, say, performance standards and other verifiable figures such as production volume or energy use.

CDM projects must always be validated, verified and certified by operational entities accredited by the Executive Board. As an institutional safeguard against baseline inflation in JI projects, however, the emission reductions must be verified by the host country itself or by an independent entity if the host country does not meet certain pre-defined eligibility requirements. This two-track approach for JI works as follows (CP, 2001b: 13). If a JI host country, in accordance with Protocol Articles 5 and 7, has a national system to estimate and register emissions and annually submits a national inventory report, a host Party may verify reductions as being additional. If a host Party does not meet these eligibility requirements, verification shall occur by an independent entity accredited by the so-called

---

[6] For instance, if the policy in a JI host country is a relative standard that requires a certain quantity of CO_2 per unit of output or energy, the baseline emissions for the host firm can be calculated by multiplying this standard with its expected production volume. If the host firm emits less CO_2 than this baseline figure because a JI project is implemented, emission reductions are achieved for which the investor can obtain ERUs.

Article 6 Supervisory Committee. The "slow" track for JI projects was created to maximize the environmental integrity of the project where the host Party has a weak emission registration system. The "fast" track for JI projects was created to keep the cost of implementing a project low for host countries with reliable emission registration systems. The emission reductions must be verified carefully in order to prevent the host country from transferring too many credits to investors and from running the risk of being in non-compliance.

Gaming (baseline inflation) affects the level of projects' baselines, free-riding (crediting non-additional projects) affects the number of projects credited, whereas leakage (indirect environmental effects) affects the projects' overall environmental impact (Ellis & Bosi, 1999). The main institutional issue with respect to JI and CDM baselines, then, is to find a methodology that balances between maximizing the environmental integrity of a project and minimizing its transaction costs. For example, aiming at maximizing the environmental integrity by collecting as much information as possible about a project's reference case may inhibit a project's cost-effectiveness. Flexible procedures for baselines in order to minimize transaction costs may result in (certified) emission reduction units that in reality do not take place. Although a strict third party verification procedure could reduce the latter effect, it would increase transaction costs again, which would probably be shifted to the project developers. This shows that baseline determination — although the projects will certainly lower the costs of climate policy — still incorporates some trade-off between cost reductions and environmental integrity.

4.4.2. Ex Post Corrections of the Micro-Baseline

An institutional opportunity to improve the effectiveness of project-based emissions trading is the application of ex post baseline corrections. This means that baselines are adjusted during the project if the actual circumstances deviate significantly from the ex ante baseline assumptions. The advantage is that it increases the likelihood that generated credits are based on real emission reductions. The disadvantage is that it raises transaction costs by magnifying uncertainty about the amount of credits the project will generate.

In many existing (AIJ) projects, additionality is based on a static (ex ante) baselines that do not change during the implementation of the project. This means that its environmental effectiveness is affected in the case that the ex ante assumptions and actual developments diverge. Estimates of the associated baseline uncertainty range from 25% to as much as 60% (Begg & Parkinson, 2001). Therefore, in response to the alleged baseline problem of low or uncertain environmental effectiveness, some authors have proposed to allow for ex post baseline corrections (e.g. Begg et al., 1999; Ellis & Bosi, 1999). Dynamic baselines can be adjusted during the project if the actual circumstances deviate significantly from the ex ante baseline assumptions, for example, those on economic growth, energy use or available technology. Such adjustable baselines would enhance the environmental effectiveness of GHG mitigation projects, since it increases the likelihood that generated credits are based on real emission reductions (Jepma et al., 1998). The baseline parameters could be updated, for instance, annually (e.g. Ellis, 1999a: 24) or after some years (e.g. Begg et al., 1999: 180).

However, the choice between static and dynamic baselines, again, involves some trade-off between economy and environment. Dynamic baselines are more effective, but they hinder trading. It is not only more complex than a static counterfactual (Aslam, 1999), but a dynamic baseline also magnifies investment uncertainty (Michaelowa, 1998a). When static baselines are used, the baseline emissions are certain, although it is uncertain how high the project's actual emissions will turn out to be. When dynamic baselines are used, not only the measurable emissions, but also the baseline emissions themselves are subject to uncertainty (Sutter et al., 2001). The consequence is that it might scare off investors, so that there even may be no emission reduction project at all (e.g. EcoSecurities, 2000: 6). Nevertheless, Chomitz (1999) argues that risks can be reduced by tying dynamic baselines to easily observable variables, such as load factors, exchange rates, central bank interest rates and fuel prices.

To provide a simple illustration of the functioning of an ex post baseline in an emission reduction project, consider the following hypothetical example. Suppose that Germany finances a coal gasification CDM project in India to start in 2000 with a duration period of 10 years. Presume that the baseline is decided to be the estimated average of GHG emissions from Indian power plants, which will not participate in CDM projects, from 2001 to 2010. However, in 2006 the actual average emissions from Indian power plants turn out to be lower than the baseline estimate for 2006 because technological innovation in India developed faster than anticipated. In that case, the project baseline will be lowered in 2006 to the actual average emission level of Indian power plants. This improves the project's environmental integrity, but it also means that the project will generate less emission reduction credits than foreseen at its start.

There are, in fact, some real-life precedents of ex post baselines. Jepma et al. (1998), for instance, describe the Costa Rican Protected Areas Project in which 15.6 million tons of carbon equivalent is sequestered by protecting a forested area of 530,000 ha from being cut down. A buffer of 700,000 tons of carbon equivalent has been created to be able, among other things, to adjust the baseline during this forest conservation AIJ project. Another example is provided by Sutter et al. (2001) who describe the Swiss-Romanian Thermal Energy Project (STEP).

This AIJ project, which aims at reconstructing two distric heating systems, includes a provision for baseline revision if heat production changes significantly, for instance due to new consumers of heat (like new buildings) being added to the distribution network.

Despite such practical examples, the issue of ex post baseline corrections seems to be largely off the negotiating table because adjustable baselines raise transaction costs. For that reason, incorporating them in international accords would have made countries such as Japan, Canada and the Russian Federation less willing to ratify the Kyoto Protocol (in particular after the withdrawal of the US from the Protocol in March 2001). Although some argue that the absence of adjustable baselines in the Marrakesh Accords of 2001 seems to leave open this possibility, for instance, for CDM projects (JIQ, 2002: 3), the most important point for our chapter is that ex post baseline corrections simply exist as an institutional option to improve the environmental effectiveness of project-based emissions trading.

4.4.3. Standardization of the Micro-Baseline

Another institutional opportunity to improve the effectiveness of project-based emissions trading is baseline standardization. Under standardized baselines, project partners have fewer possibilities to claim more credits by inflating baseline emissions. This option turned out to be more politically acceptable than the option of baseline corrections. The reason for this is that baseline standardization not only strengthens environmental integrity, but also reduces transaction costs because it will not be necessary anymore to construct a baseline for each individual project. Moreover, an element of ex post baseline corrections can be introduced by verifying the standardized baselines after some period and adjust them on environmental grounds if necessary.

In most existing (AIJ) projects, a baseline is constructed for each individual project. The baseline of a JI project could (but need not) be verified by an independent entity if the host Party meets the eligibility requirements on reporting and accounting of emissions defined under Articles 5 and 7 of the Protocol (CP, 2001b). The baseline must be verified by such an entity (accredited by the so-called Article 6 Supervisory Committee) if the host Party does not meet these requirements. CDM projects must always be validated, verified and certified by operational entities (accredited by the Executive Board). In the 1990s, about 75% of the AIJ pilot projects applied third party baseline assessment (Schwarze, 1998).

However, in order to reduce gaming incentives and transaction costs, and soften the trade-off between cost reductions and environmental integrity, several authors have proposed to standardize baseline determination procedures (e.g. Jepma et al., 1998; Kerr, 1998; Hargrave et al., 1999a). Baseline standardization implies the development of "business-as-usual" scenarios for project categories differentiated by, for instance, region, time, project and/or technology type. These scenarios could be determined by a panel of experts. Each specific project should fit into one of the categories. The (ex ante or ex post) emission reductions of a specific project can simply be calculated by subtracting the (predicted or observed) emissions from the baseline emissions of the relevant category. This makes a third party check for each baseline much easier or even unnecessary.

Transaction costs are reduced because it will not be necessary anymore to invest time and effort in constructing a baseline for each and every project. In principle, standardization may have a negative effect on environmental effectiveness if the baselines in these categories, which remain "best guesses", are set at a level so (low or) high, that they lead to systematic (under- or) overestimating the emission reduction in projects. This bias is referred to, for instance, as "gaming at the system level" (Chomitz, 1999: 91/2) or "systematic error" (Trexler & Gibbons, 1999: 134). Therefore, an element of ex post baseline corrections could be introduced in the standardization approach. With a view to the environmental effectiveness of JI and CDM projects, the standardized baselines in the categories have to be verified after some period and adjusted if necessary, thereby responding, for instance, to technological and economic changes differing from the assumptions that form the basis of the original baseline categories.

There are three general methodologies for baseline standardization (e.g. Puhl, 1998; Hargrave et al., 1999a):

- the matrix approach;
- the benchmarking approach;
- the top-down approach.

First, the matrix approach places pre-defined default scenarios for several project categories into a matrix. The investor and the host of a JI or CDM project look up the baseline in the matrix, for example, available on the FCCC Internet homepage, to calculate the credits that will accrue from the project. There are various possibilities to define the dimensions (the rows and columns) of such a matrix. Jepma et al. (1998) propose to adopt default project/technology baselines with a possible differentiation by country or region. Hargrave et al. (1999a) suggest to define the baseline technologies not only for certain regions as well as project and sector types, but also for a specified time. The baseline for a real-life project would equal the emission level for the specified technology. Projects that introduce technologies with lower emissions than the specified baseline technology are considered to meet the additionality requirements. In this design, project participants have the possibility to demand an ad hoc adjustment of the baseline for their particular case with a view to some unique circumstances, but they have to bear the costs of such a procedure, while facing the risk of losing the appeal.

Puhl (1998) also mentions the possibility to define a standard baseline for narrow technology categories that would automatically qualify as additional, such as wind and solar generation projects (see also Trexler & Gibbons, 1999: 131). Moreover, in principle, certain project types can also be designated as non-additional in advance, thereby excluding them from the matrix, such as (certain types of) forestry projects (e.g. Cullet & Kameri-Mbote, 1998).

In all variants of the matrix approach, the matrix could be updated periodically by adding certain technologies to the categories, for instance, those which have reached a certain threshold share in a countries' technology inventory (Hargrave et al., 1999a), so that these technologies are no longer considered to be additional. In theory, the updated matrix could be used retroactively to change existing project baselines. Although the option of ex post corrections is desirable from the perspective of environmental effectiveness, it also increases credit output uncertainty, which affects its political acceptability.

Second, the benchmarking approach, as proposed by Trexler & Gibbons (1999), uses emission performance "benchmark" rates, for instance, determined on the basis of historic or projected sector-specific emission intensity trends, to calculate project emission baselines. Benchmarks define standard emission factors for a certain project in a particular host country. These factors could be derived, for instance, from default projects (Luhmann et al., 1995), recent historical country/sector/fuel averages or recent/future marginal technology as proxies of most likely investment. Like the matrix approach, all JI or CDM projects that reduce emissions below the benchmark levels would automatically qualify as additional and generate credits. However, unlike the matrix approach, benchmark baselines can be based on a mix of technologies rather than a specific technology, and benchmarks can be forward-looking based on projected technologies rather than the current capital stock (Hargrave et al., 1999a).

Benchmarks can be established on the level of either projects, regions or countries, possibly differentiated by sector, either based on projected or historic emissions. Creating benchmarks for more sectors and project types not only increases development costs, but also enhances its environmental effectiveness. Benchmarks need not be static, set as a constant over the project's lifetime, but can also be dynamic, changing periodically over the project's lifetime.

Third, top-down baselines are project-specific micro-baselines that have been derived from national or sectoral macro-baselines.[7] The micro-baselines would be

---

[7] National macro-baselines are usually referred to as top-down baselines, whereas (sub-)sectoral macro-baselines are sometimes also referred to as multi-project baselines (e.g. Ellis & Bosi, 1999).

established by allocating the aggregate macro-baseline to individual project activities. Although there is a clear analogy with emissions trading where an emission ceiling could be divided into smaller permits to be allocated to domestic emitters, top-down baselines are still baselines, which do not (necessarily) allocate absolute emission ceilings. These baselines could be moving with economic growth, for instance, if they are defined as the multiplication of some energy-efficiency standard with the production volume of firms. Under a permit trading scheme, however, the maximum allowable emission level for firms would be fixed, also in the case of economic growth.

In the case of JI, the macro-baseline could be the national assigned amount of an Annex B Party. The micro-baseline for a particular project could be the relevant national environmental policy (such as standards, taxes, benchmarking or covenants) as planned to comply with this emission ceiling. This would also reflect the requirement that a project-specific JI baseline shall take into account relevant national and/or sectoral policies (CP, 2001b: 18). If legally binding emission standards are lacking, the assigned amount could be used as a basis to calculate the GHG emissions per unit of energy for its sectors and/or technologies at which its commitment would be achieved (Jepma et al., 1998).

For instance, suppose that the commitment of some Eastern European country can only be fulfilled, as part of a set of measures, if the CO_2 emissions per unit of energy produced in the power sector would be reduced by 15% on average. In this hypothetical example, -15% would constitute the micro-baseline for JI projects in the power sector only. If such a project would generate an emission reduction of 20%, 5% (considered to be additional) would then be credited. Although the top-down approach can be applied in JI projects, the political will and administrative capacity for the approach may be lacking in the host countries. Rather, which should not be forgotten, project-based emissions trading "(…) is a device for avoiding the difficulties of setting the sectoral or national caps" (Chomitz, 1999: 24).

The top-down approach is even more problematic for CDM projects. In the case of the CDM, the macro-baseline cannot be the assigned amount, since developing countries do not have emission ceilings under the Kyoto Protocol. Instead, it is possible that developing countries construct non-binding simulated targets, or even adopt internationally binding "growth targets" (e.g. Baker & Barrett, 1999), for instance, measured in GHG emissions per unit of output rather than in terms of absolute emissions. However, there is a considerable risk that the macro-baseline will be set higher than business-as-usual predictions, leading to baseline inflation on both macro- and micro-level.

Although top-down baselines for developing countries would reduce transaction costs and thus increase the level of investment activity, they also imply capacity building and upfront costs of macro-baseline determination and allocation. Furthermore, emitters do not necessarily face sanctions if they exceed non-binding top-down baselines. Although it is possible to decide that CERs can only be used by the buyer if all sources under the macro-baseline have met their micro-baselines (individually or aggregate), this rule would require domestic penalties on sources that produce emissions above baseline levels, which in effect would convert the micro-baselines into binding targets.

Following Jepma (1999b), the different baseline methodologies could be summarized as follows: the project-based approach is micro-based and focuses on what would have happened without the project, the standardized matrix and benchmark approaches are meso-based and focus on what could have happened without the project, and the top-down approach is macro-based and focuses on what should have happened without the project. From the perspective of environmental effectiveness, baseline standardization is a desirable institutional opportunity, since it reduces baseline inflation.

Standardizing baselines is politically more acceptable than adjusting baselines because the former reduces transaction costs, whereas the latter raises transaction costs. For this reason, contrary to ex post corrections, baseline standardization appeared, for instance, in the text of the Marrakesh Accords of 2001. The CDM Executive Board shall develop and recommend to the CoP the "(…) appropriate level of standardization of [baseline] methodologies (…). Standardization should be conservative in order to prevent any overestimation of reductions in anthropogenic emissions" (CP, 2001b: 46).

Within the set of standardization options, the top-down baseline approach is preferable for projects in JI host countries that should develop environmental policy for domestic sources anyway to comply with their assigned amounts. Top-down baselines are not feasible for CDM projects in developing countries without such policy or in JI host countries where environmental policy has not yet matured: in those cases, the matrix approach (either or not with benchmarking) is a transparent and cost-effective alternative. Permit trading, however, remains to be the superior institutional arrangement because the emission transfers occur under the emission ceilings of individual polluters. Nevertheless, the options of baseline standardization and baseline corrections show that there are institutional opportunities for project-based emissions trading to narrow the effectiveness gap with a tradeable emission rights system.

4.5. Conclusion

Neoclassical analysis demonstrates that permit trading is environmentally effective because emission sources trade under absolute emission ceilings. The effectiveness is uncertain for instruments without emission ceilings, such as credit trading or emission reduction projects, as explained in earlier chapters. In this chapter, we have largely confirmed, but also nuanced this traditional view in environmental economics on the basis of institutional considerations.

Just like Liebowitz & Margolis (2000) clarified the survival of the QWERTY-keyboard by contesting the superiority of its alternatives, it helps to explain the inclination of policy makers to add sub-optimal designs to the existing environmental policy framework by nuancing the effectiveness advantages of permit trading as well as the presumed ineffectiveness of credit-based approaches. The same type of arguments applies as in the technological lock-in literature (discussed in the previous chapter), namely an absence of consensus, relative fitness problems and uncertainty. This caused the superiority of permit trading to be contested in politics.

First, the most important effectiveness problem of permit trading (and government trading) was considered to be that of hot air trading. A country (or firm) has hot air if its business-as-usual emissions remain below its official emission ceiling. Emission rights can then be traded and used to cover emissions that might have remained unused without emissions trading. Many emissions trading "blueprints" assume that emission ceilings are set lower than business-as-usual emissions (e.g. Baumol & Oates, 1988; Anderson et al., 1999). However, decision makers had information that this assumption, with some uncertainty, may not be met under the Kyoto Protocol, for instance, regarding the Russian Federation.

The emission ceilings are still respected when the hot air is traded, so that effectiveness is achieved in its formal interpretation. However, hot air trading does disturb effectiveness in (what might be called) its ethical interpretation because it can make overall emissions higher with than without emissions trading. Without hot air trading, the actual emissions of all emission sources together could have been lower than the overall target. Permit trading was considered environmentally "fit" in some (formal), but not all (ethical) respects, which helps to explain the institutional lock-in. The absence of consensus about whether hot air trading is an environmental problem or not made permit trading look suspicious instead of superior. It actually became an institutional barrier to get emissions trading functioning, in particular when the EU proposed, after the targets were negotiated, to limit hot air by restricting trading.

Second, politicians were confronted with a growing literature about the institutional opportunities to improve the effectiveness of credit-based approaches, like JI and the CDM. Project-based emissions trading faces the baseline problem of estimating future emissions at the project site in the absence of the project. Because the baseline is a counterfactual (that can be set too high), effectiveness is uncertain, but ex post baseline corrections and baseline standardization make such projects less inferior than some contend. Correcting the baseline ex post means that the baseline is adjusted during the project if the actual circumstances deviate significantly from the ex ante baseline assumptions. This environmental option turned out to be politically unacceptable because it raises transaction costs by magnifying uncertainty about the amount of credits that the project will generate. Standardizing baselines means that business-as-usual scenarios are developed for several project types and regions, so that project partners have fewer possibilities to claim more credits by inflating baseline emissions. This appeared to be politically acceptable because it also reduces transaction costs as it will not be necessary anymore to construct a baseline for each individual project.

Our analysis shows that formal and informal institutional factors, including equity concerns, gave rise to some kind of relative fitness problem by making permit trading, which is environmentally superior considering its use of emission ceilings, look ineffective in some respects, for instance, considering the allocation issue of hot air trading. Moreover, project-based flexibility was perceived to be more institutionally "fit" than contended in the neoclassical economic hierarchy of market-based climate policy when it became clear that project baselines could be enhanced and that baselines could be derived from existing environmental regulation (if present). These issues, which were surrounded by uncertainty and ambiguity, made decision makers reluctant to switch to permit trading. In addition, this superior alternative would involve relatively high set-up costs. The transaction costs to set up and to operate within such markets will receive full attention in the next chapter.

Chapter 5

Transaction Costs of Market-Based Climate Policy

5.1. Introduction

After studying efficiency and effectiveness in the previous chapters, this chapter examines the transaction costs of different types of market-based climate policy instruments and provides an assessment of the empirical literature on this traditional new institutional economics topic, while extending the analysis with a political transaction cost comparison.

According to Kerr & Maré (1997) and Krutilla (1999), many simulation studies that calculate the efficiency gains of emissions trading ignore the possible effect of transaction costs. Two more or less recent examples of studies that ignore such costs are the modeling exercises by Sijm et al. (2000) and Ciorba et al. (2001). However, several authors emphasize that transaction costs play a key role in the success of a tradeable permit or credit scheme (e.g. Hahn & Hester, 1989; Tietenberg et al., 1999). Neoclassical economists usually argue that permit trading has lower transaction costs than credit-based approaches because credit transfers require pre-approval. However, they do not systematically compare the political transaction costs to set up those market instruments. This chapter not only studies these set-up costs, but also nuances the transaction cost advantages of permit trading and the presumed disadvantages of project-based trading on the basis of institutional considerations, just like Liebowitz & Margolis (2000) clarified the survival of the QWERTY-keyboard by questioning the superiority of its alternatives. Our comparative analysis of (political) transaction costs helps to explain why environmental regulation is path dependent and why politicians are tempted to add sub-optimal designs to extant policy.

This chapter is organized as follows. Section 5.2 defines the concept of transaction costs. Section 5.3 considers the asymmetrical "model versus muddle" assumption behind some transaction cost studies that compare permit trading with credit-based instruments. Section 5.4 examines and nuances the transaction cost advantages of tradeable emission rights, for instance, by considering institutional barriers that arise from incremental design and set-up costs, and gives an overview of empirical transaction cost figures in existing permit trading markets. Section 5.5 analyzes the institutional opportunities to lower the transaction costs of project-based emissions trading, including JI and the CDM, for instance, by considering baseline standardization and multilateral funds, and considers empirical evidence of transaction costs in AIJ pilot phase projects. Section 5.6 discusses the methodological problems of comparing the transaction costs between different types of flexible instruments as well as of comparing different types of transaction costs. Finally, Section 5.7 presents the conclusion.

5.2. Definition of Transaction Costs

Already in 1969 Arrow wrote: "The identification of transaction costs in different contexts and under different systems of resource allocation should be a major item on the research agenda of (...) the theory of resource allocation in general" (Arrow, 1969: 48). In those years, a new branch of institutional economic research emerged, referred to as transaction cost economics (TCE), which is usually associated with Coase (1960) and Williamson (1975). Although it took some time for the literature on economic instruments for environmental protection to recognize its relevance, there is now a growing awareness of the importance of transaction costs (e.g. Bressers & Huitema, 1999), which is largely triggered by some difficulties that economists observed to get emissions trading accepted and running smoothly.

Although emissions trading lowers the costs of climate change mitigation, transaction costs may reduce its cost-effectiveness. In this chapter we distinguish between two types of transaction costs:

- market transaction costs;
- political transaction costs.

First, from a new institutional economics perspective that builds upon neoclassical assumptions, Stavins (1995) defines transaction costs as the difference between the buying and selling price of a commodity in a given market. According to Stavins, transaction costs are generally ubiquitous in market economies, since parties to transfers (for instance, in property rights such as tradeable emission permits) must find one another, communicate and exchange information. Furubotn & Richter (1997) prefer to refer to this type of costs as market transaction costs, whereas we use the term ex post transaction costs introduced by Vollebergh (1994) to reflect the idea that these are the costs of transferring property rights between parties in a market after (ex post) this market has been set up by politicians.[1]

Following Coase (1960), Dudek & Wiener (1996) explain that market transaction costs typically consist of search costs, negotiation costs, approval costs, monitoring costs, enforcement costs and insurance costs. The transaction costs of monitoring emissions and enforcing the environmental policy are usually borne by the government. Mullins & Baron (1997) subdivide transaction costs into direct costs (e.g. the money spent to initiate and complete a trade) and opportunity costs (e.g. the loss of time and resources through delay and managerial attention). Furubotn & Richter (1997) subdivide transaction costs into fixed costs and variable costs of which only the latter depend on the number or volume of transactions.

As transaction costs increase, the price received by sellers is depressed relative to the price paid by purchasers (Stavins, 1995). Trade will be profitable only if the exchange rate adjusted credit or permit prices differ more than the transaction costs incurred of transferring the credit or permit (e.g. Hinchy et al., 1998). Transaction costs, which are likely to be highest in the initial phases of a pollution market, may greatly reduce the cost savings that are potentially achievable by reducing the number of trades that are made (e.g. Jackson, 1995; Pearce, 1995; Mullins & Baron, 1997). Due to transaction costs, the degree of utilization of market-based climate policy will be reduced (e.g. Michaelowa & Stronzik, 2002).

Second, from a neo-institutional economics perspective, transaction costs do not only originate when property rights are transferred between parties in a market that is functioning, but transaction costs are also incurred, in a broader sense, when these rights are created (or protected) through political, administrative or judicial decisions (e.g. Krutilla, 1999; Allen, 2000). This reflects the idea that "(...) property rights themselves are costly (sometimes too costly) to impose (...)" (Cole, 2000: 306). Therefore, according to North (1990: 28, 61), transaction costs not only consist of the costs to protect property rights and enforce agreements, but also of the costs to define those rights. In Chapter 3 we have referred to these costs as set-up costs. They are the costs involved in establishing or changing an institutional arrangement, which may require the creation or alteration of property rights. Furubotn & Richter (1997) as well as North (1990) refer to these costs as political transaction costs, whereas Vollebergh (1994) uses the term ex ante transaction costs to reflect the idea that these are the costs of setting up a market before (ex ante) this market is functioning.[2] Haddad & Palmisano (2001: 442) use the term development costs and Banuri et al. (2001: 52) classify them as

---
[1] However, Vollebergh (1994) used the term ex post transaction costs in a stricter sense to refer to the administrative costs of monitoring and enforcement.
[2] However, Vollebergh (1994) used the term ex ante transaction costs in a stricter sense to refer to the costs of obtaining information in the phase of designing economic instruments for environmental policy.

implementation costs, which include the costs of making changes in existing rules and regulations. In general, these costs rise as complexity increases.

The IPCC acknowledges that political transaction costs are usually not fully covered in the (more neoclassical than institutional) economic analyses of environmental policy instruments because they are "(...) different to those costs conventionally considered as transaction costs" (Banuri et al., 2001: 52). Nevertheless, Dixit (1996) and Williamson (1997a, 1999), for instance, acknowledge that the transaction cost approach can be fruitfully applied to public administration and politics, respectively. Haddad & Palmisano (2001: 442) as well as Janssen (2000) are one of the few authors (see also Eckersley, 1993) to explicitly recognize the costs of setting up market-based climate policy instruments.

Haddad & Palmisano (2001: 441) argue that permit trading has relatively high "development costs" because the associated emission ceilings are difficult to change, assuming that they are based on inalienable property rights. However, this assumption can be criticized because a tradeable permit is not so much permanent, private property right, but rather an authorization (or hybrid property right), which can be terminated or limited by the government, with an emission ceiling that declines each year (e.g. Tietenberg, 2002: 5). Where we do not fully agree with Haddad and Palmisano's argument, Janssen (2000) does not even explain what particular factors contribute to the set-up costs. In the next chapters, on the contrary, it is our purpose to analyze and discuss these factors in detail. Examples of set-up costs from the perspective of government and administration in the field of policy preparation are the costs of gathering and processing information, the costs of developing the required legal framework, the costs of (re)allocating property rights, and the costs of dealing with lobbying efforts and cultural resistance.[3]

In each of the following sections, we will first consider market transaction costs before we assess the level of political transaction costs. We start, however, by analyzing some weak spots in the argument that permit trading has superior transaction cost properties compared to project-based emissions trading.

5.3. Model Versus Muddle?

One element of the theoretical superiority of permit trading, next to its efficiency and effectiveness, is that it is thought to have lower transaction costs than the other

---

[3] Interestingly, although Dixit (1996) gives a broad definition of transaction costs and even formulates a transaction cost politics (TCP) perspective, he does not look at political transaction costs as such by "(...) taking for granted the existence of a governance structure that assigns initial rights and enforces (...) agreements to trade these rights" (Dixit, 1996: 37). Also Estache & Martimort (1999), who write about politics, transaction costs and the design of regulatory institutions, do not recognize the existence of political transaction costs.

flexible instruments. Neoclassical economists defend this opinion by arguing that transfers in a system of tradeable emission rights will be automatically registered and only have to be checked at the end of the year, while project-based approaches, such as JI and the CDM under the Kyoto Protocol, require advance approval of every single trade because of the baseline problem (discussed in the previous chapter) (e.g. Hahn & Stavins, 1999; Tietenberg et al., 1999; Vrolijk & Grubb, 2000).

Transaction cost theory would then expect that decision makers choose the instrument with lowest transaction costs, but politicians have actually started by setting up credit trading schemes and hesitated to implement permit trading, at least for some time (as in the case of the EU). It appears, however, that the transaction cost advantages of permit trading and the presumed disadvantages of project-based trading are less straightforward than the aforementioned traditional analyses have suggested if more institutional factors are brought into the analysis. Just like Liebowitz & Margolis (2000) clarified the survival of the QWERTY-keyboard by questioning the superiority of its alternatives, we largely confirm but also nuance the traditional hierarchy of flexible instruments for climate policy and extend the analysis by considering the political transaction costs of setting up such markets, which helps to explain why the superior alternative of permit trading is not readily accepted and implemented.

When writing about the cost-saving potential of emissions trading, Ingham (1992: 117) stresses that it is methodologically wrong to compare the costs of poorly designed emission standards with the costs of a perfectly designed tradeable permits. This argument can be extended to credit-based approaches. It is incorrect to compare the transaction costs of poorly designed (project-based) credit trading schemes with the transaction costs of perfectly designed permit trading systems. However, this is what often happens in the aforementioned studies: the traditional view that permit trading does not require advance approval of every single transaction contrary to emission reduction projects is, to some extent, a "model versus muddle" comparison.

The underlying asymmetric assumption often seems to be that environmental policy (including clear emission targets for firms as well as reliable monitoring and effective enforcement mechanisms) is well developed in the case of permit trading, but underdeveloped in the case of credit-based approaches. Moreover, it is assumed that politicians (want to and) succeed in implementing a full-scale permit trading from the start. These assumptions can be criticized and relaxed. Because "frictionless ideals are useful mainly for reference purposes" (Williamson, 1979: 261), we will take the traditional (neoclassical) argument as a reference point and introduce some "muddle" elements into its analysis of permit markets as well as some "model" elements into its analysis of credit markets. We will also introduce the impact of politics by modeling the differences in the political "muddle" between the market-based climate policy instruments necessary to establish them.

We emphasize that the traditional view in environmental economics does not take into account several institutional barriers that could raise transaction costs under permit trading, such as a design with a small number of traders, nor does it consider the institutional opportunities to lower transaction costs for emission reduction (JI and CDM) projects, such as baseline standardization (Woerdman, 2001c). Next to considering their political transaction costs, we will provide an overview and analysis of the empirical evidence of transaction costs both from existing (non-GHG) emissions trading markets (e.g. Stavins, 1995; Tietenberg et al., 1999) and from the pilot phase for AIJ projects (e.g. Fichtner et al., 1999; Michaelowa & Stronzik, 2002).

5.4. Transaction Costs of Tradeable Emission Rights

The transaction costs of permit trading will be lower than for credit-based approaches such as JI and CDM, since the latter face a baseline problem that requires formal approval of each transfer contrary to the former (e.g. Tietenberg et al., 1999). However, this claim loses some validity if the underlying assumption of full-scale firm-to-firm trading in a perfect world is relaxed and if the set-up costs of such a scheme are taken into account. They act as institutional barriers raising the (market and political) transaction costs of emissions trading. By considering these costs, we also introduce and model the political "muddle" when establishing permit markets. Furthermore, it is instructive to look at empirical data on market transaction costs in already existing (domestic) permit trading markets, such as the US SO_2 allowance trading scheme.

5.4.1. Incremental Design, Set-Up Costs and Thin Markets

The idea that tradeable emission rights have lower market transaction costs than project-based emissions trading assumes that the neoclassical (international) firm-to-firm trading blueprint is implemented in reality. This blueprint presupposes, among other things, perfect competition, a downstream emissions trading system with many participants, the absence of market power and perfect international enforcement in the case of non-compliance of nation states. However, relatively few markets meet the assumptions of perfect competition (e.g. Helpman & Krugman, 1989; Stavins, 1995). This is also the perception among several politicians, which makes them doubt whether the transaction costs of permit trading will be as low as models presume. Decision makers had information that transaction costs would increase if governments, such as the FCCC Parties under the Kyoto Protocol, set various rules on emissions trading to cope with the complex problems of an imperfect world. An absence of theoretical and societal consensus, relative fitness problems and uncertainty made them reluctant to switch to permit trading, which contributed to a (temporary) institutional lock-in.

In addition, some authors underline the possibility, which is also recognized in the path dependence approach, that institutions may actually come about that raise transaction costs (Nooteboom, 2000: 100), for instance, because some institutional arrangements are primarily established to meet policy goals other than cost-effectiveness (such as environmental integrity or equity). It is thus too simple to refer to such cost-raising institutions as "government failure" by only considering the efficiency criterion as Estache & Martimort (1999: 3) do. Other criteria play a role in politics as well (e.g. Dixit, 1996: 147; Fisher et al., 1996: 405). Moreover, in the eyes of decision makers, implementing market-based climate policy could require some trade-off to be made between the economy, in this case transaction costs, and the environment. Creating institutions to make permit trading superior in one respect, such as effectiveness, could (but need not) affect its superiority in some other respects, such as transaction costs. Despite such relative fitness problems, transaction costs appear to be an important driver of the political process. The existence of multiple criteria, trade-offs, relative fitness problems and the political impact of transaction costs is demonstrated in the following examples (that partly draw from earlier chapters). They confirm that rules governing the trading system, for instance, to strengthen effectiveness or equity, can have a "dramatic effect" on transaction costs (e.g. Mullins & Baron, 1997: 31).

Zhang (1998c), for instance, proposes to tackle the "hot air" problem by imposing a transaction tax on emissions trading with countries, particularly in Central and Eastern Europe, whose negotiated emissions budget is probably larger than their business-as-usual emissions. This proposal knowingly raises transaction costs, but accepts this to strengthen environmental integrity. For the first reason, though, this proposal never gained wide political attention. Another example is the (rejected) proposal of the EU to place a quantitative restriction on the use of the Kyoto Mechanisms, implying that 50% of the Kyoto commitments should be achieved domestically (SBSTA/SBI, 2000), for instance, to restrict "hot air" trading (and for other reasons, including equity, which will be highlighted in the last few chapters of this book). Such a ceiling on trade would have strengthened environmental integrity, among other things, but it would also have raised transaction costs as it requires a pre-approval of each trade to make sure that a transaction does not fall behind the national threshold (Zhang, 2000a: 323). For this reason, other industrialized countries considered the EU proposal to be politically unacceptable.

In a similar fashion, governments could decide, for instance, that international emissions trading must satisfy various rules on liability, risk insurance and compliance in order to ensure effective enforcement (e.g. Jepma et al., 1998). An example, already decided upon in the context of the Kyoto Protocol, is the requirement that each Annex B Party shall maintain a commitment period reserve, which should not drop below 90% of its assigned amount (or 100% of five times its most recently reviewed inventory whichever is lowest), to restrict the discretion to oversell (CP, 2001b: 54). An Annex B country is allowed to trade more units only if and to the extent that its reviewed emissions are lower than 90% of its assigned amount. Checking whether this is the case raises transaction costs, but only moderately since emissions have to be reviewed anyway. Another example, that is not (yet) established, is buyer liability which not only strengthens effectiveness if buyers have a stronger willingness to comply than sellers and if the enforcement system is weak (Nentjes & Klaassen, 2004), but also considerably raises transaction costs because the permit buyer has to check whether the seller is in compliance as the former will be held liable for the non-compliance of the latter. Nevertheless, because the transaction costs of this environmental option are high and, for instance, those of the commitment period reserve are low, the latter was, in fact, adopted by politicians contrary to the former.

Considering the dynamics of the political process on climate change, Heller (1998: 114) points out that "(…) transactional costs of political mobilization are associated with displacing embedded policies (…)". Only a few authors acknowledge that any welfare assessment of permit trading needs to be adjusted by taking into account the transaction costs thrown up by the political process itself (e.g. Heyes & Dijkstra, 1999), such as the time-consuming lobbying process of negotiating an acceptable permit allocation that, for instance, requires a choice between grandfathering and auctioning (Woerdman, 2000b). The transaction costs of a permit trading scheme rise if they are defined in a broader sense to incorporate the costs of establishing such a regulatory regime in which pollution rights are (re)distributed (Krutilla, 1999). These political transaction costs to set up the institutional arrangement are likely to increase as the (re)distribution of property rights deviates more from the status quo (e.g. Rolph, 1983; Welch, 1983; Krutilla, 1999), for instance, because this would intensify rent-seeking activities by lobby groups. A consideration of set-up costs reflects the neo-institutional economic viewpoint that transaction costs are all costs of human interaction over time (North, 1997: 149).

It has been indicated before that permit trading has relatively high set-up costs (as will be demonstrated in detail in several of the next chapters) because there are specific legal problems and issues of cultural resistance that arise from its explicit (re)allocation of property rights. Not just auctioned emission rights, but also grandfathered permits imply a large deviation from the status quo because of the implied wealth transfer (Welch, 1983). Contrary to permit trading, as will be demonstrated in subsequent sections, credit trading and JI are institutional arrangements with relatively low set-up costs because they can incrementally use existing environmental policy (as the baseline) from which to calculate the (tradeable) emission reductions.

Various proponents of permit trading create the impression that if the "cost barrier" of setting up the scheme is taken, permit trading will have lower market transaction costs than the project-based flexible instruments. However, a low-cost trading scheme cannot simply be presumed, since its efficiency and transaction cost properties critically depend on its design. On the one hand, transaction costs can indeed be low if transfers are automatically registered and checked at the end of the year (e.g. Tietenberg et al., 1999). On the other hand, transaction costs may rise if governments do not implement the downstream trading "blueprint" in which both large and small emitters receive permits. Instead, which is ignored in many transaction cost studies that treat emissions trading either as an ideal model or as a black box (e.g. Michaelowa & Stronzik, 2002), these countries can design and implement domestic permit trading systems with a limited number of participants (e.g. Zhang, 1998c). Politicians may choose such a limited design, for instance, to deal with uncertainties and complexities in an incremental fashion. The EU, for instance, intended to follow, what they call, a "step-by-step approach" when they were making plans to develop an emissions trading system (COM, 2000a: 10).

Transaction costs could increase in a permit trading system with a limited number of participants, albeit not necessarily. Nentjes et al. (1995: 55) write that the theoretical possibility of the presence of (high) transaction costs seems to be irrelevant *when* the market for tradeable carbon permits is designed in such a way that the number of participants is large. Consequently, *if* the permit market is designed for a relatively small number of traders, transaction costs may increase. In general, since transaction costs will decrease as the number of traders increases (e.g. Tietenberg, 1992; Stavins, 1995), transaction costs will increase as the number of traders decreases (e.g. Heister et al., 1992; Pearce, 1995). However, there are important exceptions to this general rule. First, transaction costs do not have to increase when there are fewer traders if this also means that transactions become larger, which lowers the transaction costs per tonne of carbon traded. Second, search and bargaining costs can be kept low if the small number of traders already know each other and communicate regularly, which was, for instance, the case in the US lead phasedown program (Kerr & Maré, 1997).

Instead of a large international market of interlinked domestic permit trading schemes implemented in each Annex B Party, a fragmented carbon market is emerging that consists of a growing number of domestic and regional permit trading schemes which are not yet interlinked, each with different designs and trading rules. If the domestic schemes that emerge are finally connected, but continue to evolve in this fragmented way, as Rosenzweig et al. (2002: 36) expect for the short term, it will result in higher transaction costs than assumed in the ideal model due to the associated differences and complexities. Rosenzweig et al. (2002: 35) also present some kind of lock-in argument. They contend that changing the domestic schemes that now emerge to make them compatible is likely to be difficult because it could affect the interests and competitive positions of those firms that have a stake in the existing, sub-optimal design (where only a few firms, or even no firms at all, are regulated by means of tradeable permits under obligatory emission ceilings).

Each domestic market itself contains a limited number of participants. For instance, in the domestic, mandatory permit trading scheme for CO_2 emissions in Denmark that operated from 2001 to 2003, only electricity producers participated, whereas in the domestic, voluntary permit and credit trading scheme for GHG emissions in the UK that is operational between 2002 and 2007, also other companies which already have (relative) emission or energy targets are allowed to participate if they want. To facilitate administrative oversight, as discussed in earlier chapters, decision makers could restrict permit receivers to fossil fuel producers (upstream system), who will pass on their permit costs in a mark-up on the fuel price for both small emitters (such as households and car drivers) and large emitters (such as utilities and industrial sources). Another option is to allocate permits to fossil fuel producers and large emitters (hybrid system). To facilitate incremental change and learning, politicians can also distribute permits exclusively to large emitters, such as the electricity sector (as initially proposed by the EU (COM, 2000a)), while small emitters are regulated via taxes or standards (mixed system). According to Michaelowa (1998b), an upstream system, for example, is likely to suffer from relatively high transaction costs because the number of traders will be small (compared to a downstream system). However, to judge the effect on transaction costs of upstream and other systems with a limited amount of potential traders, a distinction has to be made between search and bargaining costs, incurred by those who trade, on the one hand, and monitoring and enforcement costs, incurred by the government, on the other hand.

On the one hand, if there are less potential traders, it may be more difficult (than in a downstream system) to find a suitable trading partner, which raises search costs. Nevertheless, information facilities that are both easily accessible and reliable (such as a clearinghouse) reduce this potential problem (e.g. Tietenberg, 1999). However, transaction costs also depend, to some extent, on the "thickness" of the market concerning the amount of trades that occur in the market (e.g. Liski, 1999). In a thick market, many traders are active and trades occur regularly, whereas in a thin market, only a few trades occur. If there are less potential traders, transactions are likely to occur less frequently (than in a downstream system), which makes the market "thinner". This could increase price uncertainty on the market as information which is relevant for the traders and their transactions does not dissipate speedily through the market, which complicates the bargaining process between buyers and sellers and adds to their information and negotiating costs. This could be partly offset by auctioning (a part of) the permits to give a price signal to the market (although the same problem returns if these prices would show a large variety, which depends on the design of the auction, e.g. Lyon, 1982).

On the other hand, checking compliance for only a limited number of traders saves administrative monitoring and enforcement costs for the government. This is a transaction cost advantage of upstream as well as hybrid and mixed schemes over a downstream system if the latter directly includes and monitors small end-users, such as individual motorists (Bohm, 1999). Nevertheless, in the previous chapter it was demonstrated that monitoring can also be organized upstream in a downstream trading scheme, so that administrative costs can also be reduced if many sources are directly included in the tradeable permit system. The dominant perception among policy makers, however, was that "(...) the number of participants in the trading system might become too large and pose problems for the monitoring and enforcement of the rules" (Oberthür & Ott, 1999: 196).

If politicians would start with an upstream, hybrid or mixed system and experiences are satisfactory, they may extend the trading system to other sectors. By imposing the flexible instrument on more entities, the government creates positive network externalities that result in lower market transaction costs for the users, such as lower search costs and lower costs of exchanging information. The more entities are subject to a particular flexible instrument, the easier it is for them to communicate and trade with each other when they use similar emission reduction entitlements. This would also increase the scope for efficiency gains.

Government trading, however, is another story. In principle, it depends on information mechanisms whether government-to-government emissions trading will have higher transaction costs than international private emissions trading. Nevertheless, the transaction costs of government-to-government emissions trading are expected to be higher than those in the case of international firm-to-firm trading because firms would have more and better information (for instance, on their marginal abatement costs) than governments to achieve cost-effective emission trades (e.g. Tietenberg, 1992). This does not mean that private sources have perfect information about the costs of abatement options: they do not (e.g. de Savornin Lohman, 1994). Moreover, the transaction costs per unit of emissions traded decrease as the quantity of trade in the intergovernmental deal becomes larger (Boom & Nentjes, 2000).

Next to the problem of market power in a government trading market (e.g. Gusbin et al., 1999), it is feared that political considerations or issue linkages may distort the presupposed economically rational market behavior of governments. For instance, a government could refuse to enact an efficient trade with a country whose non-economic policy or ideological views are perceived to be objectionable, or it could be inclined to enact a relatively expensive trade (compared to cheaper possible deals in other countries) in order to intensify general trading relations with certain politically favored countries. Nevertheless, international trading between private (or the so-called "legal") entities under the responsibility of the Parties under the Kyoto Protocol has explicitly been acknowledged as a policy option in the Marrakesh Accords of 2001. To make this possible, domestic permit trading schemes must be developed and could eventually be connected (under conditions sketched in Chapter 2) to create an international market (Zhang & Nentjes, 1999).

The fact that permit transfers can be checked at the end of the year (and do not have to be checked for each transaction) depresses market transaction costs, but these costs could become higher in a "thin" market when politicians decide to start with a small number of traders to facilitate incremental change and administrative learning. The political transaction costs of permit trading are relatively high because they largely replace existing environmental policy by explicitly (re)allocating property rights, whereas credit-based approaches have lower set-up costs as they build upon existing environmental policy. Decision makers knew that the institutional set-up of a tradeable emission rights system would be a tremendous task (Oberthür & Ott, 1999: 204).

These political transaction costs of permit trading, as well as the ambiguity and uncertainty about its magnitude, posed an obstacle for policy makers to switch to this superior alternative. The implication, as demonstrated in the next chapters, is also that legal problems and cultural frictions are larger in the case of permit trading than in the case of credit-based flexibility options. These factors are institutional barriers that make the transaction costs of emissions trading potentially higher than various economists have claimed in their models, but we have also seen that there are design opportunities to lower these costs, for instance, by creating adequate information facilities. Nevertheless, the fact that its superiority was contested, both by decision makers and in the literature, made some contribution to the institutional lock-in by giving political priority to start with (sub-optimal) emission reduction projects.

5.4.2. Empirical Evidence of Transaction Costs in Permit Trading Markets

Just like in every other market (e.g. Masten, 1996), transaction costs appear to be "common" in permit trading markets (Stavins, 1995: 144). In the market for lead permits during the lead phasedown from 1982 to 1987 in the US, Kerr & Maré (1997) estimate that transaction costs resulted in an efficiency loss in the order of 10%. Fisher et al. (1996) state that a source of indirect evidence of the prevalence of transaction costs in these early permit markets comes from a bias toward internal trading (within firms) as opposed to external trading (among firms). However, these early markets contained specific trade restrictions and regulatory uncertainties, increasing transaction costs, that are absent in the more recent US SO_2 allowance trading market where trading is unrestricted and regulatory property rights are clearly defined and protected (e.g. Tietenberg, 1999).

According to Klaassen & Nentjes (1997: 395), brokerage fees in the US SO_2 allowance trading market, which give some indication of the magnitude of transaction costs that result from searching and negotiating, are about 5% of the transaction value. Because brokerage fees fell from about $1.75 per allowance in 1994 to about $1.00 per allowance in 1996, it can be calculated on the basis of data gathered by Ellerman et al. (1997: 32–33) that transaction costs in this market have dropped to about 3% (1.5% for each side of the transaction). Based on figures provided by Joskow et al. (1998), it appears that the average commission figure per allowance per trade in 1996 was less than 2% of the prevailing spot price for SO_2 allowances. Probably because the transaction volume increased further, both Brockmann et al. (1999: 90) and Hargrave et al. (1999b: 11) note that transaction costs have decreased to approximately 1% of each trade (according to brokers active in the SO_2 market, as they say). Conrad & Kohn (1996) conclude that transaction costs have not significantly affected the trading and price of SO_2 allowances.

However, when drawing a parallel between the transaction costs in the US SO_2 emissions trading market and those in a possible future GHG emissions trading system, one has to realize that the former is a national scheme with many participants, while the latter — if agreed upon — could eventually be an international scheme that requires some additional trading rules to ensure environmental integrity and compliance. In addition, transaction costs will decline as the number of potential traders and the number of transactions per source increase and vice versa (Stavins, 1995). The amount of participants, in its turn, depends on market design. This means that transaction costs could rise relative to the international firm-to-firm trading blueprint in the case of government trading under or in the case of small, for instance, upstream, domestic trading schemes (e.g. Michaelowa, 1998b).

Finally, Jepma & Munasinghe (1998: 306) underline that the supposition of low transactions costs with many potential traders and transactions only applies to the final stage of a full-grown market rather than to the time-consuming process leading up to it. However, as far as we know, no quantitative assessments exist of the complete range of political transaction costs to set up such markets. There are, however, a few studies that have tried to find indicators to be able to calculate some of those costs. For instance, Versteege & Vos (1995) estimate that the preparation time of a tradeable permit scheme for SO_2 and NO_x emissions for the energy-intensive sectors amounts to 7 years. Assuming that per year 10 man-years are devoted to its preparation, those set-up costs that arise from policy preparation are a few million euros. However, the optimistic assumption of 10 man-years can well be challenged and there are more (direct and opportunity) costs involved when setting up a permit trading system than just labor costs. Furthermore, "objective" empirical cost assessments are only one input in the "subjective" perceptions that decision makers have about the political transaction costs of permit trading.

5.5. Transaction Costs of Project-Based Emissions Trading

In general, Williamson (1997b: 7) writes that "(…) transaction cost economics, always and everywhere, is an exercise in comparative institutional analysis — where the relevant comparisons are between feasible alternatives (…)". In our case of comparing market-based climate policy instruments, some economists argue, as pointed out before, that baseline-and-credit trading schemes entail higher transaction costs than cap-and-trade systems because credit transfers require advance approval of every single trade, while transfers in a permit trading system will be automatically registered and checked at the end of the year (e.g. Tietenberg, 1992; Mullins & Baron, 1997; Hahn & Stavins, 1999; Tietenberg et al., 1999; Sijm et al., 2000). In the context of the Kyoto Protocol, this view culminates in the formulation that "(…) project-based mechanisms, CDM and JI, will always have higher transaction costs than emissions trading by their very nature" (Vrolijk & Grubb, 2000: 9).

We have already indicated that this economic hierarchy of climate policy instruments is, to some extent, a "model versus muddle" comparison which asymmetrically assumes that environmental policy is well developed in the case of permit trading, but underdeveloped in the case of credit-based approaches. Many of those studies suppose that institutional arrangements to lower market transaction costs for project-based emissions trading are absent. In the next subsections we will not only nuance this (implicit) assumption behind various traditional economic studies of market-based environmental regulation, but we will also provide an overview and analysis of the empirical evidence of market transaction costs in the AIJ pilot phase.

5.5.1. Baseline Standardization, Capacity Building and Multilateral Funds

In reality, institutions are a "mixed bag" of factors that lower and factors that raise transaction costs (North, 1990: 63). Therefore, "design matters" not just for permit trading, but also for emission reduction projects. There are some institutional opportunities to lower the market transaction costs of project-based emissions trading. Many examples in fact emerged in the context of the Kyoto Protocol. "The transaction costs of (...) Joint Implementation can be significantly reduced through conscious attention to critical design elements" (Dudek & Wiener, 1996: 3). To some extent, the same argument applies to the CDM as "(...) barriers to an efficient functioning of a CDM fund can be overcome by designing it properly" (Michaelowa & Dutschke, 1998: 36). This made politicians feel less convinced about the economic inferiority of such (sub-optimal) instruments, which contributed to their institutional lock-in.

Although market transaction costs for CDM projects are relatively high because they must always be validated, verified and certified by operational entities accredited by the Executive Board, a two-track system was created for JI projects in the Marrakesh Accords of 2001 (CP, 2001b: 13). If a JI host country not only has a national system to estimate and register emissions and removals, but also annually submits a national inventory report, a host Party may verify reductions as being additional. If a host Party does not meet these eligibility requirements, verification shall occur by an independent entity accredited by the so-called Article 6 Supervisory Committee. In other words: if the host Party has a weak emission registration system, an institutional "slow" track is necessary to maximize environmental integrity. However, if a host Party has a reliable emission registration system, environmental integrity is stronger so that an institutional "fast" track can be taken which keeps JI market transaction costs relatively low. This is also the reason why the two-track system appeared to be politically acceptable in the first place.

Another point is that the transaction costs for project-based emissions trading, including JI and the CDM, will decrease over time as a result of learning effects (Michaelowa, 1995; Puhl, 1998), an evolutionary factor that is underestimated in the static Coasian transaction cost framework (Langlois, 1994; Nooteboom, 2000). Next to institutional arrangements that are designed to stimulate learning, it will be explained that standardizing (baseline) procedures, strengthening capacity building and developing multilateral funds are among the main institutional options to lower these market transaction costs, although they tend to raise set-up costs.

The first option that is expected to reduce market transaction costs is the standardization of micro-baseline determination procedures (e.g. Jepma et al., 1998; Hargrave et al., 1999a), which was already discussed in the previous chapter in the context of preventing baseline inflation. Baseline standardization implies the development of business-as-usual scenarios for project categories differentiated by, for instance, region, time, project and/or technology type. These scenarios could be determined by a panel of experts. Each specific project should fit into one of the categories. The (ex ante or ex post) emission reductions of a specific project can simply be calculated by subtracting the (predicted or observed) emissions from the baseline emissions of the relevant category. This makes a third party check for each baseline much easier or even unnecessary. Transaction costs are reduced because it will not be necessary anymore to invest time and effort in constructing a baseline for each and every project. It remains necessary to verify whether the investor and host have properly calculated the emission reductions, but this will be less time-consuming than verifying case-by-case baselines.

A practical example of standardization is the matrix approach, which places pre-defined default scenarios for several project categories into a matrix. The investor and the host of a JI or CDM project look up the baseline in the matrix, for example, available on the FCCC Internet homepage, to calculate the credits which will accrue from the project. Although a matrix is likely to work efficiently once it has been established, it should not be forgotten that it still takes expertise, time and money to develop the (standardized) baselines for each cell in the matrix to begin with. However, because it reduces transaction costs, baseline standardization appeared on the political agenda, for instance, in the text of the Marrakesh Accords of 2001 (CP, 2001b: 46).

Additional options to reduce the transaction costs for project-based emissions trading are to standardize the emission abatement reporting procedure, to develop standard contracts for project partners, to strengthen the institutional capacity in the host countries and/or to set up information exchange and trade facilities, such as a clearinghouse (Dudek & Wiener, 1996; Michaelowa & Dutschke, 1998). The United Nations Industrial Development Organization (UNIDO), for instance, has provided manuals on guidelines for technology transfer as well as procedures for project accreditation and verification with the explicit aim to reduce transaction costs for CDM projects (ENB, 1999). Moreover, capacity building in host countries can make it easier, for instance, for potential project partners to find one another and for governments to monitor the emissions.

Moreover, instead of a bilateral approach, a unilateral approach can be taken in which the host engages in self-financing of the emission reductions. This is expected to lower transaction costs because a host that is also the investor will be more familiar with local conditions than a foreign investor (Black-Arbelaez et al., 2000; JIQ, 2001a). The disadvantage is that it requires substantial host country project development and financing capacities. If these are not sufficiently available in particular (developing) countries, transaction costs can be lowered via a multilateral approach by clustering several emission reduction projects in a portfolio. These are implemented by specialized intermediaries transferring the funds to individual subprojects, which can be particularly relevant for small-scale (CDM) projects (e.g. Michaelowa & Dutschke, 1998; Ghosh, 1999). Without multilateral funds, small projects are less attractive than large projects because their start-up costs and operational costs are more or less similar in absolute terms (e.g. EcoSecurities, 2000).

In practice, for instance, the Executive Board of the World Bank has established and approved the so-called Prototype Carbon Fund (PCF) in order to reduce transaction costs (JIQ, 1999). The PCF is a pilot activity, operational since 2000 and scheduled to terminate in 2012, which facilitates learning-by-doing with investments restricted to a maximum of $180 million. It is a mutual fund in which (private and public) investors pool capital to be invested in GHG emission reduction projects in co-operation with potential host countries for AIJ, JI and the CDM. The credits acquired through these projects will be returned to the investors. In 2001 the multilateral PCF was ready to invest $145 million in JI and CDM projects on behalf of 6 countries and 17 companies (JIQ, 2001b).

Although several authors, like Trexler & Kosloff (1998) or Woerdman & van der Gaast (2001), have calculated that developing countries seem to offer the largest low-cost potential, it is also clear that the informational, institutional and infrastructural constraints are higher in such countries than in countries with economies in transition (e.g. Karani, 1997). Under the Kyoto Protocol, this generally makes the transaction costs for CDM projects higher than for JI Article 6 projects (Sokona & Nanasta, 2000), also given the adaptation tax and the relatively strong sustainability requirements under CDM Article 12. However, with respect to potentially lowering the transaction costs of emission reduction projects implemented in and in co-operation with developing countries, several authors (e.g. Aslam, 1999; Dutschke & Michaelowa, 1999) mention the possibility of designing the CDM as either a simple project exchange, a clearinghouse (similar to a broker) or a multilateral fund (similar to the PCF) in which the credits, initially accruing to the CDM, are distributed to the investors according to their share.

If projects are to be compared systematically with permit trading markets under the assumption of well-designed environmental policy, similar ideal "model" circumstances could be assumed for such projects as well (where some economists asymmetrically tend to assume "muddle" circumstances). If the Kyoto Protocol will be effectively implemented in Central and Eastern Europe, the transaction costs for JI projects, for instance, could be lower than several researchers have predicted, as we will explain below, even when some institutional shortcomings are taken into account.

Various studies have argued that JI has a baseline problem (in essence similar to the CDM) assuming that a choice has to be made between several seemingly "reasonable" baselines for each individual project (e.g. SEVEn/JIN, 1997). However, the host country of a JI project has committed itself to an assigned amount (contrary to a CDM host country), which implies that a Central or Eastern European government has to define environmental policy targets for its domestic emitters. If it has done so, the JI baseline to calculate the additional emission reductions could be derived from the defined environmental policy for the host firm or sector involved. This would also be in line with the requirement formulated in the Marrakesh Accords of 2001 that the project-specific JI baseline shall take into account relevant national and/or sectoral policies (CP, 2001b: 18).

For instance, if the policy in a JI host country is a performance standard that requires a certain quantity of CO_2 per unit of output or energy, the baseline emissions for the host firm can be calculated by multiplying this standard with its expected production volume or energy use. If the host firm emits less CO_2 than this baseline figure because a JI project is implemented, emission reductions are achieved for which the investor can obtain ERUs. If the host invests in the project itself (self-financing of emission reductions), we speak of credit trading (or unilateral JI). This would not necessarily require a pre-approval of each transaction: the scheme can be designed in such a way that compliance is checked at the end of the year, similar to permit trading schemes. This also means that the transaction costs of both credit trading and ("fast" track) JI will not diverge as much from permit trading as traditional environmental economics literature suggests.

Importantly, the fact that credit trading and JI can use existing environmental policy as the baseline from which to calculate the (tradeable) emission reductions also means that they have relatively low set-up costs compared to instruments that instead require an explicit (re)allocation of property rights. Nevertheless, the political transaction costs of establishing credit trading in industrialized countries are likely to be lower than those of establishing JI in countries with economies in transition where environmental policy and emission monitoring are still in their infancy. The political and market transaction costs of the CDM are higher because environmental policies and institutional capacities are less developed in developing countries than in Annex B countries. But once these flexible instruments have been set up, their market transaction costs can be lowered through the institutional opportunities offered by baseline standardization, capacity building and multilateral funds.

This strengthened the perception among various decision makers, not only that consensus was lacking on the economic disadvantages of project-based mechanisms, but also that the presumed inferiority of such instruments could be contested by reducing their "muddle" on the basis of institutional enhancements. This made them more optimistic towards these sub-optimal regulatory tools than neoclassical economic models would expect, which contributed to their (temporary) institutional lock-in. The transaction cost gap between permit trading and credit-based approaches can be reduced, but not resolved. This is also confirmed by empirical data.

5.5.2. Empirical Evidence of Transaction Costs in AIJ Projects

The impact of transaction costs is that they raise costs for the emission traders involved, which lowers the trading volume or even prevents transactions from occurring (Michaelowa & Stronzik, 2002: 11). It would be interesting to know the exact level of such costs, but the availability and quality of transaction cost data regarding project-based emissions trading are not always satisfactory.

From an empirical point of view, Palmisano (1996) claims that transaction costs in early project-based credit markets in the US governing air pollution control (notably the so-called "offset", "bubble" and "netting" policies in which an emission growth at one source could be compensated by an emission decline at another source) have not prevented trading. Although the evidence is "rather mixed" (Jepma & Munasinghe, 1998: 306), there is "abundant anecdotal evidence" indicating the prevalence of significant transaction costs in some of these early US credit trading programs according to the IPCC (Fisher et al., 1996: 423). From a theoretical point of view, Palmisano (1996) expects that transaction costs in a market for carbon credits will be lower than in those (non-carbon) markets, given the potentially larger financial magnitude of carbon trades. In addition, it must be stressed that the early US emissions trading programs of the past contained specific trade restrictions and regulatory uncertainties that increased transaction costs, but which do not have to be copied to a future carbon trading market (e.g. Ingham, 1992; Tietenberg, 1999).

Interestingly, without considering permit market transaction costs, Haites (2000) presents a trading model under the Kyoto Protocol in which the reference case assumes transaction costs of 25% for the CDM and 15% for JI. He also performs two sensitivity analyses in which transaction costs are assumed to be 50% for the CDM and 35% for JI as well as 10% for both JI and the CDM, respectively. These quantitative figures reflect the perception of many economists about the magnitude of transaction costs of credit-based approaches and can therefore be used as a theoretical reference point for the (incomplete) empirical figures found in the studies discussed below.

Empirical data of transaction costs for emission reduction projects to combat climate change can be found under the Kyoto Protocol in the pilot phase for Activities Implemented Jointly (AIJ), which started in 1995 and, although initially scheduled to be completed earlier, has been extended after 2000 by the FCCC Parties because of positive learning experiences (e.g. CP, 2001b: 46). The aim of the pilot phase is to gain experience with the potential environmental, institutional and cost-effectiveness aspects of GHG emission reduction projects. The typical characteristic of this pilot phase (contrary to JI and the CDM) is the absence of crediting, since Parties are not allowed to use the reductions achieved through AIJ projects for the fulfillment of commitments under the FCCC. The AIJ projects give some indication of the level of JI and CDM transaction costs.

In the late 1990s, some authors have written about the transaction costs in one or a few AIJ projects. In 1997 the Nordic Council of Ministers assessed the transaction costs of 10 AIJ projects in Eastern Europe. Total transaction costs, including JI specific transaction costs such as baseline determination and GHG emission reduction monitoring, ranged from 12 to 19% of the total initial investment in energy sector projects, and from 15 to 30% in smaller and more complex industrial sector projects. The transaction costs for the JI acceptance procedure ranged from 1 to 8% (JIQ, 1996, 1997). In a case study, Fichtner et al. (1999) calculated the transaction costs for six selected AIJ projects from all over the world, which appeared to range from 1 to 15% of the total project costs. Furthermore, countries that have concentrated their AIJ investments in the same country, region or sector, have been able to reduce transaction costs (Schwarze, 1998; Ellis, 1999b). In the early years of the 21st century, more extensive research was done on the transaction costs in AIJ projects. Michaelowa & Stronzik (2002) and Fichtner et al. (2003) are one of the few authors who provide an extensive overview of empirical data on AIJ transaction costs in more than 50 projects.

Fichtner et al. (2003) have considered 64 (out of 144) projects that contained detailed information on both production and transaction costs. From this sample, they conclude that transaction costs range from 7 to more than 100% of production costs with most projects lying between 14 and 89% (Fichtner et al., 2003). In an average project, 50% of the transaction costs are technical assistance costs, 36% are administration costs, 12% are the costs for follow-up projects and 2% are reporting costs. The authors underline that their figures are higher than those reported in other studies, but they also acknowledge the poor data quality in (their sample of) AIJ projects.

Michaelowa and Stronzik come up with (somewhat) lower figures by calculating the transaction costs of 51 Swedish AIJ projects carried out in the Baltic states because this is "(…) the only AIJ program with a consistent reporting of transaction costs (…)" (Michaelowa & Stronzik, 2002: 16). They do not list figures from other AIJ projects because these projects differ too much in their definition of transaction costs, as they say. The average transaction costs (resulting from technical assistance and administration) are about 20% of total project costs for energy-efficiency projects and about 14% for renewable energy projects. Moreover, taking into account the starting dates of the projects, the authors indicate that these costs have declined over time (from about 17% in 1994 to 13% in 1998 for energy-efficiency projects and from about 18% in 1993 to 14% in 1998 for renewable energy projects). Michaelowa & Stronzik (2002) show that the transaction costs of renewable energy projects are smaller than those of energy-efficiency projects, primarily because the former are larger in terms of emission reductions generated.

However, when drawing a parallel between AIJ on the one hand and JI and CDM on the other hand, one has to keep in mind that the aforementioned studies are limited to less than half of the existing pilot phase projects. Moreover, since no credits may accrue from AIJ which deprives investors from the essential incentive to participate, the number of projects is likely to be smaller under the pilot phase than under JI or CDM as the first commitment period approaches. This delays possible learning effects, thereby preventing transaction costs to decline further. In addition, in the AIJ pilot phase baseline determination has not been standardized, while the establishment of standardized baselines, if developed and agreed upon, would substantially lower transaction costs for such projects, as we have seen in the previous subsection. Therefore, one could argue that the transaction costs of AIJ projects are likely to be an upper bound for the transaction costs of JI and CDM projects.

When comparing the empirical data on transaction costs found in this section with those in the previous section, it becomes clear that the transaction costs for credit-based approaches are higher than those for permit trading, although there are several institutional opportunities to bring them down. However, there are some methodological problems when making this comparison.

5.6. Methodological Problems of Comparing Transaction Costs

One might argue that it is not strange to obtain ambiguous (empirical) results from comparing the transaction costs in more or less different markets because each market has its own typical transaction costs (e.g. some particular markets may have high search costs, whereas others may rather have high approval costs). Unfortunately, different studies do not rarely focus (implicitly) on different types of transaction costs (like search costs versus approval costs) or they do not (sufficiently) define the type of transaction costs they analyze, which makes them difficult to compare in a systematic fashion. Although it is instructive to consider empirical data on transaction costs of both permit trading and emission reduction projects, one must be aware of the methodological problem of comparability with regard to the cost components involved.

5.6.1. Comparing AIJ Transaction Costs with Permit Trading Transaction Costs

In the case of permit trading, each firm or industry pays for the measurement and registration of its own emissions, but firms' transaction costs usually do not include the transaction costs of monitoring and enforcement which are borne by the responsible government. Although including the latter cost components would clearly raise the total costs of permit transfers (Fisher et al., 1996), transaction costs in permit trading markets are generally expressed as a percentage of the transaction value with respect to the costs that private entities incur (such as search costs, negotiation costs and insurance costs).

In order to facilitate a comparison between permit trading and existing or emerging project-based credit markets such as the AIJ pilot phase, some argue that the transaction costs in emission reduction projects have to be expressed as a percentage of the total investment, yielding relatively high AIJ transaction cost figures. However, in several cases (notably in the so-called simulation projects as discussed in the previous chapter), the amount invested to generate a reduction of GHG emissions in an AIJ project is only a percentage, say 10%, of the capital supplied by the investor. This is, for instance, the case in simulation projects (Woerdman & van der Gaast, 2001). Those AIJ projects did not primarily focus on cost-effectiveness, but rather added a market-based climate component — such as baseline determination or emission reduction measuring — to already ongoing projects (mostly fuel switching and energy-efficiency projects) to obtain institutional experience by simulating as if they had been set up as JI (or CDM) projects. Consequently, it would be doubtful to relate the transaction costs of the total investment to the 10% GHG emission reduction component of the investment. Therefore, it is also possible, and probably more relevant, to express the transaction costs for those AIJ projects not in terms of the investment-related component (such as the transaction costs associated with obtaining a construction license), but rather in terms of the AIJ/JI/CDM-related component (such as the transaction costs associated with paying the consultant responsible for monitoring and verification). Obviously, the transaction costs related to the value of the GHG emission reductions of the project are lower than those related to the total investment.

Indeed, some studies presented above have considered such AIJ specific transaction costs, which would presumably facilitate a more adequate comparison with transaction cost figures in permit markets. In that case, based on the scarce empirical data available, transaction cost percentages for both permit and credit trades seem to lie somewhere within a range of about 1–10% of the transaction value. Nevertheless, the early credit-based emissions trading systems in the US as well as the experimental and international market for AIJ projects (where credits can not be used for compliance purposes) clearly experienced higher transaction costs than the well-established and domestic SO_2 allowance trading market in the US.

However, the fact that the transaction costs in the US were higher in the early credit trading programs than in the current American permit trading program for

SO_2 emissions and the fact that international AIJ projects have higher transaction costs than domestic SO_2 allowance trades in the US do not necessarily imply that (project-based) credit trading programs have higher transaction costs than permit trading programs *in general*. The transaction costs in those two types of programs crucially depend on their specific design characteristics. First, comparing those programs with their respective predecessors is methodologically wrong because the trading program under the Kyoto Protocol will be different, both in geographical scope, sector coverage, emission type, institutional arrangements and trading rules. Second, the trade restrictions and regulatory uncertainties that increased transaction costs both in the early US credit trading programs of the past and in the present AIJ program do not have to be copied to the design of permit and credit markets under the Kyoto Mechanisms (e.g. Ingham, 1992).[4] Third, there are some empirical data which have confirmed that transaction costs could be low, not only for permit trading (e.g. Hargrave et al., 1999b), but also for, mainly large-scale, climate change mitigation projects (e.g. Fichtner et al., 1999; Michaelowa & Stronzik, 2002).

5.6.2. Comparing Market Transaction Costs with Political Transaction Costs

We have criticized the idea that market transaction costs for permit trading are always (much) lower than those for credit-based approaches. Moreover, we have introduced an institutional economics perspective by incorporating political transaction costs (or set-up costs) when comparing market-based climate policy arrangements, including the Kyoto Mechanisms, which further nuances the traditional neoclassical viewpoint in environmental economics. The methodological problem of comparing market transaction costs with political transaction costs is that both have their roots in different economic research traditions.

The neoclassical tradition focuses on the efficiency of equilibrium outcomes in which the fittest will survive (or the fitter under incomplete information), assuming rational and cost-minimizing actors with given preferences, operating in an institutional vacuum or operating within given institutions. The new institutional tradition (behind the concept of market transaction costs) builds upon neoclassical analysis by assuming cost minimization, but it also focuses on the efficiency of processes in the context of institutions and recognizes that costs may occur when property rights are transferred. The neo-institutional tradition (behind the concept of political transaction costs) also recognizes that costs may occur when property rights are established by analyzing the emergence and disappearance of (formal)

---

[4] They will, in fact, not be copied: one example is the "fast" track project cycle created for JI projects in host countries with reliable emission registration systems.

institutions, not only in terms of efficiency, but also in terms of (changing) cultures and perceptions, that are outcomes of a historical and continuing path-dependent process where boundedly rational actors form preferences and show routine (as well as learning) behavior (e.g. Allen, 2000; Nooteboom, 2000).

However, to nuance the traditional economic hierarchy of market-based climate policy, flexible instruments like the Kyoto Mechanisms must be compared with each other by adding a new dimension or perspective. For this purpose, market transaction costs should not so much be compared with political transaction costs for each instrument, but the instruments should rather be compared with each other in terms of both market and political transaction costs. To provide a real-life institutional example, this is done for the Kyoto Mechanisms in Table 5.1.

Although it would have been possible to construct a more detailed table based on the nuances outlined in this chapter (for instance, by making a distinction between "fast" track and "slow" track JI projects), it is not desirable — albeit important in the verbatim analysis above — to incorporate them in a visualization as it would result in a cluttered and unsurveyable collection of qualifications. Therefore, the table has been deliberately kept simple by only making a distinction between permit trading, credit trading/JI and the CDM and by only using the values "high", "medium" and "low" to make clear classifications of associated transaction costs. The advantage of such an approach is that the table stays transparent, the disadvantage is that the table is no more than a rough reflection of the much richer and differentiated analyses we have tried to give in the earlier sections.

The first row of Table 5.1 reflects the traditional (neoclassical) view in environmental economics that the market transaction costs in credit-based arrangements, in particular for the CDM, will be higher than in permit trading systems, as discussed before. This view asymmetrically assumes, to some extent, a "model versus muddle" situation in which environmental policy is well developed in the case of permit trading, but underdeveloped in the case of credit-based approaches.

The second row of this table reflects our refinement of this view, which is taken as a starting point, by introducing some "muddle" elements in the permit trading model and some "model" elements in the credit-based approaches. The market transaction costs for permit trading are relatively low if transactions can occur freely with annual checks, but they could become higher either when politicians decide in an incremental fashion to start with a "thin" market of limited scope or when information is incomplete, whereas these costs can be lowered for credit-based approaches through the institutional opportunities offered, for instance, by baseline standardization, capacity building and multilateral funds. Moreover, the market transaction costs for JI projects can be kept at reasonable levels in the "fast" track project cycle which is allowed when the JI host Party has a reliable

Table 5.1: Market and political transaction costs of the Kyoto Mechanisms.

Type of transaction costs	Theoretical framework	Permit trading	Credit trading + JI	CDM
Market transaction costs	Neoclassical economics (traditional view)	Low	High	High
	New institutional economics (this chapter)	Medium	Medium	High
Political transaction costs	Neo-institutional economics (this chapter)	High	Low	Medium

emission registration system. In addition, both credit trading and permit trading do not necessarily require a pre-approval of each transaction and compliance can be checked at the end of the year, which results in moderate transaction costs. Despite the opportunities to lower market (and political) transaction costs for CDM projects in developing countries, they will still be relatively high because environmental policies and institutional capacities remain to be less developed and because the sustainability requirements of Article 12 are stronger than for JI Article 6 projects within Annex B countries.

The third row of Table 5.1 reflects our assertion that the political transaction costs of permit trading are relatively high because they largely replace existing environmental policy by explicitly (re)allocating property rights, whereas credit-based approaches have lower set-up costs as they can build upon existing environmental policy. In general, North (1990: 51) suggests that political transaction costs are higher (and more difficult to measure) than market transaction costs when he writes that "(…) markets [that approximate] the Coase zero transaction cost conditions (…) are scarce enough in the economic world and even scarcer in the political world". If this is also true for the Kyoto Mechanisms, the third row will dominate political developments (rather than the first or the second), which helps to explain that credit-based approaches are more easy to implement than permit trading schemes. Where neoclassical economists expect governments of Annex B countries to opt for permit trading because of its low market transaction costs when the market is established, our institutional economics approach rather expects that these politicians are tempted to opt for credit trading/JI because of the low political transaction costs of establishing this market — assuming that their decision making behavior is guided more by political transaction costs (which they perceive to be higher for themselves) than market transaction costs (which are probably lower for legal entities, such as firms).

Contrary to the path dependence approach, TCE cannot explain the survival of sub-optimal institutions (e.g. Nooteboom, 2000). Rather, TCE assumes "(...) that the organization of hierarchies is the result of a minimization of transaction costs" (Estache & Martimort, 1999: i). In terms of market transaction costs, the traditional view is that permit trading is optimal, whereas credit trading, JI and the CDM are sub-optimal. Yet the latter do persist in political reality, although the former is recently moving center stage. TCE is only able to explain their "survival", to use the evolutionary term, if the neo-institutional concept of political transaction costs is introduced, which appeared to be relatively high in the case of permit trading. However, the problem is that political transaction costs in the form of perceived set-up costs, that include legal and cultural barriers, are difficult or even impossible to quantify. The implication is that political transaction costs (which may also change over time) can be better dealt with in a path dependence analysis than in a (static) neo-classical or new institutional economic framework (e.g. Magnusson & Ottosson, 1997: 3).

As part of (or next to) the perception of such (political) transaction costs to set up market-based climate policy instruments, like the Kyoto Mechanisms, detailed research is required of the role of laws and culture in relation to networks and learning (e.g. Magnusson & Ottosson, 1996: 354; North, 1997: 151). This will be done in the next chapters. The introduction of such formal and informal institutional factors in this chapter has made clear that there are barriers and opportunities which were not considered, or insufficiently appreciated, in the traditional analyses of flexible instruments for environmental policy. Our institutional economics approach has demonstrated that there is ambiguity and uncertainty on the superior transaction cost properties of permit trading. Nevertheless, this instrument still turned out to be relatively fit from this perspective, also when some institutional "muddle" was taken into account, but not from the perspective of political transaction costs. In the eyes of policy makers, this particular view gave rise to some relative fitness problem associated with permit trading. The fact that the neoclassical superiority of permit trading was contested on such institutional grounds has contributed to the lock-in of inefficient environmental policy.

5.7. Conclusion

Traditional studies based on neoclassical economics argue that tradeable emission rights have superior transaction cost properties over project-based emissions trading because permit transactions can occur freely with annual checks, whereas credit transfers require pre-approval (e.g. Hahn & Stavins, 1999). However, two basic institutional considerations change this picture. First, these studies often asymmetrically (and implicitly) assume a "model versus muddle" situation in which environmental policy is well developed in the case of permit trading, but underdeveloped in the case of credit-based instruments. Therefore, we have introduced some "muddle" elements in the permit trading model and some "model" elements in the analysis of credit approaches. Second, these studies have not systematically compared the political transaction costs to set up those market instruments. Therefore, we have tried to make a comparative assessment of such set-up costs by analyzing the political "muddle" necessary to establish them. This nuances the transaction cost advantages of permit trading and helps to explain why some politicians are tempted to opt for credit markets, just like Liebowitz & Margolis (2000) clarified the survival of the QWERTY-keyboard by questioning the superiority of its alternatives.

There are several institutional barriers that raise market transaction costs for permit trading. Trading rules and market designs could be established for reasons other than efficiency (like environmental integrity or administrative transparency). An example is the EU proposal to place a quantitative ceiling on the use of the Kyoto Mechanisms to restrict hot air trading, which would have required a pre-approval of each trade to make sure that a transaction does not fall behind that ceiling. Because it increases transaction costs, however, this proposal was rejected. Another example is that transaction costs could become higher when politicians incrementally decide to start with a limited number of participants to facilitate learning. This would make the permit market "thin" and prices more uncertain, which increases the information and bargaining costs of buyers and sellers. Nevertheless, information facilities (such as a clearinghouse) could reduce this problem and checking compliance for only a limited number of traders saves monitoring and enforcement costs for the government (although administrative costs can also be kept low in a downstream system if monitoring is organized upstream, as argued before).

At least as important is that there are several institutional opportunities to lower the market transaction costs for credit-based instruments. First, credit trading does not necessarily require a pre-approval of each transaction and compliance can be checked at the end of the year, which results in moderate transaction costs. Second, next to learning effects, the transaction costs of project-based trading can be lowered by standardizing (baseline) procedures, strengthening capacity building and developing multilateral funds (although they do tend to raise set-up costs). Another example, in the context of the Kyoto Protocol, is the so-called "fast" institutional track created for JI projects. If the host Party has a reliable emission registration system, it may verify reductions as being additional. Despite the aforementioned opportunities to lower transaction costs for CDM projects in developing countries, they will still be relatively high because environmental policies and institutional capacities remain to be less developed and because the sustainability requirements under Article 12 are stronger than those under Article 6 for JI projects.

An overview of empirical studies clearly shows that the transaction costs in permit markets are usually a few percentages of the transaction value, for instance, in the well-developed US SO_2 allowance trading scheme, whereas these costs are (much) higher in credit markets, for instance, in the early US credit trading programs and in projects under the international AIJ pilot phase. However, we emphasize that these data are difficult to compare, not only because each market has its own typical transaction costs and because different studies often focus (implicitly) on different types of transaction costs, but also because the trade restrictions and regulatory uncertainties that increased transaction costs in the US credit markets of the past and in the experimental AIJ program do not have to be copied to the design of credit-based approaches in present or future market-based climate policy. Nevertheless, as more information on high transaction costs became available from AIJ projects, an increasing number of policy makers began to consider setting up tradeable emission right schemes.

However, the political transaction costs of permit trading are relatively high because they largely replace existing environmental policy by explicitly (re)allocating property rights, whereas credit-based approaches have lower set-up costs because they can use existing environmental policy (as the baseline) from which to calculate the tradeable emission reductions. Permit trading deviates more from the status quo than credit-based approaches, which gives the former higher political transaction costs (e.g. Welch, 1983; Krutilla, 1999). Politicians were well aware of this (e.g. Oberthür & Ott, 1999: 204), which gave rise to some kind relative fitness problem: permit trading was considered superior in some (market transaction cost), but not in all (political transaction cost) respects. An absence of consensus about the superiority of permit trading, and uncertainty about its (high) level of set-up costs, contributed to the institutional lock-in of other, sub-optimal designs.

North (1990: 51) suggests that political transaction costs are generally higher (and more difficult to measure) than market transaction costs. If this was also the perception among policy makers, for instance, in the context of the Kyoto Protocol, then political transaction costs have played a (major) role in governmental decision making, which helps to explain why politicians are tempted to add sub-optimal designs to extant policy instead of readily accepting permit trading. The formal and informal institutional barriers that contribute to these political transaction costs, including equity, are analyzed in the next chapters. To understand the nature of these barriers, we will make a step from new institutional economics to neo-institutional (law and) economics.

PART III
INSTITUTIONAL LAW AND ECONOMICS

Chapter 6

WTO Subsidization Law and Distortions of Market-Based Climate Policy

6.1. Introduction

This chapter specifies the formal constraints to implementing market-based climate policy by formulating the economic and legal conditions, both in terms of efficiency and equity, under which international differences in the domestic allocation of tradeable emission rights lead to competitive distortions and actionable subsidies under the rules of the World Trade Organization (WTO).

Where we have mainly focused on nuancing the economic hierarchy of market-based climate policy on the basis of the new institutional economics in the previous chapters, we will now shift our attention to explaining the political hierarchy of this type of policy by considering its formal institutional barriers. Permit trading is the superior alternative, but politicians tend to make existing environmental policy more flexible by incrementally using sub-optimal credit-based approaches that implicitly allocate emission entitlements. Under permit trading, they would have to (re)allocate tradeable emission rights explicitly, which makes the allocation problem and its consequences more visible.

The legal problem of allocating permits is that some governments could grandfather their permits while others auction them, which may distort competition and lead to inadmissible subsidization according to WTO law under certain conditions (e.g. Petsonk, 1999; Werksman, 1999b). To find these conditions, we will use neoclassical as well as neo-institutional law and economics perspectives (e.g. Medema et al., 2000). These two perspectives capture the legal relevance of both efficiency and (perceived) equity considerations when allocating permits (e.g. Woerdman, 2004a,b). The idea that permit trading is "completely neutral" with respect to equity because permit allocation does not affect efficiency, as Ciorba et al. (2001: 8) contend, turns out to be wrong.

This chapter is organized as follows. Section 6.2 defines the concept of competitive distortions. Section 6.3 specifies the economic conditions, both in terms of efficiency and equity, under which international differences in domestic permit allocation procedures may lead to competitive distortions. Section 6.4 uses this economic framework to specify the legal conditions under which grandfathering could be seen as a subsidy under WTO law. Section 6.5 extends our law and economics analysis by taking a neo-institutional approach that focuses on perceptions in politics to find out whether politicians see grandfathering as a subsidy under WTO law and, subsequently, whether they desire an international harmonization of permit allocation. Finally, Section 6.6 presents the conclusion.

6.2. Definition of Competitive Distortions

Although frequently used by governments and firms, the notion of competitive distortion is not common in economic theory. In this chapter, we distinguish between two different definitions (or: interpretations) of the competitive distortion concept:

- distortion of efficient competition;
- distortion of fair competition.

These definitions can be traced back to the analysis of environmental regulation by van der Laan & Nentjes (2001) who make a distinction between competitive distortion as an inefficiency in the allocation of resources and competitive distortion as an inequity of firms' starting conditions. The first definition is based upon neoclassical economics which focuses on the efficiency of outcomes in an equilibrium setting. A competitive distortion arises if an environmental policy instrument entails a price deviation from the welfare optimum under perfect competition, thereby reducing the efficiency of (inter)national trade. The second definition is based upon neo-institutional (law and) economics which focuses on equity in a path-dependent setting. A competitive distortion then arises if the introduction of an environmental policy instrument leads to unequal changes of the competitive relations among comparable firms, thereby reducing the (perceived) fairness of (inter)national competition.

The latter view does not so much originate from 17th century mercantilist theory that aims to eliminate comparative disadvantages and cost differences among firms and nations, as van der Laan & Nentjes (2001: 137) suspect, but rather from 20th century neo-institutional economics that aims to include legal and cultural institutions into economic analyses. The equity interpretation of competitive distortion reflects the (Aristotelian) proportionality-based fairness principle according to which burdens and benefits should be distributed in proportion to the contributions of the subjects (see Banuri et al., 1996: 86).

In the next section we will apply, specify and modify the aforementioned distinction made by van der Laan & Nentjes (2001) to analyze the issue of competitive distortions in relation to permit allocation from an economic point of view. To do so, we will develop a neoclassical and a neo-institutional approach to the economic theory of competitive distortions, which enables us to shed more light on the question whether grandfathering distorts competition and on the subsequent question, to be discussed later on, whether grandfathering will be seen as a subsidy under WTO law.

6.3. Economic Analysis of Permit Allocation and Competitive Distortions

Many neoclassical economists realize that permit allocation is the "largest impediment" to permit trading (e.g. Ellerman, 1998: 1). A major issue, also in the context of the Kyoto Protocol, is whether international differences in domestic permit allocation procedures will lead to competitive distortions. Some argue that a country which grandfathers permits would give its firms a competitive edge against countries which either auction their permits or use carbon taxes (e.g. Romstad, 1998; Anderson et al., 1999; Hourcade & Le Pesant, 1999). Generally, their analyses suggest that grandfathering permits to sector x in country A and auctioning permits to an identical sector y in country B leads to a competitive advantage for the former. The question is whether this is correct. If this is correct, then grandfathering could be seen as a subsidy under WTO law. This, in turn, could imply that governments should not be left free to choose any domestic permit allocation rule in the case of international firm-to-firm trading. Allocation formulae could then be harmonized across the participating nations and domestic observance of such international rules could be made part of the eligibility criteria to join the permit trading system.

In this chapter, we focus on permit trading, not only because allocation decisions remain implicit under credit trading, but also because there is (de facto) no choice between grandfathering and auctioning under credit-based approaches. Under such approaches, emissions are basically allocated for free, credits accrue bottom-up and national emission budgets, like the assigned amounts under the Kyoto Protocol, do not have to be divided into emission caps for polluters before trading can begin. Permit trading, however, entails an explicit choice between grandfathering or auctioning property rights that have to be allocated top-down in the form of emission ceilings before trading can begin.

In this section, we will analyze the issue of permit allocation by using and specifying the neoclassical efficiency approach to competitive distortions, both under perfect and imperfect competition, as well as the neo-institutional equity approach to competitive distortions.

6.3.1. Perfect Competition, Efficiency and Opportunity Costs

Following neoclassical welfare economics, the concept of competitive distortion can be interpreted as an inefficiency or price distortion. A competitive distortion is then defined as any measure which entails a price deviation from the welfare optimum under perfect competition, thereby reducing the efficiency of (inter)national trade. To address the competitive distortion issue from an efficiency perspective, an analytical distinction is made between (a) perfect and imperfect competition in the market as well as between (b) opportunity costs and the financial position of firms. In perfectly competitive markets, international differences in domestic permit allocation procedures will not lead to competitive distortions according to economic theory. To follow the argument, it is crucial to understand the concept of opportunity costs.

Auctioning entails costs for firms. They have to buy the permits in an auction, which means that they pay for their emissions as well as for their subsequent emission reductions. A government can alleviate competitiveness concerns by recycling the revenues to lower other pre-existing distortionary taxes, such as taxes on labor and capital (Zhang, 1999b). Under grandfathering, firms receive their permits for free, which means that they do not have to pay for their emissions, but only for their subsequent emission reductions. Only in the presence of distortionary taxes, grandfathering is less efficient than auctioning, because grandfathering does not generate revenues that can be recycled to lower such taxes (Goulder et al., 1999; see also Burtraw et al., 2001).

However, not only auctioning, but also grandfathering entails costs for firms, which has been overlooked by several early as well as more recent economic studies on permit allocation (e.g. Lyon, 1982; Romstad, 1998). The essential point is that grandfathered permits have an opportunity cost when they are used for covering the emissions of the permit owner (Nentjes et al., 1995; Grafton & Devlin, 1996; Koutstaal, 1997; Hargrave, 2000). The opportunity cost is the revenue foregone by not selling the permits but using them in producing output. This opportunity cost, which is equal to the price for which the permit can be sold, must be included in the product price. Instead of using permits to cover emissions, the firm could have sold the permits. The revenue foregone is a cost to the firm, comparable with the "interest foregone" on own capital. Hence, grandfathered firms do not have a cost advantage over auctioned firms abroad (or over domestic newcomers), just because they received permits for free. Although there are no price distortions when comparing grandfathering with auctioning in perfectly competitive markets, grandfathering does imply a transfer of wealth to firms, since they receive an input which has a certain market value. Therefore, grandfathering permits could be viewed as granting a subsidy to the firm. However, this subsidy is a capital gift to the firm with the character of a lump sum subsidy (Jensen & Rasmussen, 1998; Hargrave et al., 1999b). In efficiency terms, a lump sum subsidy is not distorting in the product market, since it does not affect marginal emission reduction costs and it does not alter the output and price decisions of firms.[1]

Some believe that grandfathering emission permits implies a competitive distortion because it would have the same distorting effect as granting tax exemptions to certain sectors or firms (e.g. energy-intensive sectors or the export industry). However, grandfathered permits are not the same as tax exemptions precisely because of the opportunity costs of the former. Indeed, a tax exemption is inefficient because it induces different prices per unit of emission for different firms (e.g. Hoel, 1997). This can be best explained by regarding GHG emissions as an input in the production of a certain output. Different prices for the same input involve differences in marginal productivity and therefore entails an inefficient allocation of emissions among firms or sectors. Tax exemptions do not have opportunity costs (since you cannot sell them), but rather imply that emissions are an input without a price. However, grandfathered permits have opportunity costs and therefore entail a price. Contrary to tax exemptions, efficiency is not distorted by using grandfathering (if competition is perfect).[2]

6.3.2. Imperfect Competition, Inefficiency and Financial Positions

The arguments above assume perfectly competitive markets. Although the opportunity costs of grandfathering are equal to the financial costs of auctioning, international permit allocation differences will affect the financial position of

---

[1] A simple real-life example of opportunity costs is a young farmer who inherits the agricultural land of his father. Although he did not have to buy the land, he still has to include the opportunity costs of the land in the price of, say, the corn or milk he produces, because he could have sold the land instead of using it. Of course, he does have a financial advantage over a young competitor who had to buy the land from another farmer, but (in case of perfect competition) the opportunity costs of the land, in principle, prevent him from setting lower prices so that competition is not distorted.

[2] Some even turn (environmental) economic theory upside down and believe that not so much tax exemptions, but even the carbon taxes or auctioned carbon permits themselves are "distortionary" for firms. However, Bohm (1999) clearly explains that the initial permit volume reflects a global environmental concern induced by a process of international negotiations and that the tax or auction price, which reflects this concern, emerges as a corrective rather than a distortionary levy just like any other environmental "tax" reflecting similar concerns would.

firms. This aspect becomes relevant if competition is imperfect (and if the concept of competitive distortion is interpreted in terms of equity, which will be discussed in the next subsection). The so-called "new" positive (international) trade theory (e.g. Helpman & Krugman, 1989), which considers the economic impact of imperfect markets, learns that competitive distortions may arise in imperfectly competitive permit, capital or product markets. In general, imperfect markets do exist in reality (e.g. Spencer, 1990; Bovenberg, 1993), also in the context of environmental policy (e.g. Ulph, 1999), but an imperfect permit market can be avoided if politicians succeed in creating a thick market with many (small) traders as discussed in the previous chapters.

A grandfathered firm does not have to buy its permits contrary to an auctioned competitor. A grandfathered firm only has to pay for its emission reductions and not for its emissions, so that it has a lower cash outflow than an identical firm which has to buy its permits.[3] In other words, a grandfathered firm initially buys the permits from itself (opportunity costs), while an auctioned firm buys the permits from the government or the public (cash outflow). If a grandfathered firm receives its permits for free, it obtains a non-distortionary windfall profit (e.g. Romstad, 1998; Bohm, 1999). Since grandfathering implies a capital gift to the firm, a grandfathered firm has more financial resources, or own capital, than an auctioned firm, which (ceteris paribus) gives the former a stronger financial position than the latter.

Where a firm in one country receives grandfathered permits and a firm in another country has to acquire auctioned permits, a competitive distortion could arise under imperfect competition in the following three exceptional circumstances. Grandfathering gives a financial advantage:

- to an inefficient firm;
- to a firm that engages in predatory pricing on an imperfect product market;
- to a firm while auctioned firms borrow money on an imperfect capital market.

First, the fact that grandfathering gives an undertaking a stronger financial position than its foreign auctioned competitor may reinforce existing inefficiencies. Consider the case of an inefficient firm (with negative profit) which is grandfathered. Due to the financial advantages of grandfathering it is able to remain in business for a longer period than it would under auctioning (e.g. Malueg, 1990; Crane et al., 1998; Hargrave et al., 1999b). However, instead of assuming that the firm will make permanent losses (which presupposes perfect foresight), it is also possible that the firm only makes temporary losses. If the losses are permanent, the financial advantages of grandfathering unnecessarily delay

---

[3] To be more precise, a grandfathered firm only has to pay for its emissions in so far as it needs extra permits above its grandfathered quantity.

the inevitable fall of the inefficient firm, but if the losses are temporary, grandfathering helps the firm to stay in business, whereas an auctioned firm financed with borrowed money might be forced into bankruptcy. In the real world of imperfect foresight, a firm does not know whether the losses it makes will be permanent or are only temporary: it can hope for a better future (and possibly rightly so). Whether it is better from an efficiency perspective that a profit maximizing firm which makes losses leaves the industry or stays in business can only be concluded afterwards. This means that permit allocation differences can strengthen existing inefficiencies, but certainly do not cause an inefficiency here.

Second, a competitive distortion may arise if the grandfathered firm in one country starts a price war with the auctioned competitor in another country on an imperfect product market according to the so-called deep purse theory or theory of predatory pricing (Koutstaal, 1997). Grandfathering implies a capital gift to the firm which increases its resistance against periods of losses which use up the firm's capital. In a price war the grandfathered firm will lower its prices, which forces its rival to follow him to a level where both firms will incur losses. The grandfathered firm can outlast the auctioned firm (or entrant) in a price war because of its larger capital reserve.[4] After elimination of its competitor, it acquires a monopoly position and raises the price to the monopoly level. This particular outcome is an inefficiency that can be traced back to differences in permit allocation between governments. However, according to Nentjes et al. (1995), predatory pricing is unlikely to occur in an emissions trading market, not only because it is a risky and expensive strategy as emphasized in the theory of industrial organization, but also because energy-intensive firms usually do not compete on monopolistic markets and the additional capital requirements to buy emission permits are usually a small part (no more than a few percentages) of the total capital requirements. Furthermore, although some industries have expressed the perception that their competitors abroad use predatory pricing (e.g. Michaelowa et al. 1998), a dominant firm that starts a price war to push aside its competitor abroad could be prosecuted by the authorities that enforce antitrust policy in industrialized countries.[5] Not the difference in financial positions between firms itself (for instance as a result of permit allocation), but the abuse of this difference by

---

[4] A similar approach, albeit from a business economics perspective, would be to consider grandfathered permits as a form of off-balance sheet financing. While free allowances have an opportunity cost, they have no accounting cost, so that the pressure to earn a return on assets with financial costs may not apply equally to grandfathered permits compared with auctioned permits.

[5] In markets in general, empirical evidence of predatory pricing has actually been found, for example in the case of the large multiproduct firm *AKZO* which tried to eliminate the smaller firm *ECS* in the British benzoyl peroxide market and in the case of *Tetra Pak* which tried to eliminate its smaller competitors in the Italian market for non-aseptic cartons (Hildebrand, 1998: 78–79). However, these cases also underline that there are legal opportunities to punish firms that abuse their market power by means of predatory pricing. In a WTO context, the "predator" could be accused of dumping.

a firm to engage in predatory pricing leads to inefficiencies, which can be countered by means of legal action.

Third, in the case of an imperfect capital market, the findings of Koutstaal (1997) as well as Grafton & Devlin (1996) imply that a grandfathered firm has a competitive advantage if the auctioned competitor abroad needs to borrow money to buy the permits. The interest to borrow money could exceed the interest on own capital due to the imperfect capital market. The permit expenditures of the auctioned firm are higher than the opportunity costs of the grandfathered firm due to the interest it has to pay for its loans. However, the practical relevance of this argument is negligible. Capital markets almost always work imperfectly in reality, but the interest difference under consideration is small compared with total production costs. Moreover, the loans will be short term in a liquid permit market and (contrary to the interest on own capital) the interest charges for loans are tax deductible.

It is shown above that, if the assumption of perfect markets is dropped, competition could only be distorted as a result of differences in permit allocation between countries in some exceptional cases. This underlines that inefficiency is not a strong argument to see such differences (grandfathering versus auctioning) as a distortion of international competition. Equity is a different story, as will be explained and elaborated upon in the next subsection.

6.3.3. Fair Competition, Equity and Level Playing Field

The implication of the Coase theorem for market-based climate policy is roughly that the efficiency of permit trading is unaffected by the initial allocation of pollution rights when transaction costs are negligible (e.g. Coase, 1960). Although some economists believe that this should make permit trading and allocation neutral to any equity consideration (e.g. Ciorba et al., 2001: 8), others recognize that an efficient outcome is not necessarily (perceived as) an equitable one (e.g. Devlin & Grafton, 1998: 42).

Although there is no single interpretation of the concept of equity or fairness, neither in philosophical theory nor in political practice, an unfair competitive distortion in the context of environmental policy usually refers to a distortion of the "level playing field" for firms, whereas the associated inequity is primarily defined, or perceived, in terms of an inequality or asymmetry (e.g. Jepma & van der Gaast, 1999; Yamin & Lefevere, 2000; Jepma, 2002). This raises the question of how to define level playing field within the concept of competitive distortion as inequity.

Hargrave et al. (1999b: 11) mention what they call the "proverbial" level playing field, but they only relate it to the (possibly different) levels of emission targets for firms in different countries, not to the (possibly different) methods of allocating these targets. Where they do consider the methods of grandfathering and auctioning, they also analyze the issue of fair competition. However, despite their extensive and solid analysis, Hargrave et al. (1999b: 6) only provide a strictly neoclassical economic definition and elaboration of fair competition by interpreting it as efficient competition that requires marginal cost equalization.[6] This basically reflects the orthodox economic view that equity is the same as efficiency, or that the market is always fair, which Rose & Kverndokk (1999: 371) refer to as "market justice". The problem is that, if applied in a world where law and politics matter, they are not able to capture those legal and political interpretations of rules and perceptions on fair competition that consider an equitable allocation to be different from an efficient one. It will be demonstrated in the next sections that the latter interpretations are a substantial part of legal and political reality.

Contrary to Hargrave et al. (1999b), van der Laan & Nentjes (2001) do not define equity in terms of efficiency and give a static interpretation of the inequity view of competitive distortions. In the context of environmental policy in general, they contend that a level playing field can be interpreted as meaning equal starting conditions. Fair competition would exist in the case of a uniform legislative regime or equally stringent environmental legislation for comparable producers. An unfair competitive distortion would then emerge if there are "(...) differences in cost conditions created by differences in national legislation that arise out of divergent national preferences (...)" (van der Laan & Nentjes, 2001: 138). Although van der Laan & Nentjes (2001) focus on international differences in the stringency of emission standards (target level), the focus in this chapter is on international differences in permit allocation (instrument level). Moreover, where the latter authors analyze differences in costs for similar industries in different countries, we concentrate upon differences in the financial positions of such firms. The consequence is that their interpretation of equity needs to be either adjusted or extended.

When permit trading is introduced into existing environmental policy, firms will compare their financial situation with the situation of their competitors both before and after permit allocation. Also Hargrave et al. (1999b: 53), despite their narrow interpretation of equity as efficiency, acknowledge this when they relate unfair competition to unequal cost increases as a result of the introduction of regulation. In our theoretical framework, this means that equity will be judged in a dynamic, path-dependent process and not in a static, isolated setting. Firms will not demand

---

[6] According to these authors, "(...) fair competition [requires that] competition among firms is based on true economic costs of production and is not distorted by subsidies. (...) If firms in the same sector but in other countries are facing lower marginal cost increases from climate policy, those facing the higher costs can claim unfair competition" (Hargrave et al., 1999b: 6). Viguier (2000) takes a similar view.

that permit allocation makes their (absolute) financial positions equal. Instead, they will demand that they do not lose more in (relative) financial terms than their competitors (or that their competitors gain more) as a result of permit allocation alone. This "dynamic" approach to the concept of unfair competitive distortion does not so much require equality of firms' starting conditions, but rather equality of changes in their competitive relations.

Consequently, the problem of international differences in domestic permit allocation procedures is the inequality of the changes in firms' financial positions resulting from the mere process of permit allocation. The inequity view does not so much think of competition being distorted because firms face different laws, but rather because these different laws have different financial consequences for firms that compete on a market. The level playing field then refers to the competitive relations between firms. This implies that an unfair competitive distortion would arise if the allocation of permits leads to unequal changes of firms' relative financial positions.

The level playing field approach neither objects to the fact that the competitive positions of firms can be unequal because they have different market shares, nor to the fact that their relations may change, both before and after permit allocation, because of their economic activities and strategies. Rather, the level playing field or financial position approach contends that the competitive positions of firms are not allowed to change because of the political process of permit allocation itself. This means that the level playing field is maintained if permit allocation leaves the relative financial positions of firms (their competitive relations) unaltered. An unfair competitive distortion is then thought to occur if the allocation of permits leads to an unequal absolute or proportional (re)distribution of financial burdens or benefits for firms relative to the pre-allocative status quo. In less plain words: a competitive distortion as inequity would arise if the allocation of permits itself leads to unequal changes of firms' relative financial positions.[7]

We have demonstrated that a grandfathered firm in one country has a lower cash outflow and hence more financial resources than an identical firm in another country which has to buy its permits. Therefore, such international differences in domestic permit allocation procedures lead to an unequal distribution of burdens for comparable firms which changes their competitive relations disproportionally, so that competition is distorted, by definition, according to the level playing field approach. The inefficiency interpretation would not call this distributional inequality a competitive distortion, since efficiency is not negatively affected, but rather a financial advantage for the grandfathered firm.

---

[7] Our approach builds upon the work by Rolph (1983) and Welch (1983) who provided empirical evidence (already in the early 1980s) for their hypothesis that establishing property rights is only politically feasible if it does not involve substantial redistributions of wealth, but rather maintains the economic status quo by more or less preserving the existing economic relationships among firms.

Next to the grandfathering versus auctioning issue, there is the issue of permit trading versus credit trading that raises similar concerns in terms of competitive distortions in an international context. If politicians in one country develop permit trading for internationally competing and energy-intensive sectors, whereas the government of another country creates credit trading for similar sectors, competing firms will have to operate under different domestic environmental regimes. Competition is then distorted in its efficiency interpretation (Nentjes, 2000). When the economy grows in a permit trading scheme, the demand for permits under the emission ceiling will rise, so that environmental scarcity under the absolute emission ceiling is fully reflected in the permit price. This price will also be a part of the product price. However, in a credit trading scheme, the supply of emission credits also rises when the economy grows, since firms do not have an absolute emission ceiling, so that the environmental scarcity will not be fully reflected in a price for every unit of emission. This not only makes credit trading inefficient, but it also makes products cheaper compared to firms with absolute caps. The implication is that not only efficiency, but also the level playing field is distorted, because both systems have different financial consequences for the firms involved.

6.4. Legal Analysis of Permit Allocation and WTO Subsidies Law

The WTO, established in 1994, constitutes the formal institutional foundation of the multilateral trading system and marks the boundaries of how governments are allowed to design and implement domestic legislation that affects trade. The general objective of the WTO is trade liberalization, but not all trade is covered by its principles, such as trade in capital. Rather, the WTO rules cover the trading in goods and services, building mainly upon the 1947 General Agreement on Tariffs and Trade (GATT) and the 1994 General Agreement on Trade in Services (GATS), respectively. The basic right (and obligation) for its Members is to ensure trade without discrimination between domestic and foreign goods, services and service suppliers in favor of its own ("national treatment principle") or between other Members ("most-favored nation principle"). In principle, the dispute settlement procedure of the WTO is binding upon its members, demanding (a) a change in the measure that led to the complaint, (b) a compensation payment for trade damages incurred through the distorting measure or (c) a retaliation by the adversely affected party (e.g. Vaughan, 1999).

Several studies have analyzed the potential (in)compatibilities of an international emissions trading scheme with WTO rules (e.g. Michaelowa et al., 1998; Parker, 1998; Petsonk, 1999; Werksman, 1999b). In this chapter, we pose the question under what conditions tradeable permits are to be considered as (so-called "actionable") subsidies. Many authors agree that the trading of permits itself is not covered by WTO rules, because tradeable emission permits are neither goods nor services, but rather government-issued licenses that entitle the holder to carry out a regulated activity within its territory (e.g. Petsonk, 1999; Werksman, 1999b). The allocation of permits, however, will be covered by WTO rules, because this is likely to affect the competitive relationship between products and services governed by WTO disciplines (e.g. Petsonk, 1999; Werksman, 1999b).

On the one hand, some literature emphasizes that permit allocation represents the establishment and distribution of property rights over GHG emissions, which in itself lies outside the mandate of the WTO and is not likely to raise concerns as long as it does not discriminate against imported products or against foreign firms who want to operate in the Member State (e.g. Zhang, 1998c). Even if permit allocation would be inconsistent with certain WTO principles, it could still be argued that this is irrelevant, for instance when the governments with tradeable emission right systems agree to leave the choice between grandfathering and auctioning to each individual government. This would be in line with the 1969 Vienna Convention on the Law of Treaties, which implies that WTO Members may voluntarily decide that certain WTO rules do not apply between themselves (e.g. Rutgeerts, 1999).

On the other hand, however, decision makers faced an intense academic and political debate whether or not grandfathering is to be considered an "actionable subsidy" under WTO law, for instance if permits are grandfathered to legal entity x in country A while a competitor y in country B has to buy its permits by means of an auction. To formulate the conditions under which this is the case, we will analyze permit allocation in relation to the legal concepts of actionable and non-actionable subsidies in the WTO Agreement on Subsidies and Countervailing Measures (ASCM) below, based on the economic theory outlined above.

6.4.1. Permit Allocation and Actionable Subsidies

In the economic analysis of the previous section it was shown that grandfathered permits, which have the character of a non-distorting lump sum subsidy, imply the same (opportunity) costs as auctioned permits, so that grandfathering and auctioning do not differ in terms of efficiency. Nevertheless, it was also demonstrated that grandfathering gives the favored firm an advantage — in terms of a capital gift or windfall profit — relative to its auctioned competitors abroad, so that grandfathering and auctioning have different consequences for firms in terms of their financial positions and level playing field (equity) as well as for efficiency under imperfect competition.

Therefore, the relevance of the actionable subsidies issue largely depends on (a) whether the WTO ASCM has defined and/or interpreted, and (b) whether a WTO panel — in the case of a dispute — will regard, an actionable subsidy either as a lump sum payment that does not affect the efficiency or (opportunity) costs of firms, or as a financial advantage that affects the financial positions of and thus the level playing field (fair competition) among firms. More broadly, the issue depends on whether the WTO will perceive international permit allocation dissimilarities only in terms of efficiency or also in terms of equity.

According to Article 1 [definition] and Article 2 [specificity] of the WTO ASCM, a "subsidy" is deemed to exist (i) if there is a financial contribution by a government which confers a benefit to a specific enterprise/industry or group of enterprises/industries within its territory and involves a (potential) direct transfer of funds or liabilities, or (ii) if government revenue that is otherwise due is foregone or not collected. An example of the waiving of government revenue otherwise due are fiscal incentives such as tax credits, although it should be stressed that the exemption of an exported product from duties or taxes borne by a like product when destined for domestic consumption, or the remission of such duties or taxes in amounts not in excess of those which have accrued, shall not be deemed to be a subsidy according to the GATT Article XVI and the ASCM Annexes.

According to Article 5 [adverse effects] and Article 6 [serious prejudice] of the ASCM, an "actionable subsidy" is deemed to exist if the subsidy causes adverse effects or serious prejudice to the interests of other WTO Members. An example of an adverse effect is injury to the domestic industry of another Member (ASCM Article 5(a)). An example of serious prejudice is a significant price undercutting by the subsidized product (ASCM Article 6.3(c)) or a clear and permanent increase in the world market share of the subsidizing Member (ASCM Articles 6.3(d) and 6.4).

The analysis of whether grandfathering is subsidization falls back on the two different interpretations of competitive distortion outlined before. On the one hand, it could be argued that the grandfathered permits do not seriously affect or prejudice the interests of (the auctioned industry of) the foreign country, simply because both industries face the same opportunity costs and allocative or financial differences do not affect the efficiency of the emissions trading regime (e.g. Hargrave et al., 1999b). On the other hand, it could be argued that the ASCM does not neglect the financial aspects of a subsidy. Indeed, grandfathered permits could be seen as a financial contribution and benefit to a specific firm or sector, because they imply a capital gift for the favored recipient. Furthermore, one could claim that grandfathered permits are a direct transfer of funds or government revenue foregone, because the government would otherwise have collected revenues in the alternative of auctioning and grandfathering means that the auction revenue otherwise due is given to the polluters (see also Welch, 1983: 168; Burtraw et al. 2001).[8] It could thus be contended that grandfathered permits in one country and auctioned permits in another seriously affects or prejudices the interests of (the domestic industry of) the latter country, because it unequally changes the financial positions of their industries, thereby affecting the level playing field.

Grandfathering is not an actionable subsidy in the efficiency interpretation of competitive distortions. There is no "benefit" because of its opportunity costs and the auctioned industry abroad is not injured as grandfathering does not affect efficiency. Moreover, grandfathering does not lead to price undercutting if the market is perfectly competitive. However, grandfathering could be seen as an actionable subsidy in the equity or level playing field interpretation of competitive distortions (or if competition is imperfect), because grandfathered firms have a stronger financial position than their auctioned competitors abroad. This means that grandfathered firms have a benefit, which could be seen as to affect the interests of the domestic industry in the country that uses auctioning. Moreover, grandfathering could imply that revenue otherwise due is foregone, because the government would otherwise have collected revenues in the alternative of auctioning.

In addition, there are some legal WTO texts that emphasize the importance of a level playing field in which firms are able to compete fairly or equitably. For instance, a subsidy which confers a benefit to a specific enterprise or industry "(…) could enable domestically produced products to compete unfairly (…)" (Werksman, 1999b: 258–259). Interestingly, although permits are not "services", the GATS provides that a government is not allowed to treat foreign-like services or service suppliers less favorable than its own, where treatment shall be considered to be less favorable "(…) if it modifies the conditions of competition (…)" in favor of its own services or service suppliers (GATS Article XVII:2). In a similar reasoning, it could be maintained that grandfathering permits favor firms compared to foreign competitors with auctioned permits and modifies the conditions of competition due to the associated different financial consequences. Furthermore, past GATT/WTO panels have found that the violation of the "national treatment principle" includes measures affecting the so-called competitive opportunities of enterprises (Vaughan, 1999) and that foreign new entrants must be able to compete on a level playing field with an established domestic-like product (Werksman, 1999b). More generally, in 1994 the Committee on Trade and Environment explicitly referred to (among other things) the equitable nature of the multilateral trading system (Rutgeerts, 1999: 61).

Most economists agree that auctioning reduces the political acceptation of a tradeable emission rights system, because it poses a financial burden on

---

[8] In addition, some authors argue that an emissions trading scheme will involve the dislocation of vested interests (e.g. Werksman, 1999b: 261).

the energy-intensive industry, whereas grandfathering promotes the political acceptation, because emission sources only have to pay for their additional emission reductions and not for their emissions (e.g. Koutstaal & Nentjes, 1995; Dijkstra, 1998). However, the formal constraints posed by the WTO made politicians think that they could not exclude the possibility that grandfathering would be considered an actionable subsidy. Moreover, the analysis bears some kind of relative fitness problem, because grandfathered permits are only an actionable subsidy in some (equity), but not in all (efficiency) respects. This ambiguity and uncertainty made governments hesitant to switch to permit trading, which made some contribution to the institutional lock-in of building sub-optimal, credit-based approaches upon extant environmental policy.

The key findings of our law and economics analysis are summarized in Table 6.1. Grandfathering could become a WTO issue if some governments (for instance, some Parties to the Kyoto Protocol) decide to auction permits, whereas others prefer to hand them out for free. Whether grandfathering will be seen as an actionable subsidy basically depends on whether the WTO will interpret it in terms of efficiency or in terms of equity. Although grandfathering is not an actionable subsidy in the efficiency interpretation of competitive distortions because its opportunity costs are equal to the financial costs of auctioning, it could be seen as an actionable subsidy in the equity interpretation of competitive distortions (or if competition is imperfect) because grandfathered firms have a stronger financial position than their auctioned competitors abroad. Moreover, grandfathering is likely to be seen as a subsidy if the WTO does not accept "opportunity costs" as costs similar to those of auctioned permits in the first place.

6.4.2. Permit Allocation and Non-Actionable Subsidies

Even if grandfathering will be considered to constitute a subsidy, which depends on the (efficiency or equity) perspective taken, it may still be allowed if it is seen as a non-actionable subsidy on the basis of ASCM Article 8.2(c). This section of the law allows assistance to promote the adaptation of existing facilities (which have been in operation for at least 2 years) to new environmental requirements imposed by law and/or regulations which result in greater constraints and financial burden on firms, provided (among other things) that the assistance (i) is a one-time non-recurring measure and (ii) is limited to 20% of the cost of adaptation.

The former provision could imply that the financial burden on firms is legitimately relieved by allowing a one-time allocation of grandfathered permits. The latter provision is even more debatable, partly because there are several possible ways to define and calculate the "costs of adaptation". It could imply that only a small number of permits may be grandfathered, arguably, equal to 20%

Table 6.1: Conditions for competitive distortions and actionable subsidies.

Context	International GHG emissions trading between firms Grandfathered permits for legal entity x in country A Auctioned permits for comparable legal entity y in country B		
Economic concept	Competitive distortion		
Legal concept	Actionable subsidy (WTO law)		
Approach	Distortion of efficiency (inefficiency view)		Distortion of level playing field (inequity view)
Competition distorted/subsidy actionable	No	Yes	Yes
Theory	x and y have similar (opportunity) costs	x has a stronger financial position than y	x has a stronger financial position than y
Conditions	Macro-focus Perfect competition	Micro-focus Imperfect competition	Micro-focus Equity is (not equality but) equal changes
Details	x is not favored (within the meaning of WTO ASCM Articles 1–6), because x must include the opportunity costs of using the gratis permits (lump sum subsidy) in the product price	x is favored (within the meaning of WTO ASCM Articles 1–6), among other things, because x outlasts y if a price war occurs and because y pays a higher interest rate if it has to borrow money to buy the permits	x is favored (within the meaning of WTO ASCM Articles 1–6), because the gratis permits for x are a capital gift, whereas y has to buy its permits, so that the mere process of permit allocation entails an unequal (and thus unfair) change of their financial positions and competitive relations

Key: x, legal entity (say, a firm) with grandfathered permits in country A; y, comparable legal entity (say, a comparable firm) with auctioned permits in country B; WTO, World Trade Organization; ASCM, WTO Agreement on Subsidies and Countervailing Measures.

of either the total required amount of permits, the opportunity costs of the permits or the emission reduction costs. In any case, the next chapter will make clear that WTO subsidies law is relatively stringent in comparison with EC state aid law (e.g. Vikhlyaev, 2001), not only with respect to labeling grandfathering as a form of subsidization, but also with respect to exempting grandfathering from the legal objections against subsidization.

A WTO dispute could also arise over several other issues associated with permit allocation, which increases legal complexity and thereby adds to the switching costs of permit trading. Although we will not discuss them at length, an example is the application of the general exceptions, for instance those provided in GATT Article XX (e.g. Rutgeerts, 1999; Werksman, 1999b). Roughly, this section of the law allows, among other things, the adoption and enforcement of measures related to or necessary for the protection of the environment as long as they are not arbitrarily or unjustifiably discriminating between countries where the same conditions prevail. Not surprisingly, the discussion in particular centers on the interpretation of the word "necessary". WTO/GATT jurisprudence (for instance of US Gasoline panel 6.24) suggests that a measure is necessary if alternatives, which are less GATT inconsistent or less trade restrictive (or which are GATT consistent), are not available to the WTO Member or cannot be reasonably expected to be employed (see Rutgeerts, 1999; Werksman, 1999b).

In a comparison of grandfathering and auctioning, the interpretation of the latter jurisprudence depends mainly on whether one is willing to accept the opportunity cost argument or the level playing field argument. According to the first argument, both grandfathering and auctioning are alternatives with the same (opportunity) costs, which by definition makes them equally GATT (in)consistent. According to the second argument, both grandfathering and auctioning are alternatives that have different effects on the financial position and level playing field of firms. In that case, one could argue that grandfathering should be viewed as an actionable subsidy that is (more) GATT inconsistent in comparison with auctioned permits. However, a Member can still try to refer to the broad and vague provision of "reasonable expectations" by claiming that it cannot be reasonably expected to employ either auctioning (e.g. for reasons of political acceptability) or internationally prescribed domestic allocation rules (e.g. for reasons of state sovereignty).

6.5. Political Analysis of Perceptions on Subsidization

Ellerman (1998: 1), for instance, writes that "(...) the allocation of rights raise[s] fundamental issues of equity that lie pre-eminently in the political realm". Despite the early negligence as a result of an exclusive (neoclassical) orientation on efficiency, the economic literature on emissions trading has shown, since the mid-1990s, an increasing interest in the distributional problems of permit trading (e.g. Heller, 1998; Endres, 1999; Kerr, 1999). There are several reasons for this.

First, the negotiation process concerning the Kyoto Mechanisms moved from agenda building to decision making and policy implementation. To some extent, it could be argued that economists have contributed to the emergence of emissions trading on the international political agenda. Despite their focus on efficiency, economists are now confronted with, and hence write about, real-world distributional problems in the decision and implementation process of market-based climate policy. Second, whereas economists seem to have prevailed during the agenda-formation process, non-economists are becoming increasingly involved in the scientific debate on emissions trading, such as lawyers and political scientists, now that the political process moves towards implementing the flexible instruments. Third, policy makers have become institutionalized in the "policy-science" debate, for instance in the IPCC process, which has provided practical questions and perspectives — for instance on permit allocation — to the scientific community, including economists.

From a neo-institutional perspective, it does not so much matter in politics whether international differences in domestic permit allocation can "objectively" lead to competitive distortions according to some economic theory or equity principle. Rather, the agents involved act "subjectively" on the basis of their perceptions of such issues. In the context of competitiveness, for instance, Golub ascertains: "Regardless of whether the available evidence proves or disproves a conclusive relationship between economic performance and environmental regulation, the important point (...) is that the possible or perceived loss of competitiveness constitutes as much a political as an empirical matter, and figures prominently in industry's resistance (...)" (Golub, 1998: 8).[9]

Therefore, this section takes another neo-institutional economics perspective by analyzing perceptions to supplement the law and economics analysis of competitive distortions. We want to know whether or not actors perceive grandfathering as an actionable subsidy under WTO law and, subsequently, whether or not they prefer an international harmonization of permit allocation.

6.5.1. Perceptions in Political Negotiations on Permit Allocation

Perceptions matter in international climate change politics and these perceptions may change over time (e.g. Fermann, 1997: 192). The fact is that various actors,

---

[9] Likewise, Rowlands (1998) suggests that the EU policy for ozone layer protection has been shaped to a large extent by a perceived negative relationship between economic competitiveness and environmental regulations.

at least for some time, had the perception that international permit allocation dissimilarities may lead to competitive distortions (e.g. Woerdman, 2000b). This posed a barrier to the international political acceptance of permit trading. Not only will interest groups use the level playing field argument in their lobby for protection (e.g. Faure, 1998; van der Laan & Nentjes, 2001), but concerns about competitive distortions with regard to emissions trading have actually been raised in the international negotiations, for instance by developing countries (e.g. SBSTA/SBI, 1999: 28) and the EU (e.g. COM, 1999a, 2000a) in the context of the Kyoto Protocol. This was also expressed on the political agenda. At CoP4 in 1998, the FCCC Parties drew up a work program, the Buenos Aires Plan of Action, containing several controversies to be decided upon, such as the desirability of "non-distortion of competition" in the context of international emissions trading (BAPA, 1998: Decision 7/CP.4, Article 17, issue 15). This means that, apart from its conditional — or arguably questionable — economic relevance, the competitive distortion issue has become politically relevant by playing a role in the international negotiation process. Not only efficient competition, but also fair competition appears to be an important consideration for politicians when it comes to allocating permits.

On the one hand, this should not lead to "paranoid" views about the legal incompatibility of flexible instruments as the WTO is not a self-enforcing legal regime, but relies largely on the willingness of one or more members to initiate a complaint (e.g. ENB, 2000). Moreover, it is uncertain whether a WTO panel, in case of a possible dispute concerning permit allocation, will regard an actionable subsidy in terms of macro-level efficiency and opportunity costs or in terms of micro-level financial effects and fair competition.

On the other hand, this uncertainty did make politicians fear the (high) political transaction costs of permit trading. The uncertainty was enhanced by information that the WTO would not only consider the direct and de jure effects of permit allocation, but also the indirect and de facto impact on trade (e.g. Werksman, 1999b). In addition, Parker (1998: 3) not only claims that certain parts of the WTO Agreements, such as ASCM Article 2 [specificity], are intentionally vague, but he also explains that several determinations (such as specificity, injury or serious prejudice) are highly fact-specific and are usually decided upon in practice on a case-by-case basis without clear guidelines. While recognizing the uncertainty resulting from the preliminary nature of the flexible instruments at that time, and the continuously evolving mandate of the WTO itself, Kim (2000) not only underlines that potential WTO conflicts do exist, but also finds that the likelihood of a dispute is high because of the far-reaching economic impact of (market-based) climate policy.

When anticipating the (impredictable) outcome of a hypothetical WTO dispute, decision makers realize that the WTO is not so much an economic organization, but rather a legal regime as well as a political institution (e.g. Parker, 1998). Vikhlyaev (2001: 19) stresses that the WTO dispute settlement mechanism is a "quasi-judicial" system in a political environment. From a neo-institutional perspective, this implies, among other things, that perceptions and power are likely to play a role in the panel's rulings.

With regard to perceptions, those who support the neoclassical economic view that an efficient market is a fair market (because marginal costs are equalized), such as the US, Canada and Australia, are likely to perceive the aforementioned international permit allocation dissimilarities as unproblematic. These actors will tend to view emissions trading as inherently fair (see also Hargrave et al., 1999b: 65) or even inherently WTO compatible (e.g. Petsonk, 1999: 22). The argument is that the WTO supports free trading and that emissions trading is a form of free trading, while the principle of state sovereignty precludes international interference with domestic permit allocation. Instead, those who support the neo-institutional economic view that equity is not equal to efficiency, such as the EU and several lobby groups (in particular those who will be worse off under the proposed allocation scheme), are likely to perceive international permit allocation dissimilarities as problematic. They are more inclined to view such differences as potential competitive distortions (see also Hourcade & Le Pesant, 1999: 9). The argument is that grandfathering leads to an unfair financial favor in comparison with auctioned foreign competitors.

With regard to power, it could be hypothesized that the US, who have been among the strongest defenders of international permit trading without restrictions, will be able to exert the most political pressure within the WTO. First, despite their withdrawal from the Kyoto Protocol, the US remains to be the largest GHG emitter in absolute terms. Second, Bush (2002) supports credit trading in GHG emission reductions and voluntary domestic permit trading initiatives are still being developed within the US, while it cannot be ruled out that the US joins the Protocol again in the future (for instance in the case that the Democrats win the elections). Third, the Americans have been more influential than the Europeans during the FCCC negotiations in the past with respect to defining and elaborating policies and measures (e.g. Ringius, 1999: 13).

Neither from an economic, nor from a legal or political perspective it can be said with certainty whether or not the WTO will regard upon grandfathered permits as an actionable subsidy when a competitor abroad has to buy its permits. There are some indications that the WTO could take the financial and fairness effects of permit allocation into consideration, but efficiency or state sovereignty considerations may well prevail. It is, however, probably too simple and insufficient to assume WTO compatibility of permit trading solely on the basis of the economic concept of opportunity costs. The legal ambiguity whether or not grandfathering will be considered an actionable subsidy adds to the switching costs of permit trading.

6.5.2. International Harmonization of Permit Allocation Rules

Because grandfathered firms in one country, ceteris paribus, have a stronger financial position than their auctioned competitors in another country, the industry is likely to lobby in favor of grandfathering. Some conclude that public choice theory would expect governments to mirror this powerful interest group preference, leading to a wide application of grandfathering. When permit trading is accepted, grandfathering is indeed the most used form of allocating the emission rights. However, one should not forget that most industries do not want to have their emissions capped in the first place and usually prefer to make existing environmental policy more flexible by using credit-based approaches. Moreover, next to the preferences of target groups, governments will take other considerations into account in their decision on permit allocation. It is possible that some governments will decide to use auctioning partially (or even exclusively) for specific sectors or sources, for example in order to enhance efficiency by lowering other pre-existing distortionary taxes, or to achieve equity by establishing property rights for polluters on the basis of a wealth transfer to the public (see also Welch, 1983; Grafton & Devlin, 1996). Some form of auctioning has, in fact, been considered by some governments, including Norway and the Netherlands (e.g. Commissie Vogtländer, 2002; Harrison & Radov, 2002).

The perception that international permit allocation dissimilarities may lead to distortions of competition has lead to an intense political debate — in particular in the EU — on the desirability of international harmonization of permit allocation rules (e.g. Golub, 1998; COM, 2000a). It is striking, in this respect, that the literature in favor of harmonization (e.g. Lefevere & Yamin, 1999) is predominantly European instead of American, for instance. A part of the explanation is that Americans are more market oriented and have more experience with the instrument of emissions trading than Europeans. Because a fundamental issue is the extent to which national sovereignty, as it currently exists, should be maintained (Rutgeerts, 1999), another part of the explanation is that the EU is more willing and used to transferring, sharing or pooling state sovereignty by means of harmonizing (environmental) legislation than countries outside their community (e.g. Vikhlyaev, 2001).

Under the Kyoto Protocol, the FCCC Parties already agreed to make coordination voluntary in Article 2. The CoP may decide to coordinate policies and measures if it considers coordination to be beneficial, thereby (among other things) striving to minimize adverse effects, including effects on international trade and social, environmental and economic impacts on other Parties (Article 2.3) as well as taking into account different national circumstances and potential effects (Article 2.4). Such coordination, if desirable, would be politically difficult, because it requires an additional intergovernmental decision. This explains that a coordination of permit allocation methods, so far, has not emerged in international agreements that elaborate the Kyoto Protocol.

The situation is different for the EU. There, the question whether or not to set international rules for domestic permit allocation shifted around the turn of the century to a discussion *to what extent* common rules should be formulated and what allocation aspects should be harmonized (Kerr, 1999; Yamin & Lefevere, 2000; for an early discussion see Koutstaal, 1997). Not only is EC law more likely to interpret grandfathering as a form of subsidization than WTO rules, as it refers more frequently and explicitly to financial effects and equity consequences than WTO rules, but the European Commission also feared that competitive distortions and state aid could arise under EC Article 87 when Member States would be left free to choose their allocation method (e.g. COM, 2000a). Therefore, grandfathering became the harmonized rule to allocate the permits in Europe. Article 10 of the EU Directive on emissions trading states that every Member State allocates at least 95% of its allowances free of charge during the period 2005–2007 and that at least 90% of the allowances is allocated for free in the period 2008–2012. Outside Europe, the harmonization issue received less attention, in particular after the US had withdrawn from the Kyoto Protocol.

The desirability of harmonization depends on what law and economics perspective one takes. Those who plead against harmonization use the opportunity cost argument and define competitive distortions in terms of inefficiencies (e.g. Hargrave et al., 1999b; Zhang, 1999b; Zhang & Nentjes, 1999). They argue that firms with grandfathered permits have no cost advantage over firms with auctioned permits because they face the opportunity costs of using the permits (although they do not face the expenditure of buying the permits). This implies that international permit allocation dissimilarities do not distort competition in its efficiency meaning (in perfectly competitive markets), so that no ground for harmonization would exist. However, this does not mean that these authors are against any form of international rules for emissions trading. They are actually in favor of establishing eligibility criteria which prescribe the minimum conditions (notably with respect to domestic permit definition, monitoring, reporting and enforcement) that countries have to satisfy before they are allowed to trade emissions internationally.[10] Nevertheless, those against harmonization rather emphasize that

---

[10] Also note that an efficiency orientation does not preclude harmonization if competition is imperfect, although neoclassical economists prefer to speak of coordination, by specifying which type of instrument should be allowed (see also Hoel, 1997).

international emissions trading equalizes marginal costs and lowers total costs of reducing emissions.

Those who plead in favor of harmonization use the level playing field argument and define competitive distortions in terms of inequity (e.g. Jepma et al., 1998; Lefevere & Yamin, 1999). They also tend to confide less in the assumption that the markets for permits, products and capital will be perfectly competitive. The main argument is that a grandfathered firm has a financial advantage over an auctioned competitor abroad, because the latter has to buy its permits, whereas the former receives them for free as a capital gift. This implies that the mere allocation of permits changes the level playing field and thus distorts competition according to the equity approach, whereas it could also distort efficiency in some exceptional cases under imperfect competition, which would require international harmonization of domestic permit allocation rules. However, this does not necessarily mean that these authors want to centralize all permit allocation decisions. Rather, they would like to formulate basic rules for domestic permit allocation, for instance whether or not — and on what basis (e.g. historical emissions or energy efficiency) — to grandfather to certain sectors (like the electricity sector). Such rules could be made part of the eligibility criteria that governments must meet to join the international permit trading system.

Contrary to what some believe (e.g. Petsonk, 1999; Zhang, 1999b), state sovereignty is maintained (arguably at a "lower" level) in the case of harmonization if these governments voluntarily accept such international rules on domestic permit allocation as part of the eligibility criteria. According to Paterson (1997), the "common" definition of sovereignty is that states successfully claim the monopoly of the legitimate use of physical force within a given territory (internal aspect) and states mutually recognize each other's territorial monopolies (external aspect). In this line of reasoning, states that agree to harmonize permit allocation do not undermine their sovereignty, but reduce their autonomy in order to achieve other goals that they might pursue, such as the prevention of unfair competitive distortions. Moreover, the sovereignty of these states remains crucial to enforce and implement such harmonized rules as part of their (inter)national climate policy.

6.6. Conclusion

Permit trading is economically efficient, but politically controversial. Permit allocation is the largest impediment to implementing this superior alternative (e.g. Ellerman, 1998). Neoclassical economists find this hard to understand, because permit allocation does not affect efficiency which should make permit trading "neutral" to any equity consideration (e.g. Ciorba et al., 2001: 8). However, this chapter demonstrates that permit allocation leads to inadmissible subsidization under WTO law if the concept of competitive distortion is interpreted in equity instead of efficiency terms. We will explore this formal institution by using neoclassical as well as neo-institutional law and economics perspectives (e.g. Medema et al., 2000). Although the Coase theorem implies that trade-offs between efficiency and equity can be avoided when allocating emission rights, we show that it does not mean that choices between an efficiency and equity view on the allocation of emission rights can also be avoided, once applied to specific and concrete laws.

Politicians feared that competition may be distorted when some governments would grandfather their permits, while others would auction them. In a neoclassical interpretation, (efficient) competition is not distorted, because not only auctioning, but also grandfathering entails costs for firms (e.g. Nentjes et al., 1995). Emission rights allocated for free have opportunity costs when they are used to cover the emissions of the permit owner. A firm with gratis emission rights has to include these costs (equal to the permit price) in the product price if it does not want to go bankrupt in the longer term. This means that it cannot ask lower product prices than its competitor with auctioned rights abroad. However, in a neo-institutional interpretation, (fair) competition is distorted, because grandfathered permits are a capital gift, which implies that a company with gratis permits has more financial resources than a comparable foreign firm with auctioned allowances. Although the lump sum subsidy of grandfathering leaves efficiency unaffected, it does imply a financial advantage that affects equity. The permit allocation itself leads to unequal changes of the competitive relations among competing firms. The "level playing field" is said to be distorted.

These neoclassical and neo-institutional economic perspectives are crucial to understand the potential legal barrier of grandfathering under the WTO. Equity appears to have some legal relevance even if it does not affect efficiency. According to the WTO Agreement on Subsidies and Countervailing Measures (ASCM), a "subsidy" exists if there is a financial contribution by a government which confers a benefit to a specific enterprise/industry and involves a (potential) direct transfer of funds, or if government revenue that is otherwise due is foregone or not collected. An "actionable subsidy" exists if the subsidy causes adverse effects or serious prejudice to the interests of other WTO Members, such as injury to their domestic industry.

Grandfathering is not an actionable subsidy in the efficiency interpretation of competitive distortions. There is no benefit because of its opportunity costs and the auctioned industry abroad is not injured as grandfathering does not affect efficiency. However, grandfathering could be seen as an actionable subsidy in the equity interpretation of competitive distortions, because grandfathered firms have a stronger financial position than their auctioned competitors abroad. This means that grandfathered firms have a benefit, which could be seen as to affect the interests of the domestic industry in the country that uses auctioning. Moreover, grandfathering could imply that revenue otherwise due is foregone, because the government would otherwise have collected revenues in the alternative of auctioning. However, even if grandfathering is regarded as a subsidy, it could be exempted by labeling it "non-actionable", which is possible under the ASCM (among other things) if the assistance is a one-time non-recurring measure that promotes the adaptation of existing facilities to new environmental requirements.

This law and economics analysis can be used as an input to another neo-institutional perspective, namely one that focuses on perceptions in politics. On the one hand, there was a perceived legal ambiguity about whether grandfathering constitutes an actionable subsidy, even within the WTO itself (e.g. Vaughan, 1999). Concerns about competitive distortions and subsidization with regard to emissions trading have actually been raised in international climate negotiations, for instance by the EU. Furthermore, some (European) authors perceived an international harmonization of permit allocation rules as necessary to avoid a distortion of fair competition. On the other hand, many governments and authors did not perceive grandfathering as distorting by following the opportunity cost argument. In fact, politicians never agreed upon an equity-driven international harmonization of allocation methods, for instance because they perceived it as unnecessary from an efficiency point of view or as undesirable from a state sovereignty point of view.

Although grandfathering is financially more attractive and acceptable to firms than auctioning, it is unclear whether grandfathering is also readily acceptable to governments because it could constitute an actionable subsidy under WTO law if it is seen as a distortion of fair competition. Permit allocation dissimilarities were perceived to be legally "fit" in some (efficiency), but not necessarily in all (equity) respects. This legal ambiguity added to the switching costs of permit trading, which made some contribution to the institutional lock-in of building sub-optimal, credit-based instruments upon extant environmental policy. Under the latter type of flexible instruments, the allocation of emission rights remains implicit, whereas permit trading requires an explicit choice between grandfathering and auctioning. The next chapter extends this analysis to the legal framework of the EU.

Chapter 7

EC State Aid Law and Distortions of Market-Based Climate Policy

7.1. Introduction

This chapter extends the analysis of the previous chapter to the context of the European Union (EU) by determining the economic and legal conditions, both in terms of efficiency and equity, under which differences in the domestic allocation of emission rights between Member States distort competition and violate the state aid prohibitions and the polluter pays principle under the law of the European Community (EC).[1] An empirical analysis on the basis of the state aid decisions of the European Commission in the Danish and British emissions trading cases is also provided.

Permit trading faces some unique institutional barriers that arise from its explicit (re)allocation of emission rights. This remains implicit under credit-based instruments, which contributes to the tendency of politicians to make existing environmental policy more flexible by using such incremental and sub-optimal approaches. Although economists argue that permit trading is superior, some European governments feared that competitive distortions and violations of EC state aid rules may arise if they would grandfather their permits, while others would auction them.

Similar to the previous chapter on WTO subsidies law, we will show in this chapter that grandfathering violates EC state aid law if the financial effects of grandfathering are considered from an equity perspective. To capture the legal relevance of both efficiency and equity considerations when allocating permits, we will employ both neoclassical and neo-institutional law and economics perspectives (e.g. Medema et al., 2000). Despite these perceived allocation problems, the EU adopted a Directive in 2003 that creates permit trading in

---
[1] For those who do not know the difference between the EC and the EU, it is sufficient to know when reading this chapter that, put simply, the EC represents the legal framework, whereas the EU is a broader political institution that refers to the legal as well as political co-operation among its Member States, for instance in the field of a common foreign policy in Europe.

the EU from 2005 onwards with grandfathering as the harmonized rule (e.g. Woerdman, 2004a,b). The EU thus managed to break out from inefficient standards and ineffective taxes in climate policy, but equity is still relevant to explain the legal form of permit allocation in the design of the emissions trading system in Europe.

This chapter is organized as follows. Section 7.2 analyzes the economic conditions, both in terms of efficiency and equity, under which differences in domestic permit allocation methods across the EU lead to competitive distortions. Section 7.3 uses these economic tools to specify the legal conditions under which grandfathering could be seen as state aid under EC law. Section 7.4 supplements our law and economics analysis by taking a neo-institutional approach that focuses on perceptions in politics. This is done to find out what efficiency or equity interpretation of competitive distortions has guided (a) the state aid decisions of the European Commission on grandfathering in Denmark and the United Kingdom and (b) the legal design of the EU-wide permit trading scheme. Section 7.5 discusses the possibilities of extending our theoretical framework to another legal issue concerning permit trading, namely that of the (in)compatibility between grandfathering and the polluter pays principle. Finally, Section 7.6 presents the conclusion.

7.2. Economic Analysis of Permit Allocation and Competitive Distortions

Although policy makers were confronted with different calculations, most studies estimated that the EU could lower its compliance costs, under the targets of the Kyoto Protocol, by about 30% (e.g. Capros et al., 2000; Svendsen & Vesterdal, 2003). The EU was still skeptic towards emissions trading at the end of the 1990s, but in 2003 it adopted a Directive that enables such trading in the EU from 2005 onwards. The reasons both for its initial resistance and for this remarkable attitude change will be discussed in the next chapters.

Before the Directive was drafted, the European Commission presented a Green Paper on GHG emissions trading within the EU in March 2000 (COM, 2000a). The purpose of that document was to stimulate a discussion on this topic among stakeholders, scientists and politicians. Because the responses to the Green Paper were "overwhelmingly in favor of emissions trading" (COM, 2001a: 2), the Commission presented a proposal in October 2001 for a Directive on GHG permit trading for large emitters in the EU to start in 2005. Not only in its Green Paper, but also in its proposal for a Directive, the Commission expressed its fear that competition is distorted and EC state aid law is violated if some Member States would grandfather their permits, while others would auction them (e.g. COM, 2000a: 5, 2001a: 11).[2] A number of scientists, including some lawyers, shared this view (e.g. Cozijnsen, 2000: 5). This, in turn, could imply that governments should not be left free to choose any domestic permit allocation rule. Such rules must then be harmonized.

The question, which was unclear at the time, is under what conditions grandfathering constitutes state aid. This will be discussed below (e.g. Woerdman, 2003). Our economic framework uses neoclassical and neo-institutional law and economics by making a distinction between an efficiency view and an equity view of competitive distortions. This analytical framework, that partly builds upon van der Laan & Nentjes (2001), has already been outlined in the beginning of the previous chapter. Therefore, in the following subsections we will only briefly summarize it. For a full discussion, the reader is referred to the previous chapter.

7.2.1. Competitive Distortions, Efficiency and Opportunity Costs

From a neoclassical perspective, a competitive distortion is basically seen as a distortion of efficient competition. International differences in domestic permit allocation procedures will not lead to such a distortion. The reason for this is that not only auctioning, but also grandfathering entails costs for firms: grandfathered permits have an opportunity cost when they are used for covering the emissions of the permit owner (e.g. Grafton & Devlin, 1996).

The opportunity cost is the revenue foregone by not selling the grandfathered permits but using them in producing output. This opportunity cost, which is equal to the price for which the permit can be sold, must be included in the product price if the firm does not want to go bankrupt in the longer term. Instead of using them, the firm could have sold the permits. The revenue foregone is a cost to the firm, comparable with the "interest foregone" on own capital. This means that grandfathered firms do not have a cost advantage over auctioned firms abroad (or over domestic newcomers) just because they received permits for free. Because of this, they cannot ask lower product prices than their competitors with auctioned

---

[2] In this (proposal for a) Directive, the Commission makes a distinction between permits and allowances. A permit is a non-tradeable authorization to emit greenhouse gases. This permit simply contains a description of the name and address of the polluter, the activities of the installation and the monitoring methods as well as the obligation to surrender allowances equal to the total emissions of the installation in each calendar year. An allowance is a transferable authorization to emit 1 tonne of carbon dioxide equivalent during a specified period. In this legal terminology, only allowances can be traded, but no trade is possible in permits. However, in the traditional terminology of environmental economists, this type of system is still referred to as permit trading (or allowance trading). Therefore, in this chapter we conform with the broader climate change literature on emissions trading in which permits and allowances are usually different words for the same thing, namely the entitlement to emit 1 tonne of carbon dioxide equivalent.

rights abroad. There are no price distortions when comparing grandfathering with auctioning.

However, grandfathering does imply a transfer of wealth to firms, since they receive an input which has a certain market value. Therefore, grandfathering permits could be viewed as granting a subsidy to the firm. However, this subsidy is a capital gift that has the character of a lump sum subsidy (Jensen & Rasmussen, 1998). In efficiency terms, a lump sum subsidy is not distorting in the product market because it does not affect marginal emission reduction costs. Grandfathering does not alter the output and price decisions of firms.

Some believe that grandfathering permits implies a competitive distortion because it would have the same distorting effect as granting tax exemptions to certain sectors or firms. However, grandfathered permits are different from tax exemptions, precisely because of their opportunity costs. A tax exemption is inefficient because it induces different prices per unit of emission for different firms. Tax exemptions are not capital gifts and do not have opportunity costs, but rather imply that emissions are an input without a price. However, grandfathered permits have opportunity costs and therefore entail a price. Contrary to tax exemptions, efficiency is not distorted by using grandfathering instead of auctioning.

The arguments presented above assume perfectly competitive markets. Imperfect competition is unlikely if politicians allow for the direct participation of private entities in a European (and finally international) carbon emissions trading system that creates a thick market with many traders. If politicians succeed in doing this, which is not (yet) completely realized, there are only a few exceptional cases of imperfect competition where a competitive distortion could arise. An example is a situation on an imperfect product market where a firm with grandfathered permits starts a price war with a competitor abroad that had to buy permits in an auction (Nentjes et al., 1995). Although a grandfathered firm can outlast the auctioned firm because of its larger capital reserve, it is a risky and expensive strategy that, in addition, could lead to prosecution by the EU authorities that enforce antitrust policy.

7.2.2. Competitive Distortions, Equity and Level Playing Field

From a neo-institutional perspective, a competitive distortion is basically seen as a distortion of fair competition. Grandfathered permits are a capital gift, which implies that a company with gratis permits has more financial resources than a comparable foreign firm with auctioned allowances. A grandfathered firm simply does not have to buy its permits contrary to an auctioned competitor, so that the former has a lower cash outflow. Although the lump sum subsidy of grandfathering as well as its opportunity cost leave efficiency unaffected, it does imply a financial advantage that affects equity.

In a dynamic setting, fair competition requires equality of changes in the competitive relations between firms. The level playing field approach does not reject that the competitive positions of firms can be (and remain) unequal as a result of their economic activities and strategies, but rejects that their competitive positions are changed because of the political process of permit allocation itself. Consequently, the level playing field is maintained if permit allocation leaves these competitive relations unaltered.

This means that a competitive distortion arises if the introduction of an environmental policy measure leads to unequal changes of the competitive relations among comparable firms, thereby reducing the (perceived) fairness of (inter)national competition. According to this interpretation, international differences in domestic permit allocation procedures will lead to competitive distortions. Although these distributional differences are not inefficient, the "level playing field" is said to be distorted.

7.3. Legal Analysis of Permit Allocation and EC State Aid Law

The neoclassical and neo-institutional economic perspectives discussed above can be used to guide the legal analysis of permit allocation in the EU. Its Member States faced the question whether grandfathered permits should be interpreted as a form of state aid under EC Article 87 or not.[3] To find an answer to this question, we will analyze (a) EC Article 87 on state aid, (b) Commission (and Court) decisions and reports on state aid and (c) the (revised) Community guidelines on State aid for environmental protection. In a similar vein as the WTO issue concerning actionable subsidies as discussed in the previous chapter, we will demonstrate that the state aid issue in the EU largely depends on whether EC competition law has defined and/or interpreted — and whether the European Commission (or the European Court of Justice in case of a dispute) will regard — (potential) state aid issues in terms of (a) efficiency and opportunity costs or also in terms of (b) financial effects on, and fair competition (a "level playing field") between, firms.

There are at least five similarities between EC law and WTO rules in the discussion on permit allocation. First, just as in the case of the WTO, the debate whether permits are goods or services plays a minor role and the general opinion seems to be that permits are neither goods nor services (e.g. Lefevere & Yamin, 1999; Werksman, 1999b). Second, both EC law and WTO rules require

---

[3] The Treaty of Amsterdam entered into force on 1 May 1999 and amended the numbering of the Articles of the Treaty of Rome of 1957. The current Article 87 was originally numbered Article 92 under the Treaty of Rome.

non-discrimination when allocating permits. The national treatment principle and the most-favored nation principle of the WTO as well as Article 12 of the EC Treaty and its interpretation by the Court prohibit (arbitrary) discrimination on the grounds of nationality. Third, the crucial issue of allocating permits both in an EC and WTO context is whether differences in permit allocation between states in general and whether grandfathering in particular could constitute a form of subsidization. Fourth, as will be demonstrated below, the relevance of EC Article 87(1) on state aid, similar to the WTO Subsidies Agreement, mainly depends on the distinction between efficiency (opportunity costs) and equity (financial positions). Fifth, a WTO dispute panel as well as the Commission and the Court of the EC will decide upon this matter on a case-by-case basis, thereby considering not only the direct and actual effects, but also the indirect and potential effects of permit allocation and grandfathering.

However, there are also at least three important differences between EC law and WTO rules that are relevant for permit allocation. First, EC law places a stronger emphasis on maintaining fair conditions of competition than the rules of the WTO (Vikhlyaev, 2001: 25). Rules, decisions and documents of the EC on competition and state aid refer more frequently and explicitly to financial effects and equity consequences than relevant GATT/WTO provisions.[4] Second, Article 174 of the EC Treaty recognizes the importance of the polluter pays principle, contrary to WTO practice (Vikhlyaev, 2001: 32). This means that the question whether grandfathering still means that polluters pay is likely to play a more prominent role in economic, political and legal discussions in the EU than in the context of the WTO. Third, EC law contains more possibilities for (and the EU Member States are more willing and used to) transferring, sharing or pooling state sovereignty by means of harmonizing (environmental) legislation compared to WTO rules (and compared to the international community at large) (e.g. Vikhlyaev, 2001: 23–24). The legal tendencies towards (and the political demand for) the harmonization of permit allocation are likely to be stronger inside than outside the EU.

EC Article 87 on state aid has been elaborated in various Commission documents and Court decisions. In 2001, the Commission adopted the revised Community guidelines on State aid for environmental protection (OJ, 2001). Compared with the previous guidelines (OJ, 1994), the revised ones contain, for instance, more explicit and detailed elaborations of definitions and rules, pay more attention to sustainable development, and place a stronger emphasis on cost internalization in relation to economic or market instruments. The revised guidelines also mention the flexible instruments of the Kyoto Protocol, such as international emissions trading, but their potential effects on state aid are not

---

[4] Examples are EC Article 81, COM (1998a: 79, 1999b: 84, 2000a: 7, 2000b: 75, 2001a: 11).

elaborated. Although the revised guidelines do not differ much from the old ones, they do contain some new and more detailed instructions and exemptions, which are helpful for our analysis.

The possible implications for permit allocation of these and other state aid provisions in the context of EC Article 87 will be dealt with below. After discussing the criteria for state aid, the conditions will be analyzed under which state aid is exempted from its prohibitory status.

7.3.1. Permit Allocation and State Aid Criteria

Article 87(1) on state aid as formulated in the EC Treaty determines that "(…) any aid granted by a Member State or through State resources in any form whatsoever which distorts or threatens to distort competition by favoring certain undertakings or the production of certain goods shall, in so far as it affects trade between Member States, be incompatible with the common market". Two lines of reasoning can be applied following this legal text. According to the opportunity cost argument, as was shown in the previous section, grandfathered firms have no cost advantage over auctioned firms, which would imply that grandfathering is no aid at all or constitutes aid which does not distort efficiency, trade and competition between Member States. However, according to the level playing field argument, grandfathering distorts competition and should be regarded as state aid, because grandfathered firms receive a capital gift from a Member State which financially advantages or favors those firms relative to their auctioned competitors abroad. The definition of state aid in the EC Treaty is thus insufficient to decide whether grandfathering should be seen as state aid or not.

Fortunately, this subsidization concept has been elaborated by the European Commission (e.g. COM, 1999b: 84, 2001c: 86) and by the European Court of Justice (e.g. Case E/1/98 and E/2/98 of the Flemish region versus the Commission). Both describe state aid in terms of an "advantage". This could suggest that grandfathering should be regarded as state aid, since grandfathered permits are a capital gift which implies a financial advantage for firms. Moreover, the Commission and the Court recognize that the form in which the aid is granted is irrelevant and covers all financial means (COM, 2001c: 83–84). However, it could also be argued that grandfathered firms have no cost advantage over auctioned firms by following the opportunity cost argument. To obtain more clarity, we will examine the four criteria that the European Commission (COM, 2000b) uses to determine whether a measure is to be regarded as state aid which is incompatible with the common market. A measure is considered to be state aid if it satisfies the criteria of both (a) state origin, (b) firm advantage, (c) specificity and (d) trade effect.

With respect to the first criterion of state origin, the aid must be granted by the State or through state resources. It could be claimed that grandfathering (although it is a transfer of permits) is not a genuine or direct transfer of resources, since the permits are allocated for free by the State. Nevertheless, the opposite can well be defended by stressing that these permits have market value and that the capital gift induced by grandfathering is an (in)direct transfer of state resources. Furthermore, Jepma et al. (1999) indicate (on the basis of COM, 1998a) that the state origin criterion requires a transfer of resources from the State (or in the State) receiving, actually or potentially, less revenues in order for state aid to exist. It could be argued that the State will receive less revenues in the case of grandfathering compared to either (pre-existing) taxation or auctioning because grandfathering can be interpreted as giving the (hypothetical) auction revenue to the polluters (e.g. Welch, 1983: 168).

With respect to the other three criteria, grandfathering should be seen as an advantage that affects trade by favoring specific firms and thus distorts competition according to the level playing field approach, but not according to the reasoning of the opportunity cost approach, as we have seen. Interestingly, the Commission not only mentions the desirability of a level playing field, where firms are treated on an equal footing, in the context of state aid (e.g. COM, 1998a: 79; OJ, 2001: 13), but it also describes the firm advantage criterion as a financial advantage that improves a firm's market position (e.g. COM, 1999b: 84).[5]

Furthermore, the Court finds that also a relatively small amount of aid does not exclude the possibility that trade may be affected (COM, 2001c: 88). In a similar fashion, the Commission emphasizes that when the State confers even a limited advantage on an undertaking which is active in a sector characterized by competition, there is a distortion or risk of distortion of competition (COM, 2000b: 75), which could run counter to the Commission's goal to ensure the competitive functioning of markets (OJ, 2001: 5). The Court has also specified that aid constitutes an advantage conferred on a firm by the public authorities without payment (or against a payment which corresponds only to a minimal extent to the figure at which the advantage can be valued) (COM, 2001c: 86). However, it can also be argued that grandfathering does not distort efficiency or trade because its opportunity costs will be reflected in the product price. From this perspective, grandfathered permits internalize costs as much as auctioned permits do. This aspect would then imply that grandfathering is allowed, because cost internalization is a priority objective in the Commission's policy on the control of state aid (OJ, 2001: 5).

---

[5] This view seems to be reinforced by Jans (1995: 262) who describes specificity in the context of state aid as benefits awarded to specific industries or undertakings, which have the effect of favouring their financial or competitive position in comparison with their competitors.

Consequently, neither Article 87(1) on state aid nor its elaboration by the European Commission or the European Court of Justice seems to provide a decisive answer whether grandfathering should be regarded in terms of efficiency (opportunity costs) or equity (financial advantage). In a broader sense, according to van der Laan & Nentjes (2001), it appears that European law not only contains instances of the efficiency interpretation of competitive distortions, but also of its equity interpretation. For example, the efficiency view can be found in Article 130S of the Single European Act and in some environmental directives, like the Titanium Dioxide Directive. The equity view can be found, not only in the preamble of the Treaty of Rome ("fair competition") and in Article 81 of the EC Treaty ("distortion of competition (…) apply dissimilar conditions (…) placing them at a competitive disadvantage"), but also in the preambles of some environmental directives, such as the Drinking Water Directive ("a disparity between national legislation results in (…) a distortion of the competition") and the Sulphur Directive ("unequal conditions of competition").

7.3.2. Permit Allocation and State Aid Exemptions

Not all state aid is prohibited under European competition law. Article 87(3) as well as the Community guidelines on State aid for environmental protection (OJ, 1994), which have been revised a few years ago (OJ, 2001), provide the basis for the exceptions under which state aid is to be regarded as compatible with the common market.[6] In short, state aid can be allowed if:

(1) the aid promotes the execution of an important project of common European interest;
(2) the aid remedies a serious disturbance in the economy of a Member State;
(3) the aid facilitates the development of certain economic activities or areas;
(4) the European Council decides that the aid is compatible with the common market.

First, even if grandfathering should be seen as state aid, it can nevertheless be allowed on the basis of Article 87(3)(b) if the aid is used to promote the execution of an important project of common European interest. In 1987 the Court recognized (*Glaverbel* Case 62/87) that concerted action by a number of Member States to combat environmental pollution is an example of an important project of common European interest. The question is, however, whether grandfathering literally "promotes" climate change mitigation, as this section of the law would

---

[6] The Community guidelines on State aid for environmental protection (OJ, 1994) entered into force in 1994 and expired (after two postponements) on 31 December 2000. The revised guidelines (OJ, 2001) were put into action in 2001 and will cease to be applicable on 31 December 2007.

require. On the one hand, it can be argued that grandfathering "promotes" climate change mitigation, because emissions trading becomes more politically acceptable to firms as a result of the wealth transfer grandfathering induces (e.g. Koutstaal & Nentjes, 1995). On the other hand, it can be argued that grandfathering does not "promote" climate change mitigation when emissions trading is politically unacceptable to governments, for instance if the equity perception dominates that emission sources should not obtain the right to pollute for free and therefore must pay for it by purchasing the permits from society (Grafton & Devlin, 1996). If grandfathering is seen as state aid, it is not allowed if it merely helps firms to comply with Community regulation, unless it stimulates firms to pollute less than legally required (OJ, 2001: 6). In addition, the aid must be necessary for the adoption or continuation of the project (OJ, 2001: 6, 13), which can only be defended if the political acceptance of emissions trading exclusively hinges on grandfathering.

Second, if grandfathering would be state aid, Article 87(3)(b) also allows the aid if it is used to remedy a serious disturbance in the economy of a Member State. It is clear that auctioning (as well as taxation) entails a financial burden for polluters, but it is not evident that auctioning would thus create a "disturbance" in the economy of a Member State that is "serious" enough to allow for grandfathering as the remedy against it. Internalizing the costs of pollution by means of emissions trading is not a disturbance, but rather a correction of the economy (e.g. Bohm, 1999). Auctioning is not a disturbance either: it is even more efficient than grandfathering, because the auction revenues can be recycled to lower distortionary taxes (e.g. Goulder et al., 1999). However, if grandfathering should be seen as aid, it may still be allowed provided that it is seen as a temporary second-best solution (OJ, 2001: 5). In addition, because of its financial effects, grandfathering prevents that the competitiveness of firms — deemed important by the Commission (OJ, 2001: 5) — is reduced, as long as competitors abroad are not subject to an emission cap. This argument is valid for firms that compete on international markets, for instance with (uncapped) firms in the US, but the argument is invalid for firms that (almost) only have domestic or European competitors with emission caps.

Third, if grandfathering would be state aid, it may be exempted from the state aid prohibition on the basis of Article 87(3)(c), which provides that aid may be considered compatible with the common market if it facilitates the development of certain economic activities or of certain economic areas, where such aid does not adversely affect trading conditions to an extent contrary to the common interest. On the one hand, if the opportunity costs of using grandfathered permits imply that they do not affect efficiency and hence trading conditions, it could also be argued that they have the same effect as auctioned permits on the development of certain economic activities or areas precisely because they have no cost advantage. On the other hand, it could be defended that grandfathered permits do facilitate the development of certain economic activities or areas, because they are a capital gift giving the firm a stronger financial position than under auctioning. It could also be argued that the "conditions" of trading do not so much refer to the trading itself, but rather to the level playing field or the prerequisites for fair competition. This could suggest that permit allocation differences between Member States should be seen as harming the common interest by affecting the level playing field according to the equity view of a competitive distortion.[7]

Fourth, if grandfathering would be state aid, it could be allowed, in principle, on the basis of Article 87(3)(e), which refers to the discretionary power of the European Council to decide by qualified majority — on the basis of a proposal by the Commission-that an aid measure is compatible with the common market. In itself, this provision does not help to judge the relevance of the neoclassical efficiency argument or the neo-institutional equity argument in EC law and politics. However, it underlines that the issue whether permit allocation differences between Member States are desirable, and whether grandfathering should be seen as state aid, is likely to be decided upon not just on the basis of legal indications, but also on the basis of political considerations.

In this context, the Community guidelines on State aid for environmental protection could suggest that grandfathering is only allowed temporarily (OJ, 2001). This is the case if we assume that emission permits are a form of operating licences. An emissions trading system legally requires firms to have (grandfathered or auctioned) permits that cover the emissions to be allowed to operate. In this interpretation, the provisions for operating aid apply, provided that grandfathering should be seen as aid in the first place, of course. These provisions make clear that firms may receive the aid no longer than 5 years. If the level of aid is the same during these years ("non-degressive aid"), a firm may receive no more than 50% of the extra costs necessary to meet the environmental objectives. If the level of aid decreases each year ("degressive aid"), which should be the general rule, the intensity may amount to 100% of the extra costs in the first year, but must have fallen in a linear fashion to zero by the end of the fifth year. Tax exemptions

---

[7] In practice, Article 87(3)(c) has already been used several times to allow for state aid. Examples are the energy-intensive industries in the Netherlands, Denmark, Norway, Sweden and Finland, which were (temporarily) exempted from a tax on CO_2 emissions (Jans, 1995; Baron, 1997; Heller, 1998). Some make a comparison by claiming that grandfathering will be exempted from state aid as well because it resembles a tax exemption, but we have already explained in the previous section that tax exemptions distort efficiency, whereas grandfathered permits do not distort efficiency because of their opportunity costs. This means that if grandfathering is to be regarded as state aid, it will be exempted from the state aid prohibition rules even more easily than the aforementioned tax. The Commission already takes what it calls a "flexible" approach (COM, 2001c: 83) towards inefficient tax exemptions, whereas grandfathered permits are even more efficient than the already allowed tax exemptions.

are also seen as operating aids and may only be granted, among other things, for a limited period of time with a maximum of 10 years (or 5 years in the case of energy-efficiency improvements). A temporary relief from environmental taxes may be authorized by the Commission to hedge against the risk of losing international competitiveness. This could imply that grandfathering, which could be introduced to accommodate similar competitiveness concerns, may also be deemed compatible with European law as long as it is a temporary transition (of possibly 5 or 10 years) to an auctioned scheme.

If governments would agree upon a 5 year transition period, the implication could be that the annually (re)allocated permits are allowed to be grandfathered for 100% in the first year, 80% in the second year, 60% in the third year, 40% in the fourth year and 20% in the fifth year, so that all permits are auctioned (and thus 0% is grandfathered) in the sixth year when the transition period is over. Without considering the state aid issue, the option to start with grandfathering and provide a gradual transition to an auctioned scheme in the EU is mentioned in various studies (e.g. Harrison & Radov, 2002).[8]

However, it should be noted that the connection between competitive distortions and state aid need not be established to begin with. There are two economic interpretations of the competitive distortion concept, namely one in (neoclassical) terms of efficiency and one in (neo-institutional) terms of equity (step 1). It was outlined above that these interpretations clarify under what conditions grandfathering is likely to be seen as state aid (step 2). Nevertheless, it is still possible that grandfathering is seen as problematic without turning to state aid regulation (and thus without making step 2). When permit allocation differences across the EU are not seen as a form of subsidization, they may still be perceived as to distort competition, namely in the equity view (albeit not in the efficiency view). In that case, according to Articles 94–97 of the EC Treaty, the Council (based on a proposal by the Commission) can issue the necessary directives to combat these distortions, such as harmonization measures.

7.4. Political Analysis of Perceptions on State Aid

From a law and economics perspective, it is ambiguous whether grandfathering, in the case of permit allocation dissimilarities between Member States, satisfies each of the aforementioned four criteria for state aid to exist (namely state origin, firm advantage, specificity and trade effect), and if it does, whether or not it will be exempted from the state aid injunction. We have shown that this ambiguity, which

---

[8] However, the Community guidelines only allow firms to receive aid over their extrainvestment costs necessary to reduce emissions, so that it could be argued that they may not receive aid over their entire emissions.

helps to explain why permit trading does not rank first in the political hierarchy of market-based climate policy, is a matter of perspective. Similar to the WTO issue on actionable subsidies discussed in the previous chapter, the issue of grandfathering as state aid in the EU appears to depend on whether one is willing to accept the opportunity cost argument or the level playing field argument.

Some have argued that grandfathering may be difficult to implement in Europe, because the EC rules and decisions on competition and state aid, as discussed above, refer more frequently and explicitly to fair competition, a level playing field and equity consequences than the relevant provisions which define and specify an actionable subsidy under the WTO Subsidies Agreement discussed in the previous chapter.[9] Nevertheless, others have argued that grandfathering may be easy to realize in Europe because the exemptions to the state aid injunction in EC law are legion relative to comparable WTO provisions. In line with the general remarks by Vikhlyaev (2001: 4), this suggests that EC state aid law is less stringent than WTO subsidies law, not only with respect to labeling grandfathering as a form of subsidization, but also with respect to exempting grandfathering from the legal objections against it.

Without conducting such an analysis, van der Laan & Nentjes (2001) advised to analyze decision-making processes to increase our knowledge of whether the efficiency or the equity interpretation of competitive distortions prevails in European environmental politics. Therefore, this section extends and supplements the neoclassical and neo-institutional law and economics perspectives, as performed above, with another neo-institutional approach that considers the perceptions of political actors on these competitive distortion and state aid issues. We also take an empirical approach, as several authors desire (e.g. Mackaay, 2000), by applying our law and economics theory to the decisions of the European Commission in the case of grandfathering as (exempted) state aid in the domestic carbon emissions trading systems of Denmark and the UK.

7.4.1. Perceptions in Political Negotiations on Permit Allocation

Although some authors, from an efficiency perspective, argue that the "(...) diversity of allowance allocation rules in Europe (...) should not be considered an obstacle to implementing an EU trading system" (Viguier, 2000: 6), the fact is that

---

[9] Yamin & Lefevere (2000: 30) argue: "The aim of competition provisions of EC law is to secure a level playing field for competitors wherever they are located in the EC. This notion goes beyond the narrow conception of competitiveness which economists focus on". Likewise, van der Laan & Nentjes (2001: 148) find: "The Treaty is not as one-dimensional as economists may believe or wish". In a similar fashion, Cini & McGowan (1998: 158) conclude that, in practice, multiple policy objectives are "fundamental in tempering the neo-liberal rhetoric of the state aid directorate".

some political actors at least perceive this to be a potential obstacle, as concerns about competitive distortions with regard to emissions trading have actually been raised in the international climate negotiations, for instance by the EU itself (e.g. COM, 1999a). Permit allocation in relation to competitive distortions and state aid also appeared on the internal political agenda of the EU, given the European Commission's reference to those issues. In its proposal for a Directive on GHG emissions trading in the EU, for instance, the Commission writes: "(...) it is feared that if allowances were allocated on the basis of auctioning in one Member State but allocated free in another, competition may be distorted. (...) The proposal does not spell out what would be consistent or inconsistent forms of allocation with regards to State aid as each situation will have to be examined on its merits, [but] (...) State aid scrutiny examines the possible distortions of competition (...)" (COM, 2001a: 11–12). Another example is provided by van Heukelen of the Commission, who not only writes that Member States should ensure fair competition, but also that grandfathering, although "more politically expedient" than auctioning, entails "dangers of state aid" (van Heukelen, 2000: 11).

The discretionary power of the European Council to decide — on the basis of a proposal by the Commission — that grandfathering is exempted from the state aid provisions means that a political decision will be pivotal to the issue of permit allocation differences between EU Member States. Financial and equity arguments are likely to play a role in such a decision, not only with a view to the historical relevance of the level playing field argument in European environmental legislation (Hargrave et al., 1999b: 11) and state aid policy (Cini & McGowan, 1998: 158), but also considering the continuously recurring reference made by the Commission in its Green Paper on GHG emissions trading in the EU to "fair competition" (pages 7 and 12), "conditions for equal competition" (page 14) and a "level playing field" (page 15) as well as to the relation between competitive distortions and financial (dis)advantages (page 19) for firms (COM, 2000a).[10] This could imply that a Member State will not be left free to grandfather (or auction) permits in the amount and to whom it likes, but rather that permit allocation rules will be harmonized across the EC to avoid unfair competitive distortions.

The latter conjecture is supported by the Commission's proposal for a Directive on GHG permit trading in the EU. Although in its draft proposal, the Commission "would not harmonize the method of allocation and quantities of allowances (...)" (COM, 2001b: 3), in its final proposal, the Commission (still does not want to harmonize the quantities of allowances issued, but) wants to harmonize the method of allocation, not only for the period 2005–2007, but also for the period

---

[10] However, the Green Paper does mention the concept of opportunity costs once (page 20) when discussing the issue of new entrants to the European carbon trading market (COM, 2000a).

2008–2012. The Commission proposes that "(...) from 2005 to 2007 all Member States allocate allowances to participating installations for free. (...) Without such harmonization, (...), competition may be distorted. (...) By 30 June 2006 the Commission will review the experience gained during (...) the period 2005–2007 with a view to ascertaining which harmonized method would be most appropriate (...) in the period 2008–2012" (COM, 2001a: 11).

The reason for the Commission to propose grandfathering for the period 2005–2007 is that it is a preliminary phase, in which there are no GHG emission targets for the Member States under the Kyoto Protocol against the background of an unknown allowance price (COM, 2001a: 3, 11). This changes when the first commitment period 2008–2012 of the Kyoto Protocol commences: an allowance price will develop and the Member States will be subject to legally binding targets. Although the Commission left open, in its proposal, whether to grandfather or auction during the first commitment period, it did indicate to prefer a harmonized method, since combinations of grandfathering and auctioning may distort competition and could lead to state aid (as will be ascertained on a case-by-case basis) (COM, 2001a: 11). In the draft proposal, the Member States are required to publish and submit in advance to the Commission a national allocation plan that has to "(...) include information on (...) the measures taken to ensure that allocation will be equitable and in conformity with Article 87 of the Treaty" (COM, 2001b: 29). The reference to this Article was left out from the list of criteria in the final proposal, but a text against unduly favors and discrimination was included instead: "The plan shall not discriminate between companies or sectors in such a way as to unduly favor certain undertakings or activities (...)" (COM, 2001a: 35). Moreover, reference to Article 87 is still made in other parts of the final proposal, for instance: "Member States shall ensure that decisions taken pursuant paragraph 1 [on the period 2005–2007] or 2 [on the period 2008–2012] are in conformity with the requirements of the Treaty, in particular Articles 87 and 88 thereof" (COM, 2001a: 25).

In the environmental matter of using GHG emissions trading to combat climate change, the views of the European Parliament are relevant since the co-decision procedure applies.[11] In the draft report of the Parliament (EP, 2002), several amendments are proposed to the Commission proposal for the Directive on emissions trading. For instance, the Parliament wants to incorporate the polluter pays principle in the Directive which is left out in the Commission's proposal (EP, 2002: 20). Moreover, the Parliament wants Member States to allocate 70% of the allowances by means of grandfathering on the basis of 1990 emissions and 30% by

---

[11] In this procedure, the Council of Ministers decides under a qualified majority rule (more than 2/3 of the votes), but the European Parliament can block the Council decision with a majority of votes. This means that without the consent of the Parliament, the Council decision cannot be carried out.

means of an auction for the period 2005–2007, instead of the 100% grandfathering that the Commission proposes. Part of the justification is also "(…) to allow newcomers to enter the market on fair conditions (…)" (EP, 2002: 9). Instead of postponing the decision on allocation for the period 2008–2012 as the Commission does, the Parliament wants that it should be made clear now that from 2008 all allowances will be auctioned. "Common coordinated climate-change policies must (…) ensure (…) that the internal market is not distorted" (EP, 2002: 19).

When reading the literature on harmonization, which is predominantly European, a clear distinction should be made between the literature that discusses the harmonization of emission targets (e.g. Faure, 2001) and the literature that discusses the harmonization of policy instruments to achieve those emission targets (e.g. Viguier, 2000). As part of the latter, those who plead against harmonizing permit allocation procedures primarily use the opportunity cost argument (e.g. Hargrave et al., 1999b; Zhang, 1999b), whereas those who plead in favor of harmonization mainly use the level playing field argument (e.g. Jepma et al., 1998; Lefevere & Yamin, 1999). This does not mean that the latter authors want to centralize all permit allocation decisions, but rather that they prefer Member States to negotiate and formulate basic rules for domestic permit allocation, for instance whether or not (and on what basis) to grandfather to certain sectors (e.g. the electricity sector). Other authors steer a middle course by advancing weak or limited forms of harmonization. For instance, Kerr (1999) is against regulating the way in which EU Member States allocate permits by using the opportunity cost argument, but she uses the level playing field argument (similar to Yamin & Lefevere, 2000) to plead in favor of coordinating how much permits governments allocate to each sector. Another variant, also in the context of the EC, is provided by Lefevere & Yamin (1999), who mention the possibility of a "shared competence scheme" in which efficiency and trade barrier issues are regulated at the central level (Community) while all other issues are left to individual governments (Member States). When these alternative options still had to be worked out in detail, the European Commission and the European Parliament, as demonstrated above, had already pleaded in favor of the harmonization of the method of permit allocation.

Although Viguier (2000) writes that such a harmonization of permit allocation methods is economically not justified, which can be defended by using the opportunity cost argument, the neo-institutional approach recognizes that political actors do not justify harmonization of permit allocation on economic, but on equity grounds. Only looking at the efficiency justification of harmonization, like Viguier (2000: 8) basically does, without considering the equity justification, does not lead to a full understanding of the legal and political arguments used in practice. From a normative, neoclassical economic view, the level playing field perspective on fair competition may (or may not) be "groundless" and "misleading" (Viguier, 2000: 8), but the point is that some political actors perceived this perspective to be relevant and, in fact, used it to defend harmonization. From a positive-theoretical view, therefore, equity is not "groundless" — it is just another perspective, since an "(...) efficient outcome (...) is not necessarily and equitable one" (Devlin & Grafton, 1998: 42). When considering the role of perceptions in politics, economists "(...) must recognize the multiplicity of ends being pursued by market participants and accept those ends as given" (Cordato, 1994). In a similar vein, Dixit (1996: 147) writes that when "(...) economists (...) judge the performance of a policy-making system, they should admit the legitimacy of noneconomic goals (...)".

Although some claim that harmonization is undesirable, or even impossible, by arguing that states are sovereign (e.g. Petsonk, 1999; Zhang, 1999b), we argue that a voluntary agreement by states to harmonize permit allocation does not so much undermine their sovereignty, but rather reduces their autonomy in order to achieve other goals they might pursue, such as the prevention of unfair competitive distortions. Moreover, the sovereignty of these states remains crucial to enforce and implement such harmonized rules as part of their (inter)national climate policy. If desirable, such rules could be made part of the eligibility criteria Member States must meet to join the European (or international) permit trading system.

Cini & McGowan (1998) have drawn the conclusion that politics plays a role in the state aid decisions of the Commission, both in terms of political values and perceived national interests. The impact of values suggests that not only efficiency, but also equity is likely to be considered in a decision on permit allocation, which — in itself — increases the chance that grandfathering will be seen as state aid. However, a Member State's perception of its national interests in relation to permit allocation and state aid will lie somewhere between two extremes, that is (a) the desire to protect the national sovereignty of being free to allocate the permits as domestically preferred and (b) the desire to protect the national economy and industry against Member States who are free to choose any permit allocation they like which could result in a competitive advantage for the competitors abroad.

Which (mixture of) desire(s) would finally become dominant in the Commission as well as in other EU institutions was uncertain around the turn of the century. The proposal for a Directive on permit trading by the Commission and the Parliament's reaction on it, as discussed above, pointed in the direction of harmonizing permit allocation methods in the EU. A report of the European Climate Change Programme (ECCP, 2001), in which the Commission invited many stakeholders to participate in preparing a proposal on climate change policy, not only warns against competitive distortions due to permit allocation and underlines the desirability of a level playing field, but also states that permit allocation differences do not necessarily give rise to distortions within the internal market and claims that such distortions are likely to be temporary (ECCP, 2001: 8–10). Similar to the draft proposal for a Directive on emissions trading by the Commission (COM, 2001b), the report concluded that "Member States should be allowed to choose their own initial method of allocation, subject to obtaining any appropriate State aid approvals" (ECCP, 2001: 9). Although the rejection of harmonization, and the apparent preference for sovereignty, seems to run counter to the equity interpretation of competitive distortions, the report expects a progressive evolution towards auctioning in the longer term (ECCP, 2001: 8). This expectation was later made concrete by the Parliament's proposal to apply full-scale auctioning across the EU in the first commitment period (EP, 2002).

In 2003 the EU adopted a Directive that creates permit trading in the EU, but before that time, another political indication of which perceptions would become dominant was provided by the Commission's decisions on state aid in the domestic carbon trading schemes of Denmark and the United Kingdom.

7.4.2. The Political Precedent of Emissions Trading in Denmark and the UK

Important albeit limited "test cases" for the competitive distortion and state aid issues in an EU-wide carbon trading market were the political precedents of domestic carbon trading for the power sector in Denmark and for various companies in the United Kingdom from which the power sector is excluded. Denmark was the first EU country with a domestic and obligatory permit trading scheme, which became operational in January 2001, and the domestic and voluntary UK scheme was put into action in April 2002. The European Commission reached a decision on the allocation of permits in the context of state aid, both for Denmark (COM, 2000d) and for the UK (COM, 2001d). We will first analyze the Commission decision on state aid in the Danish case and then its decision in the UK case.

Because the Danish scheme was implemented at an earlier date than the UK scheme, it was bound to set a political precedent for future thinking about — and decisions on — (differences in) permit allocation in a European carbon trading market. Nevertheless, the relevance of the Danish case was also limited, not only because its trading scheme would already end in 2003, but also because the European Commission clearly indicated that its decision in the case of Denmark does not necessarily set a legal precedent for future decisions on emissions trading schemes.

According to Act 376 of 2 June 1999 (originally Bill 235) of the Danish parliament, tradeable permits were grandfathered to electricity producers in Denmark — irrespective of whether they were Danish or foreign owned — based on their historical CO_2 emissions during the period 1994–1998 (Folketinget, 1999; COM, 2000d). The scheme ran from 2001 to 2003, but if any new entrants would arrive during this period (which was not expected), they would be allocated quotas "on the same terms" as incumbents following objective and non-discriminatory criteria. The aforementioned Act would then be amended, which would be notified again to the Commission. According to Haites et al. (2000), the Danish reallocation of permits in the case of new entrants before 2003 implies that both newcomers and incumbents would receive grandfathered instead of auctioned permits. The emission ceilings for electricity producers had the effect that they — as a group — must reduce their emissions, but the combined heat and power plants had received less stringent (business-as-usual) emission ceilings because the latter contributed more strongly to CO_2 savings in the past. The Danish parliament acknowledged that the bill had to be notified to the European Commission, among other things, on the basis of EC Article 88(3) concerning state aid. It also laid down that the bill would not come into effect before receipt of approval by the Commission. The Commission approved the scheme by means of a letter to the Danish government dated 12 April 2000 and reached two basic decisions (COM, 2000d).

First, the Commission considered grandfathering in the Danish scheme to be state aid, because the tradeable permits have a market value and because the State foregoes revenue which could derive from auctioning the permits. Albeit limited to the Danish case, it did provide some support for one of our conjectures, namely that the Commission would see grandfathering as state aid based on the state origin criterion, translating grandfathering in terms of giving the (hypothetical) auction revenue to the polluters. The Commission has interpreted grandfathering as a wealth transfer without considering its opportunity costs and indicated that a company may use its profits from permit sales to improve its competitive position.

Second, the Commission nevertheless decided to allow grandfathering following EC Article 87(3)(c) which exempts state aid to develop certain economic activities or areas. The fact that they saw grandfathering as actually developing economic activities or areas implies that they acknowledged the financial advantage of grandfathering over auctioning as in the equity interpretation. If they would have used a (strict) efficiency perspective, grandfathering would not be seen as to have a cost advantage over auctioning because of its opportunity costs, in which view it does not develop economic activities or areas more or less than auctioning would. In short, the eight reasons for the Commission's exemption were that the Danish scheme (1) contributes to environmental protection and generates experience with emissions trading, (2) incorporates large emitters, (3) intends to participate in future international carbon trading, (4) represents emission reductions, (5) is limited to 2003, (6) does not restrict electricity imports or exports, (7) provides annual reports for transparency and (8) treats incumbents and newcomers equally

(COM, 2000d: 6, 7). These reasons to accept grandfathering despite of its state aid character were not only environmental, economic and legal, but also political in nature. The Commission approved the state aid in Denmark, among other things, not only because of the contribution to environmental protection (as emphasized in COM, 2001c: 90), but also because of its desire to gain experience with and prepare for emissions trading (reasons (1) and (3)). This provides some support for our conjecture that a Commission's state aid decision is at least partly based on political considerations and trade-offs. It also shows that there is broad room for interpreting the exemption rules which goes beyond a (neoclassical) law and economics approach. Several reasons for the exemption of grandfathering have nothing to do with the allocation per se (grandfathering versus auctioning), but rather relate to the (other and more general) characteristics of the scheme itself.

In the UK case, the Commission reached similar conclusions, in its decision of 28 November 2001, and largely used the same type of arguments as in the Danish case. Unlike the Danish scheme, the UK scheme runs from 2002 to 2007, combines permit trading with credit trading, excludes the power sector and contains absolute targets that are voluntary (e.g. DEFRA, 2002). Although this broader scheme has a more complex design than the smaller Danish scheme, the bottom line is that companies in the UK that voluntarily wish to take up absolute targets (the so-called "direct participants") receive grandfathered tradeable permits.[12] Firms with an energy-efficiency target (the so-called "unit participants") can use credit trading: according to Rosenzweig et al. (2002: 58), there are more UK companies with relative targets than with absolute targets.[13]

The Commission approved the scheme by means of a letter to the UK government (COM, 2001d). Explicitly referring to the Danish case and without mentioning the opportunity costs of free allocation, the Commission also considered grandfathering in the UK scheme to be state aid, among other things because the State foregoes revenue which could derive from auctioning the permits. Grandfathering is seen as an advantage that distorts competition with companies not having access to the scheme. Similar to the Danish case, the Commission nevertheless decided to allow grandfathering following EC Article 87(3)(c) which exempts state aid to develop certain economic activities or areas.

---

[12] This should not be confused with the subsidy that companies receive to join for which they bid in an auction. If a firm is in non-compliance or decides to drop its absolute yet voluntary target, it loses this yearly "incentive money" and the possibility to trade permits, while it must repay the subsidies it already received plus interest. Companies that were already subject to an energy tax (the "Climate Change Levy") and choose to adopt absolute or relative targets also obtain a percentage discount on this tax, which is lost if the firm is in non-compliance or decides to opt out.

[13] There is also a "gateway" which ensures that permits can be sold to the unit sector, but which controls credit sales to the direct sector: only when there has been a net flow into the unit sector will any unit sector participant be able to transfer credits to the direct sector (Rosenzweig et al., 2002: 60).

And again, the Commission also used political arguments, next to environmental, economic and legal ones, to defend the exemption.

In short, the 13 reasons for the Commission's exemption were that the UK scheme: (1) is in line with the idea that each Member State may choose the policy it wishes to comply with the Kyoto targets as long as Community provision on emissions trading are absent, (2) uses a competition-oriented instrument, (3) goes ahead before Community regulation, (4) will provide valuable learning insight for the benefit of any later initiatives, (5) is the first multilateral trading scheme in the EU, (6) achieves a net environmental benefit, (7) uses an incentive necessary to ensure voluntary participation, (8) requires companies to reduce emissions below their targets to capitalize the potential aid from free allowances, (9) recuperates the incentive in case of non-compliance, (10) is limited in time and will adapt to the requirements of an EU-wide emissions trading scheme foreseen in 2005, (11) will produce detailed annual reports, (12) undertakes to accept emissions trading based on mutual agreements with other States and (13) will elaborate and inform the Commission of non-discriminatory ways to include new entrants (COM, 2001d: 11–12). Like we have seen before, these reasons demonstrate that grandfathering is allowed as state aid by stating political desires, for instance to gain experience with emissions trading (reason (4)), and by mentioning characteristics of the scheme that have nothing to do with the allocation of permits per se (such as reasons (5), (11) and (12)). Interestingly, the Commission emphasized that the UK might have to adapt its scheme if EU-wide emissions trading would start in 2005 "(…) in order to avoid distortion of competition between allowances issued through different systems" (COM, 2001d: 12).

The presence of political arguments and the absence of the opportunity cost argument in the economic analysis of grandfathering by the Commission can be interpreted in two ways. A pessimist might see them as another example of sometimes imperfect and incomplete case-by-case decisions by the state aid directorate (see also Cini & McGowan, 1998: 143) or as another example of "infant" economic analysis in the legally-oriented state aid directorate or in EC competition law itself (see also Hildebrand, 1998: 413). An optimist might see them as an example of a Commission that was neither blind to international political developments (regarding the emerging international and/or European carbon trading market under the Kyoto Protocol) nor to national political preferences (of Denmark and the UK), and which was able to find a balance between costs and benefits and between risks and opportunities, for instance by indicating that: "The Danish CO_2 quota system has to be assessed in the light of its merits" (COM, 2000d: 6). Moreover, although their content could be criticized, the Commission decisions in the Danish and UK cases have been consistent by each time referring to the same legal provisions to characterize the grandfathering as state aid as well as to exempt the aid.

From a normative perspective, some authors would like to see that efficiency becomes the centerpiece of state aid decisions in general, albeit recognizing the political difficulty of doing so, and that the Commission at least makes explicit the non-economic objectives for which the aid may be authorized (e.g. Nicolaides & Bilal, 1999). From a positive perspective, we have found that other criteria than efficiency were used in the state aid decisions of the Commission on carbon trading in Denmark and the UK. Next to the examples discussed above, it is important to observe that the equality principle played a role in the Commission's decision on the Danish scheme (which is also part of the unnoticed level playing field argument), for instance when the Commission indicated to support the equal treatment of newcomers and incumbents in the Danish case in order to avoid competitive distortions (COM, 2000d: 5, 7). Such distortions apparently arise in the case of an unequal treatment, meaning that they interpret such distortions not in terms of efficiency, but in terms of fairness.

The political (albeit not legal) precedent created by the Commission's decisions in the Danish and UK cases could suggest that grandfathering in a European carbon trading scheme will not only be seen as state aid, but might also be exempted. This precedent is important, but also limited because of the difference of assessing domestic permit allocation in two particular Member States (of which one holds a mandatory scheme limited to the power sector and the other holds a voluntary scheme where the power sector is excluded) versus the assessment of international permit allocation dissimilarities in a possibly obligatory and EU-wide scheme with a larger number of Member States that might include electricity facilities as well as other sectors.

7.4.3. The Political Outcome of Permit Trading in the EU

Such a broad European scheme was, in fact, decided upon (EU Council, 2002). As a result, the EU adopted a Directive in 2003 that creates permit trading for large emitters in the EU from 2005 onwards with grandfathering as the harmonized rule (OJ, 2003). According to Annex I of this Directive, the permit trading system covers such installations with a rated thermal input exceeding 20 MW in the energy, metal, cement, glass, pulp and paper sectors. According to Article 3, they receive allowances that each allows the holder to emit 1 tonne of carbon dioxide equivalent during a specified period. In a so-called "non-paper", the European Commission refers explicitly to "a cap on greenhouse gas emissions" for those installations (COM, 2003: 4).

Contrary to the general design of the EU Directive on CO_2 emissions trading by choosing in favor of the superior alternative of permit trading, the specific harmonization of the allocation of emission rights, as foreseen under this Directive, is not compatible with the neoclassical economic text book model.

As we have seen, however, the European Commission feared that competitive distortions and state aid could arise under EC Article 87 when Member States would be left free to choose their allocation method. Therefore, Article 10 of the Directive requires that every Member State allocates at least 95% of its allowances free of charge during the period 2005–2007, for instance by means of grandfathering based on historical emission figures, and that at least 90% of the allowances is allocated for free in the period 2008–2012.

Because of the opportunity costs of grandfathering, harmonization on the basis of gratis allocation is not necessary according to the efficiency approach, as we have explained before. The decision to harmonize is therefore not grounded in efficiency, but in equity concerns. A neo-institutional economic approach learns that allocation free of charge distorts fair (not efficient) competition if the emission rights are auctioned in another country, because companies with free permits are financially favored above their competitors which had to buy their emission rights. Although efficiency explains the general economic design of the Directive (namely permit trading), equity explains its specific legal form (namely the harmonization of permit allocation methods).[14]

According to Article 9, each Member State shall develop a national allocation plan to the Commission on how it proposes, among other things, to allocate the allowances. The Commission may reject that plan, or any aspect thereof, on the basis that it is incompatible with the criteria listed in Annex III (or with Article 10). This Annex states that the plan "(…) shall be consistent with (…) Community legislative and policy instruments" and "(…) shall not discriminate between companies or sectors in such a way as to unduly favor certain undertakings or activities in accordance with the requirements of the Treaty, in particular Articles 87 and 88 thereof". Although reference to Article 87 was left out from the list of criteria in the proposal, it thus reappeared in the final text of the Directive.

When comparing the (draft) proposals and reports with the final Directive, it is even more interesting to see that the European Parliament made a huge shift from initially proposing 70% grandfathering in the period 2005–2007 and even 100% auctioning in the period 2008–2012 in a draft report (see EP, 2002) to finally proposing 85% grandfathering in both periods, only half a year later (see Worsley & Freedman, 2002). Where the Commission proposed to use 100% grandfathering, the final outcome of the co-decision procedure of 95% grandfathering

---

[14] From an equity point of view, the harmonization of grandfathering is also compatible with Article 97 of the EC Treaty, albeit only conceptually. An EU country that wishes to deviate by auctioning all of its allowances, which is legally not allowed under the Directive (but which is conceptually imaginable), would only cause a distortion of fair competition that is detrimental to itself. Article 96 of the EC Treaty states: "Where the Commission finds that a (…) regulation (…) is distorting the conditions of competition (…) the Council shall (…) issue the necessary directives (…)". Article 97 of the EC Treaty states: "(…) If the Member State (…) causes distortion detrimental only to itself, the provisions of Article 96 shall not apply".

in 2005–2007 and 90% grandfathering in 2008–2012 not only reflects a political compromise between the Parliament and the Commission, but also reflects a substantial attitude change on the permit allocation issue in the Parliament. This (internal) attitude change was partly triggered by the (external) threat that the Russian Federation had just made (in 2003) not to ratify the Kyoto Protocol. Various Members of the Parliament recognized that this has accelerated the co-decision procedure in the EU to stimulate Russian ratification by signaling that the EU takes climate policy and market instruments seriously and that the Russians, although the Americans have left the Kyoto Protocol, can still gain from trading emissions with the Europeans (e.g. Houlder, 2003).

The neo-institutional economic approach to permit allocation that focuses on its equity consequences is able to understand, in general, the perception in politics that different allocation methods between Member States may lead to competitive distortions and state aid. In particular, this equity perspective is able to explain, first, the decisions of the European Commission to regard (and exempt) grandfathering as state aid in the emissions trading schemes of Denmark and the UK, and second, the harmonized institutional shape of the Directive on permit trading in the EU. At the time of writing, however, it is still unclear whether the Commission will take an efficiency or equity approach to judging the national allocation plans on state aid. An example is when one government wants to allocate its allowances for free based on historical emissions, while another government wants to do this on the basis of energy efficiency, for instance by referring to Annex III that requires quantities of allowances to be consistent with the (technological) potential of activities to reduce emissions. "Normal state aid rules will apply" (COM, 2004: 11) is all the guidance that Member States got from the Commission to construct their national allocation plans.

The formal institutional constraints and legal ambiguities discussed above posed an additional barrier to implementing permit trading in Europe, which magnified the perceived switching costs of this superior alternative. The EU nevertheless succeeded in overcoming these obstacles, partly "forced" by external shocks, like the sudden threat of the Russians to withdraw from the Kyoto Protocol, and partly for reasons to be discussed in the next chapters. Although the economic advantage of grandfathering permits is now seen as legally "fit" from a state aid point of view, it may be incompatible with the polluter pays principle.

7.5. Possible Extensions of the Analysis to the Polluter Pays Principle

By choosing in favor of harmonization under the Directive on permit trading, the EU has acknowledged, as explained above, that it incorporates not only efficiency, but also equity aspects of the allocation of emission allowances into its legal decision regarding EC Article 87. It would be inconsistent if this equity aspect suddenly ceases to play a role in other legal matters related to the same allocation of emission allowances. An example is the question whether allocation of emissions free of charge, as foreseen in the Directive, is compatible with the polluter pays principle established under EC Article 174. By extending our law and economics framework to this issue, the answer could be that there is compatibility between gratis allocation and the polluter pays principle if emphasis is placed on the efficiency aspect of the distribution of emission rights, but that a consistent application of the equity approach leads to a potential legal conflict between free distribution, like grandfathering, and EC Article 174.

In the EU, the polluter pays principle demands that the costs of measures to deal with pollution should be borne by the polluter who causes the pollution (OJ, 2001: 3). However, there are "problems of interpretation" with this principle (Steenge, 1997: 122). Those who have considered the emissions trading issue basically believe that a polluter does not pay for its pollution if he obtains his emission rights for free, contrary to a polluter that has to buy the required emission rights at an auction (e.g. Nash, 2000). In our view, however, there are two possible approaches. According to the efficiency approach, a polluter also pays when he has received his allowances free of charge, namely in the form of the opportunity cost of using the allowances (next to his, direct, emission reduction costs). Instead of using the allowances, he could have sold them. According to the equity approach, not the polluter pays in the case of gratis allocation, but the "public pays", because allocating emission rights free of charge can be interpreted as a wealth transfer from the public (or government) to the polluters. In the form of emission rights that have economic value, the polluters, one could argue, receive the revenues that the government would have obtained in case of an auction.

The interesting thing now is that the Member States of the EU have chosen, as we have seen, for the harmonization of the allocation of emission allowances on a free of charge basis so as to avoid competitive distortions. This harmonization, however, is not necessary on efficiency grounds: not only companies with auctioned allowances, but also companies with allowances allocated for free make (opportunity) costs. By nevertheless choosing to harmonize, the EU apparently has interpreted the gratis allocation of emission rights not in terms of opportunity costs, but in terms of a financial advantage or wealth transfer. This advantage plays a role, as we have noted before, in the equity approach. A consistent application of this interpretation then means, however, that allocating emission rights free of charge is not compatible with the polluter pays principle under EC Article 174. Only in terms of opportunity costs this allocation method can be seen as compatible with this principle. If the EU would suddenly stress the opportunity cost of gratis emission allowances in the case of the polluter pays principle, while it did not regard upon this characteristic at all when considering the case of harmonization, it would use an inconsistent view on the allocation aspect of emissions trading.

This complex issue could remain a legal-theoretic debate behind the scenes, but it could also grow, in principle, into a political-legal conflict that would slow down (or even block) the implementation of permit trading in the Member States of the EU. It seems, however, that the political will of the European institutions to implement emissions trading based on an allocation of emission rights free of charge weighs heavier than the consistency of the legal and economic argumentations that are at the root of this system. Moreover, even if there is an incompatibility, it still does not mean that the emissions trading system could not function. It is not the first time that legal-economic inconsistencies and imperfections have been observed, for instance in political and judicial decisions regarding European competition law and individual cases of state aid in general (e.g. Cini & McGowan, 1998: 143; Hildebrand, 1998: 413), and EU policy makers have chosen for an, in principle, economically and legally solid design of trading pollution under an emission ceiling in the form of permit trading. It does mean, however, that the decision making on the Directive has been in two minds about the legal consequences of permit allocation, namely an equity view on the harmonization of the allocation of emission rights regarding EC Article 87 and an efficiency view on the polluter pays principle applied to the same allocation of emission rights regarding EC Article 174. This is inconsistent from a law and economics point of view, but more pragmatic lawyers are likely to perceive it as a handy and politically acceptable way of getting the emissions trading system functioning.

Moreover, to some extent, one could argue that at least the European Parliament has been consistent, namely in using the equity view (although they could be accused, in principle, of neglecting the efficiency view). Members of the Parliament, who initially pleaded for more auctioning, said that allocating first 90% and then 95% of the allowances free of charge would ensure the "progressive" application of the polluter pays principle and cause "less" distortions of competition, insisting that further harmonization should be considered, including auctioning, for the time after 2012 (Worsley & Freedman, 2003: 15).[15] On the one hand, it seems that a harmonized scheme that prescribes 100% auctioning would ensure the "full" application of the polluter pays principle for them, which is consistent with the equity view. On the other hand, it illustrates that in politics an allowance allocation method can be a "bit" (or "more" or "less")

---

[15] Also the European Commission explicitly stated that auctioning applies the polluter pays principle (COM, 2000a: 18), simply because polluters literally pay for their pollution by means of purchasing the permits from the government.

consistent with the polluter pays principle, whereas in a conceptual approach a particular allocation method is either consistent with this principle or not.

On such a theoretical level, our analysis also shows that the approach of various lawyers is incomplete if not incorrect. According to Nash (2000: 3, 13), for instance, the weak form of the polluter pays principle is a rule against subsidizing pollution and the strong form is a requirement to internalize the costs of pollution. Nash argues that distributing allowances free of charge as under grandfathering is inconsistent with the weak form of this normative principle, because allowances are allocated at no cost which implies a subsidy. Moreover, because inconsistency is "demonstrated" with the weak form, Nash does not check compatibility with its strong version. We are convinced that these arguments are inaccurate. First, although we agree that grandfathered permits are a subsidy, which plays a role in the equity approach, the point is that they are a lump sum subsidy which does not affect efficiency. The question of consistency with the weak form of the polluter pays principle thus depends on the (equity or efficiency) approach taken. Second, grandfathered permits do internalize the cost of pollution because of their opportunity costs. The question of consistency with the strong form thus depends on whether one recognizes and acknowledges the aspect of opportunity costs when allocating emission rights free of charge.

This indicates that our law and economics approach is, at least, capable of supplementing the existing legal literature on the issue of emissions trading and the polluter pays principle, but more research is needed to judge this. The ambiguity, however, helps to explain why (European) politicians have not readily accepted permit trading, which is most acceptable to firms in the case of grandfathering, but may run into legal conflicts with the polluter pays principle for the government, depending on the perspective taken.

7.6. Conclusion

The economically superior instrument of permit trading may not rank high in the political hierarchy of market-based climate policy. To explain this, we have discussed its formal constraints by analyzing the economic and legal conditions, both in terms of efficiency and equity, under which competitive distortions and state aid violations arise under European law when some governments would grandfather their permits, while others would auction them. EC Article 87(1) on state aid determines that "(…) any aid granted by a Member State or through State resources in any form whatsoever which distorts or threatens to distort competition by favoring certain undertakings or the production of certain goods shall, in so far as it affects trade between Member States, be incompatible with the common market". In addition, the European Commission uses four criteria to determine whether a measure is state aid, namely state origin, firm advantage, specificity and trade effect.

From a neoclassical law and economics perspective, grandfathering is not state aid, because it does not distort (efficient) competition. Trade is not affected, because grandfathering does not affect efficiency. Moreover, grandfathered firms are not favored, because they have to include the opportunity cost of their permits (which is equal to the permit price) in the product price. Therefore, the firm cannot ask lower product prices than its foreign competitor with auctioned rights. This means that grandfathered firms are not advantaged (in the sense of having lower costs), so that there is no need to harmonize permit allocation procedures.

From a neo-institutional law and economics perspective, however, grandfathering could be seen as state aid, because it distorts (fair) competition. Grandfathered permits are a capital gift, inducing a windfall profit, which gives the firm more financial resources than a comparable firm abroad with auctioned permits. The state favors specific firms by giving them a financial advantage over their auctioned competitors in another Member State. This lump sum subsidy affects trade, not in efficiency terms, but in equity terms by unequally altering the competitive relations (the "level playing field") among competing firms. Although there is not a genuine transfer of resources from the government, the state origin criterion is also satisfied if the State will receive less revenues as in the case of grandfathering, which amounts to giving the (hypothetical) auction revenue to the polluters (e.g. Welch, 1983: 168). Grandfathered firms are advantaged, due to the mere process of permit allocation, so that it is desirable to harmonize permit allocation procedures.

We have used this law and economics framework to analyze the state aid cases of permit trading in Denmark and the UK empirically. It appears that the European Commission considered grandfathering as state aid by using the state origin criterion: the State foregoes revenue which could derive from auctioning the valuable permits (COM, 2000d, 2001d). Nevertheless, the Commission exempted the aid by using environmental, economic, legal as well as political arguments: the grandfathering was allowed, among other things, by following EC Article 87(3)(c) that exempts state aid if it helps to develop certain activities or areas and by stating a political desire to gain experience with and prepare for emissions trading. Although the Commission mentioned neither the impact of opportunity costs nor the desire for a level playing field, grandfathering was interpreted as a wealth transfer which could affect the equal treatment of firms. This set a political (albeit not legal) precedent to interpret grandfathering in the EU in terms of equity.

The latter conjecture was largely supported by the Directive, adopted in 2003, that creates permit trading in the EU from 2005 onwards with grandfathering as the harmonized rule. The choice for permit trading means that efficiency has guided the general economic design of the Directive, but the choice for harmonization means that equity explains the specific legal form of the Directive. This harmonization would not have been necessary from an efficiency point of view, because grandfathered permits have opportunity costs, so that competition is not distorted when other governments auction their permits. Although the equity view made no headway in the WTO context, it was in fact used in the EU. Nevertheless, the same equity view could imply that grandfathering conflicts with the polluter pays principle under EC Article 174, arguing that the public pays as a result of the wealth transfer to the polluters. More research is needed to judge this.

Efficiency rejects, but equity explains the perception among various European policy makers that differences in permit allocation methods between countries might distort competition and could lead to state aid (e.g. COM, 2001a: 11). This legal ambiguity added to the switching costs of permit trading. While permit trading (re)allocates emission rights explicitly, politicians were tempted to build upon extant environmental policy by means of sub-optimal approaches like credit trading, that (re)allocate such rights only implicitly. Although the Europeans managed to break out by choosing permit trading, equity still explains their legal choice to harmonize grandfathering. More reasons for the (temporary) institutional lock-in and for the subsequent breakout will be discussed in the next two chapters that focus on the informal institutional barriers, including equity, to implementing market-based climate policy.

PART IV
NEO-INSTITUTIONAL ECONOMICS

Chapter 8

Theoretical Aspects of Restricting Market-Based Climate Policy

8.1. Introduction

This chapter specifies the informal constraints to implementing market-based climate policy by elaborating and criticizing various theoretical explanations of the (so-called "supplementarity") proposal of the European Union (EU) to quantitatively restrict the use of economic climate policy instruments, including equity as a cultural barrier, in the form of 16 hypotheses.

Where we have mainly used institutional law and economics to analyze the formal barriers to implementing permit trading in the two previous chapters, we will now take a neo-institutional approach by studying the informal obstacles for governments to accept economic instruments in climate policy in the next two chapters. Cultural resistance, for instance, added to their perceived switching costs of permit trading and contributed to the (temporary) institutional lock-in of incrementally building less efficient instruments, like project-based emissions trading, upon existing environmental policy.

Since the 1990s, when the United States (US) and some other countries proposed to use market-based instruments (notably permit trading) in international climate policy, the EU, for instance, expended years of political effort trying to reject or restrict this. In the context of the Kyoto Protocol, the position of the Europeans culminated in their proposal of 1999 to make the Kyoto Mechanisms supplemental to domestic action by quantitatively limiting their use. Although this (self-imposed) barrier to trade was rejected internationally in 2001, an analysis of the EU supplementarity proposal is still useful to lay bare the underlying values of why some governments wanted to limit market-based climate policy, especially permit trading. This chapter develops 16 hypotheses that could help to explain these cultural elements of an institutional lock-in (e.g. Woerdman, 2002). The next chapter will test them empirically and will explain why the Europeans managed to break out by adopting a Directive that creates permit trading in the EU from 2005 onwards.

This chapter is organized as follows. Section 8.2 gives the definition of supplementarity and its interpretation by the EU. Section 8.3 reviews the economic analyses of the EU proposal, including an identification of international gainers and losers, to find out whether the proposed trade restriction was in the economic interest of the Europeans. Section 8.4 systematically elaborates, criticizes and clusters various possible theoretical explanations of the EU proposal on supplementarity, including equity values, in the form of 16 hypotheses, which will be tested empirically in the next chapter. Finally, Section 8.5 presents the conclusion.

8.2. Definition of Supplementarity

Although the supplementarity issue has moved to the background of international environmental politics because it was solved in 2001, it was one of the largest political problems to implementing market-based climate policy at the end of the 1990s. At CoP4 of 1998 in Buenos Aires the FCCC Parties drew up a work program, the "Buenos Aires Plan of Action", in which the supplementarity issue was mentioned as one of its first elements (BAPA, 1998, Decision 7/CP.4, Annex). This work program was reinforced at CoP5 of 1999 in Bonn with a view to taking decisions on the Kyoto Mechanisms at CoP6 in The Hague in November 2000. However, this CoP6 meeting failed to produce an agreement on the institutional details of the Kyoto Mechanisms, so that a second meeting was held in Bonn under the name of CoP6 Part II. One of the main reasons why the Parties did not reach consensus at CoP6 Part I were the political differences on the issue of supplementarity (e.g. Churie et al., 2000).

Already in the pilot phase for emission reduction projects, the FCCC Parties recognized in 1995 that "(...) Activities implemented jointly under the Convention are supplemental, and should only be treated as a subsidiary means of achieving the objective of the Convention" (CoP1, 1995: 18). The Kyoto Protocol also mentions supplementarity, but does not define it in detail. According to JI Article 6.1(d), the acquisition of ERUs "(...) shall be supplemental to domestic actions for the purposes of meeting commitments under Article 3". CDM Article 12.3(b) states that Annex B Parties may use CERs only for "(...) compliance with part of their quantified emission limitation and reduction commitments (...)" and also international emissions trading under IET Article 17 "(...) shall be supplemental to domestic actions (...)".

We will not develop our own definition of supplementarity precisely because we want to investigate the definition of supplementarity offered by the EU. This definition, which was the negotiating position of the EU prior to and during CoP6 (e.g. SBSTA/SBI, 1999), roughly implied that 50% of the Kyoto commitments should be achieved domestically via a ceiling on the Kyoto Mechanisms. The EU proposal on supplementarity contained rules for buyers (demand) and rules for sellers (supply) (see EU Council, 1999). Demand and supply were to be restricted for the 5 year commitment period 2008–2012. The rules which limit the demand applied to all Kyoto Mechanisms, whereas the restriction on supply exempted units resulting from JI and CDM projects.[1]

Although the EU was the only political actor in 1999 to favor a ceiling on the use of the Kyoto Mechanisms, they were followed in 2000 by some other Parties who made comparable proposals (albeit with different formulas, percentages and coverage) to limit emissions trading in quantitative terms, notably China, India, Saudi Arabia, Senegal and the Alliance of Small Island States (SBSTA/SBI, 2000). However, the political differences among them were considerable, for instance because India favored a "quantified ceiling" on all mechanisms, whereas China only proposed a "concrete ceiling" on IET Article 17, thereby presumably excluding JI and CDM from such limitations.

The occasional climate coalition of JUSCANZ countries (Japan, the United States, Canada, Australia and New Zealand) did not want to define supplementarity, believing that each country should decide for itself how much it wants to trade. The Russians, for instance, feared that a ceiling on trade would limit their "hot air" sales. The opposition of such large countries against restrictions on emissions trading made it unlikely, already from the outset, that the EU proposal would be accepted, which is not only recognized in various studies (e.g. Barker et al., 2000), but even in a preliminary and internal draft note of the European Commission itself (EC, 1998).[2]

In fact, the story ended with a partial defeat for the EU as well as for the aforementioned non-Annex B countries. The EU was willing to give up its

---

[1] With respect to supply (sellers), the EU proposal stated that a selling Party may not transfer more than 5% of its {(baseyear emissions multiplied by 5 plus its assigned amount) divided by 2}. With respect to demand (buyers), the EU proposal stated that the purchase(s) of an Annex B Party may not exceed the higher of either 5% of its {(baseyear emissions multiplied by 5 plus its assigned amount) divided by 2}, or 50% of its {(actual annual emissions of any year between 1994 and 2002 multiplied by 5) minus its assigned amount}. Nevertheless, the EU proposal approved of more emissions trading to take place in the case (and amount) of early action. The ceiling on net acquisitions (or net transfers) could be increased to the extent than an Annex B Party achieves emission reductions larger than this ceiling in the commitment period through domestic action undertaken after 1993, if demonstrated by the Party in a (albeit undefined) "verifiable" manner and subject to an expert review process.

[2] Moreover, there was a discussion whether two Annex B countries could legally avoid the trade restriction by forming a "bubble" under Article 4. They could reallocate their targets and notify the FCCC Secretariat of their agreement. How to bring about the reallocation is not limited by Article 4, as long as the total emission ceiling of their original assigned amounts does not change. This suggests that if these two countries agree to let their legal entities transfer emissions across their national borders and change their assigned amounts correspondingly, there is no legal impediment to the underlying transfers. In a similar vein, Oberthür & Ott (1999: 149–150) found that Article 4, although its implications were not yet foreseeable at that time, might be used as an alternative to trading under Article 17, thereby avoiding any quantitative or qualitative restrictions that apply to the Kyoto Mechanisms.

proposal and accepted the unspecified requirement that domestic action shall be a "significant element" of climate policy in Annex B countries (CP, 2001a: 7). The EU made this compromise at CoP6 Part II, held in Bonn in July 2001, to prevent that some JUSCANZ countries would withdraw from the Kyoto Protocol, like the US had done a few months earlier in March 2001. This largely unexpected US decision had changed the game and the EU, who believed in the Protocol but was skeptic towards unrestricted emissions trading, rather had a market-based Protocol than no Protocol at all (e.g. Oberthür & Ott, 1999). Nevertheless, some of the environmental concerns of the EU were accommodated at CoP6 Part II by means of restrictions on the use of sinks and the requirement, among other things, that each Annex B Party shall maintain a commitment period reserve which should not drop below 90% of its assigned amount.

This international decision on supplementarity was reconfirmed in the Marrakesh Accords of CoP7, held in Marrakesh in October/November 2001, where it was determined that: "(...) use of the mechanisms shall be supplemental to domestic action and (...) domestic action shall thus constitute a significant element of the effort made by each Party included in Annex I to meet its quantified emission limitation and reduction commitments under Article 3 (...)", (CP, 2001b: 3). Moreover, although the Parties initially disagreed over the issue (Boyd et al., 2001: 5), they also decided that the eligibility to trade (which depends on compliance with the emission accounting and inventory requirement under Articles 5 and 7) does not depend on a requirement for Annex B Parties to report on supplementarity. Instead, the CoP "requests the Parties included in Annex I to provide relevant information" on supplementarity in relation to domestic action and their use of the Kyoto Mechanisms (CP, 2001b: 3). This means that reporting failures on supplementarity would not trigger a loss of eligibility to use those flexible instruments.

The supplementarity provisions in the context of the Kyoto Protocol try to prevent a situation in which the agreed emission targets would be met solely (or predominantly) by means of the Kyoto Mechanisms. The central question of this chapter is why the EU wanted to prevent such a situation and why they initially proposed to place a quantitative ceiling on trade. Our objective is to provide several positive theoretical explanations for this EU proposal on supplementarity by arranging and elaborating, as well as extending and criticizing, the clarifications found in the emissions trading literature. Although the EU proposal on supplementarity was directed against unrestricted use of all Kyoto Mechanisms, it will become clear that some reasons only apply to emissions trading, in particular permit trading, and not to the project-based flexible instruments.

Apart from some exceptions (e.g. Grimeaud, 2001), the EU proposal on supplementarity did not seem to have many supporters in the scientific community of climate change research, neither within nor outside Europe (e.g. Bohm, 1999; Petsonk, 1999; Tietenberg et al., 1999). Economists rather performed economic studies of this proposal and a few of them put forward positive explanations for the EU position (e.g. Hourcade & Le Pesant, 1999; Yamin et al., 2000). These studies and explanations will be discussed below.

8.3. Economic Analyses of the EU Proposal on Supplementarity

If one wants to know why the EU proposed to limit emissions trading and analyze the underlying cultural values, it makes sense to explore first whether it is in its economic interest to do so. If the proposal of the EU would completely follow from its economic interests, an analysis of values would not be necessary (or less relevant) to provide a positive explanation. Therefore, we will discuss and nuance the economic studies that calculate the overall economic effects of the EU proposal on supplementarity and identify the gainers and losers of this EU proposal by comparing and criticizing different economic models.

8.3.1. Overall Economic Effects of the EU Proposal on Supplementarity

How "bad" was the EU proposal in terms of overall economic effects? Economists agree that restricting the use of the Kyoto Mechanisms will raise the total costs of reducing GHG emissions in comparison with the cost-minimizing optimum (e.g. Haites, 1998; Tietenberg et al., 1999; Zhang, 2000b). This cost increase could be considerable. When comparing different studies, it turns out that the estimated cost increase in relation to a situation of unrestricted emissions trading varies in the literature from a few percentages (e.g. Rose & Stevens, 2001) to 50% (Gusbin et al., 1999) or more (Ybema et al., 1999), depending on the model and its assumptions.[3] However, the EU proposal could be put into perspective by looking at the basis of comparison in these calculations and by considering the simplifying assumptions behind the predominantly neoclassical models from which these figures originate.

First, limited trading is economically worse than full trading, but still economically better than the initial situation (or: status quo) of no trade at all. Zhang (2000b: 510), for instance, calculated that the Annex B countries

---

[3] The estimated cost increase induced by the EU proposal compared with unlimited emissions trading depends, among other things, on the shape of the marginal abatement cost (and benefit) curves, the market price for emission entitlements, the level of competition and transaction costs as well as the stringency and design of the ceiling on emissions trading (for instance, the presence or absence of an export limit next to the import limit or the interpretation of a ceiling as a percentage of either historic emission figures, assigned amounts or baseline emission projections).

(belonging to the OECD) would still be able to reduce their total abatement costs by more than 66% under the EU proposal, compared to 87% under full trading.

Second, the cost increase of a ceiling on trade is usually compared with a market optimum that is usually too optimistic (Baron, 1999b), for instance because transaction costs are assumed away. Gusbin et al. (1999) argue that transaction costs function as a ceiling on trade, since they drive a wedge between the price the supplier receives and the buyer is prepared to pay. Although the actual effect depends on the form and level of transaction costs, it might limit the potential losses associated with fixing a ceiling on the volume traded.

Third, most economic models do not consider market power, but Gusbin et al. (1999) and Ciorba et al. (2001) calculated that the Russian Federation could perhaps supply as much as 70% of the permits, which may lead to monopoly behavior and a limitation of the volume of permits supplied in an attempt to drive up their price. Market power would then have a similar effect as a ceiling on the volume of emissions traded. Ellerman (2000) stresses that even if the Russian Federation is unwilling or unable to exert market power, the export limit subsumed under the EU proposal approximates the result by restricting supply.

Although a different basis of comparison, imperfect markets and banking make the EU proposal look less "dramatic" than sometimes suggested, they do not change the fact that the EU proposal has a negative effect on overall efficiency. What, then, is the effect of the EU proposal on the permit price? The EU proposal contains a restriction on demand as well as a restriction on supply. In principle, both have opposite effects: restricting demand lowers the permit price, whereas restricting supply increases the permit price. However, one of the features of the EU proposal is that the rules which limit demand apply to all Kyoto Mechanisms, whereas the restriction on supply exempts units resulting from JI and CDM projects. This means that the demand restriction in the EU proposal is more binding than the supply restriction.

Consequently, in spite of some exceptions (e.g. Mauch et al., 1999), most authors expected that the overall result of the demand and supply restriction effects of the EU proposal, assuming perfect competition, would be a lower permit price (e.g. Ybema et al., 1999; Zhang, 2000b). This is disadvantageous for sellers and may sound attractive to permit buyers (e.g. Baron et al., 1999), but the demand restriction also forces the latter to make larger and hence more expensive domestic reductions than under unrestricted trading (Criqui et al., 1999).

8.3.2. Gainers and Losers of the EU Proposal on Supplementarity

Which countries are likely to benefit from the EU proposal on supplementarity and which are likely to be disadvantaged? Bohm (1999) claims that a ceiling on trade is not in the interest of any country, because it would make emission reductions globally more costly than possible, but we rather define country interests at the level of (regions of) countries (see van der Wurff, 1997) to find out who will gain and who will lose under a ceiling on trade. There is a consensus in the economic literature that Central and Eastern European countries as well as the developing countries will lose when a quantitative ceiling is placed on the use of the Kyoto Mechanisms (e.g. Zhang, 2000b; Metz et al., 2001; Rose & Stevens, 2001).

First, the EU proposal restricts IET and JI, which would limit the supply of hot air to about one-third of its potential magnitude and reduces the potential revenues for countries with economies in transition (e.g. Baron et al., 1999). Haites (2000) estimates that a limitation of hot air sales increases the carbon price, but strongly reduces the Annex B sellers' net revenues received (from $45.4 billion under full trading to $12.8 billion under restricted trading). Second, the EU proposal limits the demand for CDM projects, which deprives developing countries from a potential source of additional income. Zhang (2000b) calculated that a ceiling on all Kyoto Mechanisms would roughly halve the potential size of the CDM market compared to unrestricted transfers. However, there were both developing countries that opposed and supported the EU proposal, which also indicates that some follow considerations other than economic ones. For instance, from the equity perspective of rich countries "buying their way out" and "picking low hanging fruits" (e.g. Trexler & Kosloff, 1998), a reduction in CDM transfers caused by such a ceiling may be perceived as desirable. Although Yamin et al. (2000) predict a negative albeit small income effect for China and India when the CDM is restricted, it is likely that the developing countries, including China and India, will economically lose from a ceiling on the use of the Kyoto Mechanisms.

However, perhaps surprisingly, there is a lack of consensus among economists about which industrialized country will win and which one will lose under the EU proposal. Some argue that the EU loses and the US gains (e.g. Bollen et al., 1999), whereas others argue that the EU gains and the US loses (e.g. Zhang, 2000b). The main reason behind these opposing conclusions are different assumptions about the relative marginal abatement costs of these countries.

If it is assumed that the EU has higher marginal abatement costs than the US, the EU loses and the US gains. Bollen et al. (1999) calculate on the basis of the WorldScan model that a restriction on trade is not binding for the US. Because the quantitative ceiling lowers demand, the price of transferred entitlements declines, so that the US increases its imports of emission rights and undertakes less domestic action. However, the EU is constrained by the restriction and has to make larger and more expensive domestic reductions as the cost increase outweighs the advantage of lower prices on imported permits. This implies that the EU "burns its own fingers" when it proposes to restrict such trade. Also Baron (1999b) contends on the basis of a survey of several different models that the negotiating position of the EU is economically irrational, because he claims that it would be in its self-interest to buy even a larger portion of its total commitment than North America. In a survey of 10 economic models by Capros et al. (2000), most (but not all) models appear to assume that the US has lower marginal abatement costs than the EU.

However, the results change if EU demand will be lower than US demand. If it is assumed that the EU has lower marginal abatement costs than the US, the EU gains and the US loses. On the basis of the EPPA model (and by using the official and arguably too low emission projections provided by the Annex B Parties themselves), Zhang (2000b) calculates that a ceiling on trade would lead to less cost savings for the US, but more cost savings for the EU. The ceiling would reduce the sum of abatement costs and permit expenditures of the EU by as much as 40% (compared to a cost saving of no more than a few percentages under full trading). The trade restriction reduces demand and thus lowers the international market price, as explained in the previous subsection. This means that the EU proposal has the perverse effect — probably unanticipated by those who shaped the proposal with a view to strengthening domestic action — that the EU will buy more permits and will show *less* domestic action under restricted trading (than under full trading). The reason for this is that the cost savings because of the lower international price are larger than the increased costs of domestic action (Ellerman, 2000).

To make the issue even more complex, Criqui et al. (1999) have used the POLES model to calculate that an import limit of 15% would have no effect on Annex B abatement costs. Furthermore, the cost-minimizing solution for the US would be a 40% ceiling on trade and for the EU a 35% ceiling on trade. Beyond these levels, the increase in domestic costs would exceed the decrease in the market price. This would imply that both the EU *and* the US would have an interest in a ceiling, albeit a less restrictive one than preferred by the EU. The result that both the Europeans and the Americans gain by restricting trade is confirmed by Rose & Stevens (2001) who have also incorporated benefits in their analysis. Although mitigation costs rise due to the trade restriction, gross benefits increase since overall mitigation is strengthened, mainly because Russia sells less hot air.[4]

---

[4] In addition, Ybema et al. (1999) have tried to open the "black box" of the EU by identifying the gainers and losers of a trade restriction within the union. They conclude that some Member States could economically loose from a ceiling on trade, such as France and Germany, whereas others may gain, such as the United Kingdom. However, the disadvantage is that this study is restricted to the EU and does not consider the rest of the world at all, thereby neglecting the (potential) demand by the US and the supply by Russia, for instance. The latter also implies that this study is unable to anticipate the effect, as noted by Zhang (2000b), that the EU is likely to buy more permits and abate less domestically (thereby reducing its compliance costs) under a ceiling on trade compared to unrestricted trading because of the lower international market price induced by the ceiling. This effect could change the cost distribution within the EU.

The literature showed more consensus on the effect of a ceiling on trade with regard to the JUSCANZ countries other than the US. In most models, according to Capros et al. (2000), Japan is assumed to have even higher marginal abatement costs than the EU (and the US) and will lose from a ceiling on trade (e.g. Zhang, 2000b). In some models, however, Japan has relatively low marginal abatement costs and may actually gain under such a ceiling (e.g. Rose & Stevens, 2001). Few authors have modeled these effects separately for Canada, Australia and New Zealand. Because most calculate only very small effects of a ceiling on the latter three countries (e.g. Zhang, 2000b; Rose & Stevens, 2001) and because most authors expect Japan to lose, we assume that the countries of the JUSCANZ group other than the US have an economic interest in unrestricted trading.

Although the EU and the US could economically gain or lose from a ceiling on trade, depending on the level of their marginal abatement costs, it is their perception of the issue that counts according to a neo-institutional approach. Research indicating that the US might gain from a ceiling on trade was conducted in the course of 1999 and 2000, which was probably too early to be used (or to be available) when the US formulated its opinion on the EU proposal in 1999. This suggests that the US has rather acted on the basis of a market-oriented perception that full trade is good for them as shaped by and confirmed in many earlier emissions trading studies (e.g. Tietenberg, 1992; Richels et al., 1996). Similar remarks can be made for the EU. Although the EU formulated a negotiating position probably (but not necessarily) against its economic interests, the point is that the EU (later followed by India) has acted on the basis of the perception that restricting trade is rational because it serves policy goals other than efficiency. Which policy goals the EU had in mind when proposing to restrict the use of the Kyoto Mechanisms is hypothesized in the next section.

8.4. Theoretical Explanations of the EU Proposal on Supplementarity

In this section, we construct 16 hypotheses, which are tested empirically in the next chapter, that could explain the reservations of the EU against unrestricted use of the Kyoto Mechanisms. In the formulation of these hypotheses, the terms "emissions trading" and "Kyoto Mechanisms" are used interchangeably, following a reasoning that these flexible instruments under the Kyoto Protocol are all variants of the original emissions trading concept (e.g. Dales, 1968; Montgomery, 1972; Tietenberg, 1992).

Nevertheless, it will appear that some of the hypothesized reasons to restrict trading apply to all Kyoto Mechanisms (including JI and the CDM), whereas others only apply to permit (or government) trading under IET Article 17.

Although all objections against unrestricted trading are interesting to find out what drives market-skeptic governments, in particular the objections against permit trading are interesting to find why this economically superior alternative ranks low in the political hierarchy of market-based climate policy. The limitations of our theoretical analysis will also be discussed.

8.4.1. Hypotheses on Restricting the Use of the Kyoto Mechanisms

Hypothesis 1: Hot Air. According to hypothesis 1, the EU wanted to limit emissions trading in order to limit the use of "hot air" (e.g. from Russia and Ukraine), since its use would affect the environmental effectiveness of the Kyoto Protocol.

Because the assigned amounts of some Parties, notably the Russian Federation and the Ukraine, have been set higher than their actual and/or future emission levels, for instance due to their negotiating power in Kyoto in 1997, they will be able to sell emission rights that would otherwise have remained surplus in the first commitment period (e.g. Victor et al., 1998). These emission rights are then mobilized and used to cover emissions that would not have been allowed in the absence of emissions trading. This would be impossible in JI projects that involve reductions relative to actual emissions (provided that the micro-baseline is correct). Hot air trading disturbs effectiveness in its ethical interpretation, as we have explained in Chapter 4, which means that it makes overall emissions higher with than without emissions trading.

Limiting the use of "hot air" is mentioned by many economists as (one of) the main reason(s) for the EU to propose a ceiling on trade (e.g. Hourcade & Le Pesant, 1999; Petsonk, 1999; Yamin et al., 2000; Zhang, 2000b; Metz et al., 2001). The IEA has calculated that the EU proposal would limit the supply of hot air to about one-third of its potential magnitude (Baron et al., 1999). Hot air is an institutional feature that becomes an institutional barrier to get emissions trading functioning once actors view hot air as problematic and start to block the implementation of the entire scheme. It is hypothesized here that this actually happened when the EU proposed, after the targets were negotiated, to limit hot air by restricting trading.

It should be kept in mind, though, that hot air does not affect environmental effectiveness in its formal interpretation, because the official aggregate emission target is achieved also if hot air is traded. Moreover, the EU proposal reduces efficiency by limiting the possibility to trade. Furthermore, without the hot air, the US (a potential buyer) and Russia (a potential seller) might only have accepted lower emission ceilings (or even none at all), to an extent that could exceed the volume of hot air under the Kyoto Protocol (e.g. Baumert et al., 1999; Boom, 2000b).

In addition, Eastern Europeans seem to consider the tradeable hot air as a legitimate compensation for the emission reductions induced by the economic decline which resulted from the deliberately established economic transition process (Bashmakov, 1999). Finally, the EU proposal only reduces, but does not eliminate the hot air and the proposal cannot prevent the carry-over of hot air to a subsequent commitment period on the basis of Article 3.13.

Hypothesis 2: Equity. According to hypothesis 2, the EU wanted to limit emissions trading in order to stimulate domestic action with a view to equity or fairness, because Annex B Parties are responsible for the majority of historical GHG emissions and should not completely "buy their way out".

A common (albeit not the only) interpretation of equity in the context of climate change and emissions trading is the concept of (historical) responsibility (Banuri et al., 1996; Fermann, 1997). Because the Annex B Parties produced the majority of historical GHG emissions (e.g. OECD/IEA, 1996), this equity principle concludes that the industrialized countries are responsible for creating the problem of climate change and thus should also solve the problem (e.g. Sari, 1999; Neumayer, 2000). Such an interpretation of equity requires that they should not (solely) "export sacrifices" or "buy their way out" of their responsibilities by purchasing cheap credits, permits or assigned amounts from abroad, but rather that they should "clean up their own mess" by fundamentally changing their consumption and production patterns through domestic action. From this (equity) perspective, the use of emissions trading should be legally restricted (e.g. Aslam, 1999).

It is sometimes also defended, for instance by Banuri et al. (1996: 105), that the historical responsibility approach to equity is already established in the FCCC (e.g. Article 3.1) and the Kyoto Protocol (e.g. Article 17) which indicate (a) that the Parties should protect the climate system on the basis of equity in accordance with their common but differentiated responsibilities and respective capabilities, (b) that the developed countries should take the lead in combating climate change and (c) that emissions trading should be supplemental to domestic action. The historical responsibility view has been ventured frequently by developing countries in the FCCC negotiation process (e.g. Carpenter et al., 1998). Some authors mention factors which suggest that such equity considerations help to explain the EU proposal to limit emissions trading (e.g. Boom, 2000b; Zhang, 2000b).

However, neoclassical economists would argue that any allocation between domestic and cross-border emission reduction arising from unrestricted emissions trading is inherently fair (e.g. Hargrave et al., 1999b). In their view, emissions trading is always supplemental, since domestic action will take place if its marginal cost is lower than the permit price (e.g. Hourcade & Le Pesant, 1999). For instance, Zhang (2000b) projects that in a full emissions trading scenario

Annex B countries would still implement about 30% of the required emission reduction efforts by means of domestic action. It can also be objected that reducing efficiency by means of a ceiling on the Kyoto Mechanisms is in conflict with the FCCC (e.g. Article 3.3) and the Kyoto Protocol (e.g. Article 4.1), which emphasize the objective of flexibility or cost-effectiveness as well as the possibility for Parties to fulfill their commitments jointly.

Hypothesis 3: Compliance. According to hypothesis 3, the EU wanted to limit emissions trading in order to stimulate domestic action with a view to compliance, because it gets more difficult for a Party to curb emissions in a second commitment period if it has implemented its Quantified Emission Limitation or Reduction Commitment (QELRC) of the first commitment period mainly in other countries rather than initiating a trajectory of reducing emissions at home.

According to Hourcade & Le Pesant (1999), a dynamic inconsistency problem is the economic rationale behind this hypothesis: emissions trading avoids the short-term (political) costs of domestic abatement, but it may discourage the adoption of measures at home deemed necessary to reduce emissions in the long term (Michaelowa et al., 1999). This may lead to a proportionally larger increase in future costs or to a higher risk of non-compliance if the stringency of the targets in a second commitment period requires a sudden break with the trend (and lifestyles) of domestic emission growth. Limiting emissions trading then stimulates domestic action and improves the conditions for compliance in a next commitment period (e.g. Werksman, 1999a).

Although this hypothesis may reflect existing perceptions of the necessity of early domestic action, it seems to conflict with much of the conventional economic and technological literature on climate change and emissions trading. Most economists contend that emissions trading rather facilitates compliance both in the short- and long-term, because it lowers the overall costs of reducing emissions (e.g. Tietenberg, 1992; Anderson et al., 1999; Zhang & Nentjes, 1999). In this view, emissions trading can be seen as a "soft landing" (Crane et al., 1998) or "safety valve" (Werksman, 1999a) for Annex B Parties, whereas a quantitative ceiling on emissions trading, as proposed by the EU, would make compliance more difficult. Furthermore, breaking with the trend now is more costly than doing so in the future (rather than cheaper as suggested above) if prospective technological developments will make it easier and cheaper to reduce emissions and comply with the commitments.

Hypothesis 4: Technological Innovation. According to hypothesis 4, the EU wanted to limit emissions trading in order to stimulate technological innovation.

Various authors agree that emissions trading leads to less domestic action, which therefore reduces the pressure for technological innovation (e.g. Barker et al., 2000, 2001). In principle, a ceiling on trade would increase domestic action and raise overall compliance costs, which would accelerate research and development in abatement technology compared to unrestricted emissions trading. In this view, emissions trading is associated with an undesirable reduction of incentives for the industry to become "first-movers". This theory is mentioned in the literature as one of the possible reasons (e.g. Ybema et al., 1999), or even as one of the key arguments (e.g. Metz et al., 2001), behind the EU proposal to restrict emissions trading.

However, several scientists emphasize that the negative effect of emissions trading on technological development is uncertain, because the pace of technological change is difficult if not impossible to predict (e.g. Yamin et al., 2000; Metz et al., 2001). Some even claim that emissions trading has a positive effect on technological progress, although the empirical evidence is scarce (Tietenberg, 1999). Dutschke & Michaelowa (1999), for instance, argue against limiting the CDM, not only because it raises costs, but also because such a restriction would give no dynamic incentive for innovation (as it would presumably lead to a slower diffusion of clean technologies). In principle, the level of technological innovation will adapt itself to the stringency of and compliance with the commitments, as well as to the legal possibility and technical availability of options to buy reductions from abroad. Nevertheless, some authors have proposed to determine a minimum permit price to stimulate innovation, which would also secure a reasonable price for sellers, while a maximum permit price should then prevent that compliance becomes too costly for buyers (e.g. Rolph, 1983; Hourcade & Le Pesant, 1999).

Hypothesis 5: Example-Setting. According to hypothesis 5, the EU wanted to limit emissions trading in order to demonstrate the willingness to reduce emissions in industrialized countries so that developing countries are stimulated to adopt commitments in the future.

The example-setting hypothesis is mentioned by Yamin et al. (2000), Zhang (2000b) and Metz et al. (2001), among others, as one of the explanations for the EU position on supplementarity. The assumption is that developing countries refuse to consider any binding constraint unless developed countries show credibility by reducing domestic emissions. Developing countries have, in fact, ventured this opinion in the FCCC negotiations (e.g. Carpenter et al., 1998). A ceiling on emissions trading, which should force Annex B Parties to abate more at home, is then expected to alleviate the concern that emissions trading leads to reductions abroad without substantial progress of domestic abatement in the industrialized countries.

However, it could be argued that Annex B Parties still set a credible example if they jointly succeed in reducing their overall GHG emissions by at least 5%, trading or not. It can even be argued that the industrialized countries may trade as much as they want within the Annex B region as long as they attain their common target. To avoid that they solely buy cheap reductions from developing countries, the ceiling on trade could be applied to the CDM only, but this would reduce potential transfers of income and technology to the developing countries (e.g. Dutschke & Michaelowa, 1999). By restricting trade the industrialized countries do not only punish themselves, but also the developing countries. Apart from the question whether developing countries will perceive this to be fair, it could be contended that less income and less technology than would have been possible due to a restricted CDM make future targets for developing countries even less likely than under full use of the CDM.

Hypothesis 6: Liability. According to hypothesis 6, the EU wanted to limit emissions trading in order to reduce problems in the case of non-compliance, because emissions trading requires additional rules about who is responsible (buyer or seller liability) if assigned amounts, credits and/or permits have been traded which have not been backed up by real reductions.

Without any form of emissions trading, it is transparent that an Annex B Party with excess emissions is responsible for its own non-compliance under Kyoto Protocol Article 3.1. This is less clear when trading is allowed. Based on conventional practices in other areas of trade, some assume that the seller must be liable (e.g. Tietenberg et al., 1999), but this may lead to "overselling", in particular in a weak international enforcement regime where some sellers have a lower willingness to comply than buyers (Klaassen & Nentjes, 2002). Instead of reducing the potential problem of overselling via buyer liability (as proposed by the G77 and China) or shared liability (as proposed by the EU), which turned out to be politically unacceptable for the JUSCANZ countries in earlier negotiations, the seriousness (not the nature) of this problem could be decreased, as hypothesized here, by limiting the use of these flexible instruments all together (JIQ, 2000b).

However, this view must be put into perspective by acknowledging that the Kyoto Mechanisms do not aggravate non-compliance problems, but rather facilitate compliance, because they lower the overall costs of complying with the commitments (e.g. Zhang & Nentjes, 1999). The implication is that potential non-compliance problems must not be tackled by limiting emissions trading, but by designing effective enforcement mechanisms (that may not be perfectly effective under the Kyoto Protocol yet, as discussed in Chapter 4). Currently, seller liability is in place under the Kyoto regime and buyer liability turned out to be unacceptable because it raises transaction costs. Buyer liability is still preferable to a quantitative restriction on trade, though, because such a restriction crudely reduces the overall cost saving potential, whereas buyer liability increases transaction costs but leaves the efficiency untouched. Nevertheless, the political complexity of choosing among or combining the possible liability rules is recognized by various authors (e.g. Zhang, 1999a).

Hypothesis 7: Imperfect Markets. According to hypothesis 7, the EU wanted to limit emissions trading in order to limit the use of markets in environmental policy because they are likely to function imperfectly.

Emissions trading creates a market. In practice, relatively few markets meet the assumptions of perfect competition (Helpman & Krugman, 1989). A real carbon trading market may suffer from substantial transaction costs and market power if its design or functioning deviates from the neoclassical permit trading blueprint with little government oversight and many participants (e.g. Tietenberg et al., 1999). For instance, the Russian Federation is expected to contribute as much as 70% to the total supply of tradeable reductions (Gusbin et al., 1999; Ciorba et al., 2001). A monopolist may be able to exert market power in the permit market or a may use its market power on the permit market to gain power in the product market (e.g. Malueg, 1990; Westskog, 1995). In particular if only intergovernmental trading would be allowed (which was still an option at the end of the 1990s), it could be hypothesized that the EU has proposed to restrict market-based climate policy in order to limit the problems of imperfect competition associated with these flexible instruments.

Although this hypothesis may help to explain the resistance of the EU against unrestricted emissions trading, a ceiling on trade could actually enhance market power. Not only would the EU proposal increase transaction costs, because a pre-approval of each trade is required to make sure that a transaction does not fall behind the national threshold (Zhang, 2000a: 323), it could also increase market power, since the proposed demand restriction lowers the international carbon price which could stimulate the suppliers to raise the prices for which they sell their reductions. Market power can be avoided by designing a downstream trading system with many participants (e.g. Koutstaal, 1997; Nentjes, 1998; Anderson et al., 1999), although its set-up costs are likely to be relatively high. A quantitative ceiling on trade to combat transaction costs and market power is then seen as the wrong tool for a righteous goal.

Hypothesis 8: Negotiating Power Regarding the Flexible Instruments. According to hypothesis 8, the EU wanted to limit emissions trading in order to wield negotiating power with respect to elaborating the rules of the Kyoto Mechanisms (Articles 17, 12 and 6) at CoP6, because at previous CoP-meetings the EU has been able to exert influence on the level of country emission targets, but less on the choice of instruments.

Various scientists indicate that the EU has succeeded in speeding up both the climate change agenda formation process and the adoption of commitments by the industrialized countries, including the US (before they abandoned the Kyoto Protocol in March 2001), mainly because of the EU negotiating efforts to plead for a GHG emission reduction target of 15% for the industrialized world (e.g. Anderson, 1997; Ringius, 1999; Vogler, 1999). The US government initially (and strategically) signaled only to be willing to adopt a stabilization target and demanded the possibility of international emissions trading (tradeable permits and, to a lesser extent, joint implementation). The EU strongly opposed the use of market instruments (because they feared, among other things, that the US would not reduce domestic emissions), but finally had to accept their inclusion in the Protocol to obtain an emission reduction commitment by the US. Therefore, the EU cannot plead for deleting Articles 17, 12 and 6, because this would imply a renegotiation of the Protocol that could trigger non-ratification by the US.

However, from the perspective of a presumably green albeit trade-reluctant EU, a "second-best" option would then be to try to limit the application of the Kyoto Mechanisms, which would explain the inclusion of the unelaborated supplementarity provisions in the Protocol as well as their elaboration in the form of a quantitative ceiling. Furthermore, by proposing to restrict something the US eagerly wants (i.e. emissions trading), the EU has something to give away (i.e. the restriction) in return for some favorable compromises. In particular, this strategy may wield negotiating power for the EU with respect to elaborating the details of an emissions trading system, since decisions had to be made, at that time, on issues like sinks, supplementarity, hot air, eligibility, compliance, liability and non-distortion of competition (BAPA, 1998).

A critical assessment of this hypothesis could suggest that the European proposal is not so much a strategic move, but rather the revelation of a true preference for domestic action. However, even then it could be maintained that the EU proposal on supplementarity will probably have the (arguably unintended) *effect* of wielding negotiating power. Furthermore, the EU position on supplementarity is likely to be a strategic standpoint similar to its proposal of a 15% reduction prior to the Protocol negotiations in 1997, which was also seen as a bargaining position rather than a revelation of true preferences, even by the EU itself (COM, 1997: 18). Likewise, the EU proposed in 1999 to place a ceiling on cross-border emissions trading (EU Council, 1999), whereas the EU declared in 2000 that it will develop an internal emissions trading regime without mentioning anything on limiting such trade between its Member States (e.g. COM, 2000a).

Hypothesis 9: Negotiating Power Regarding the FCCC. According to hypothesis 9, the EU wanted to limit emissions trading in order to wield negotiating power with respect to the FCCC decision-making process in general, because the proposal gives the EU "something to negotiate about" which could increase the likelihood of favorable compromises at CoP6.

This hypothesis is a variant of hypothesis 8. The idea is that the EU has proposed to limit emissions trading, not only because it wants to influence the elaboration of the Kyoto Mechanisms, but rather because it seeks to influence the upcoming FCCC negotiations in general. It was already explained in discussing hypothesis 8 that by proposing to restrict something the US and some other countries eagerly desire (i.e. emissions trading), the EU has something to give away (i.e. the restriction) in return for some favorable compromises with respect to climate change issues that are not necessarily or strictly related to emissions trading alone, such as technology transfer, capacity building, compliance, the role of the GEF, national communications, agenda building, competence of FCCC institutions and the possible adoption of a second commitment period (BAPA, 1998).

Nevertheless, apart from the remarks already made under the critical assessment of hypothesis 8, it can be defended that the most controversial issues to be decided upon with respect to the overall implementation of the Kyoto Protocol are still related to emissions trading, such as sinks, supplementarity, hot air, the role of the private sector and compliance, precisely because the complexity of these issues follows from the joint execution of the agreements across national borders instead of domestic action. This would imply that the EU proposal on supplementarity is more likely to have been intended to wield negotiating power concerning the elaboration of the Kyoto Mechanisms than concerning the negotiating process in general.

Hypothesis 10: Climate Leadership. According to hypothesis 10, the EU wanted to limit emissions trading in order to not to lose the "climate leadership" to the US.

The EU has the ambition and perception of being an international environmental leader (COM, 1999a), while recognizing its difficulties in stabilizing or reducing GHG emissions, and it succeeded in performing a leadership role with respect to defining the targets of the Kyoto Protocol (e.g. Ringius, 1999). The EU can be labeled as a directional leader that is able to generate ideas for the development of the climate regime (see Gupta et al., 2000). However, the EU was not able to maintain its leadership in the field of choosing the instruments to achieve the targets, where the US appeared to be more dominant by demanding and attaining the possibility of trading emissions across national borders under Article 17 (e.g. Oberthür & Ott, 1999). It could be claimed that the EU loses its leadership if the novelty of emissions trading, to its dissatisfaction, would become the major (or the only) means of implementing the Kyoto Protocol. In this view, the EU is thought to have proposed a ceiling on trade in order not lose its (directional) climate leadership to the US.

However, it could be objected that you cannot lose something you do not have. Indeed, Gupta et al. (2000) claimed that the EU aspired but did not show credible leadership, because the majority of its Member States failed to stabilize or reduce their GHG emissions with a view to reaching the targets of their national environmental policies and the Kyoto Protocol. Furthermore, at least before the US abandoned the Kyoto Protocol, it appeared that the ability to generate ideas for the development of the climate regime had, to some extent, shifted to the US and the other JUSCANZ countries by pursuing emissions trading as one of the main innovative instruments to achieve the emission targets.

Hypothesis 11: Incrementalism. According to hypothesis 11, the EU wanted to limit emissions trading in order to prevent too radical policy changes (from standards/taxes to emissions trading).

It is widely acknowledged that the US has much more experience with emissions trading than the EU (e.g. Jepma et al., 1998; Tietenberg et al., 1999). The US started to experiment with credit trading in the 1970s, while Europe started to develop environmental policy mainly on the basis of standards and taxes. According to Hourcade & Le Pesant (1999), the fear that emissions trading will dismantle the European policy of taxes and standards partly explains the proposal by the EU to limit such trade. There was a concern about the compatibility between emissions trading and existing or future environmental policy, like the energy-efficiency requirements defined per installation under the Integrated Pollution Prevention and Control (IPPC) Directive of the EU (Yamin & Lefevere, 2000). Therefore, limiting emissions trading would give more room to and would, in this view, limit the inconsistencies with traditional and existing environmental policy in the EU, thereby facilitating an incremental and thus possibly more acceptable policy change.

Critics may wonder, however, why the EU would want to maintain its environmental policy of standards and taxes, which have not been able to place its Member States on a trajectory of decreasing GHG emissions so far (e.g. COM, 2000e). Moreover, it could be argued that the cost savings generated by emissions trading are likely to be an economic precondition for achieving the emission targets. Nevertheless, there is the political difficulty of replacing an entire system of experience and institutions related to taxation and command and control by a new system of emissions trading that would also require the renegotiation of some existing (voluntary) agreements with the industry. Albeit not efficient, a politically acceptable way to do this could be to make small steps, for instance by starting with (inter)national credit trading (e.g. Haddad & Palmisano, 2001) or by placing a ceiling on the Kyoto Mechanisms. Although the Europeans at that time also proposed to start an experimental EU-wide carbon trading scheme among large emitters by 2005 (COM, 2000a, 2001a), which reflects some willingness to take non-incremental measures, they also favored "(…) a prudent step-by-step approach in the development of emissions trading (…)" (COM, 2000a: 10), for instance by confining themselves to large fixed point sources of carbon dioxide. The hypothesis here is that similar desires for incrementalism might also have lead the EU to propose to start with limited instead of full-scale international carbon trading.

Hypothesis 12: Allocation Problems. According to hypothesis 12, the EU wanted to limit emissions trading in order to reduce the scope for allocation problems associated with emissions trading, since permit allocation gives emitters an incentive to inflate baseline emissions, triggers endless debates about which distribution is fair and could induce competitive distortions.

Hourcade & Le Pesant (1999) mention permit allocation and the associated equity and environmental problems in the context of potential explanations for the EU to propose a ceiling on emissions trading. First, if emitters know that permits will be grandfathered on the basis of historical emissions, which has been the most frequently used allocation rule in existing emissions trading systems (e.g. Rolph, 1983; Varilek & Marenzi, 2001), they have an incentive to inflate their emissions before the allocation in order to receive an emissions budget as generous as possible (e.g. Rolfe et al., 1999). Second, different political actors have different opinions and interests about which allocation of permits is fair or desirable. Grandfathering, for instance, can be based on energy-efficiency or historical emissions. Under the latter criterion, energy-efficient emission sources get less permits than heavily polluting emission sources, which is likely to be perceived as unfair by the relatively clean sources, since it would perversely reward those polluters who have been lazy towards reducing emissions in the past.

In addition, there was a fear that permit allocation could lead to competitive distortions and subsidization (under WTO or EC law) if permits are grandfathered in one country and auctioned in another. This would be no problem from an efficiency perspective that recognizes the opportunity costs of grandfathered permits, but it could be problematic from an equity point of view that stresses the financial advantage of grandfathering over auctioned competitors abroad (Woerdman, 2000a, 2003). This advantage may also distort efficiency in some cases of imperfect competition, for instance when a grandfathered firm can outlast an auctioned firm, or a potential entrant, in a price war (Nentjes et al., 1995).

It could be hypothesized that a limit on trade has been proposed by the EU because it would reduce the scale of the potential problems associated with allocating permits. It is then assumed, for instance, that a grandfathered firm has less economic possibilities to profit from its financial advantages on the international level when use of the Kyoto Mechanisms is quantitatively restricted. However, the aforementioned problems depend on the perspective taken or could be solved by other means than a trade restriction. Moreover, Crane et al. (1998) indicate that grandfathering has usually been supplemented in US practice by some measure of performance so as to not penalize emitters that have previously taken steps to reduce emissions (Rolfe et al., 1999). Furthermore, to prevent baseline inflation prior to permit allocation, an emission reduction target can be chosen on the basis of a reference year sufficiently back in time to prevent such strategic behavior (Lyon, 1982).

In addition, the equity discussions surrounding the allocation of permits may be a substantial political barrier in organizing an emissions trading scheme, but the debate need not be endless as postulated, because several systems have overcome this problem in practice, for instance via the side payment of allocating some hot air permits. Moreover, the competitive distortion issue is a matter of perspective and subsidization does not occur according to the efficiency approach. Cases of imperfection competition are thought to be exceptional in a permit trading market with many participants. Finally, if there would be a relation between allocation problems and the proposed ceiling on trade, the proposal of the EU would mean that it imposes its desire for reducing such problems also on other countries, such as the US, Japan or China, which could be seen as unfair or as a violation of state sovereignty.

Hypothesis 13: Macro-economic and Secondary Benefits. According to hypothesis 13, the EU wanted to limit emissions trading in order to stimulate the economy, because domestic action (despite its higher micro-economic costs) has more positive macro-economic impacts than emissions trading (e.g. in terms of an increase in domestic employment and reduction of non-CO_2 air pollution (secondary benefits)).

Although most economists stress that emissions trading lowers overall compliance costs relative to domestic action (e.g. Bohm, 1999), some also indicate that domestic action could have certain secondary (or: ancillary) benefits which emissions trading does not have (e.g. Barker et al., 2000, 2001). A distinction has to be made between secondary environmental effects and secondary macro-economic effects. First, these authors argue that domestic action (instead of buying reduction units from abroad) can lead to lower local pollution, less traffic congestions, a reduction of non-GHG air pollution (such as SO_2) and more technological innovation. Second, domestic action has both positive and negative macro-economic effects. Domestic action has economic costs when buying permits from abroad would have been cheaper than reducing emissions at home, but domestic action also has economic benefits as national expenditures increase when reduction technologies are bought from domestic industries, which could raise competitiveness and domestic employment.

According to Pearce et al. (1996: 218), the total of these secondary environmental and economic benefits — depending on the (assumptions about) local circumstances — could theoretically offset 30% to even 100% of abatement costs in the US and the EU. Nilsson & Huhtala (2000) contend that secondary benefits could thus alter the balance of costs and benefits of emissions trading in favor of domestic action. Yamin et al. (2000) and Metz et al. (2001) indicate that limiting the use of the Kyoto Mechanisms is sometimes justified on the basis of the argument that domestic action would yield such secondary benefits.

However, some suggest that it is strange to defend domestic GHG emission reduction on the basis of its positive side effects on other policy areas instead of its positive effects on the primary goal of combating climate change itself (e.g. Pearce et al., 1996). Moreover, the overall cost savings of emissions trading could still (and are even likely to) outweigh the secondary benefits of domestic action. Another peculiarity of the hypothesis is that it only considers the buyers' point of view. It says that potential buyers of reductions from abroad will enjoy secondary benefits when they reduce emissions themselves. However, the other side of the coin is that the potential seller is then deprived of secondary benefits. A seller would have obtained such benefits if a buyer would have purchased emission entitlements, which would have necessitated the seller (unless it has hot air) to reduce its emissions in order to create these entitlements. The emission reductions in the seller's country, for instance an Eastern European or developing nation, may also generate local secondary benefits.

From this perspective, it could even follow that the EU proposal is inequitable, because the consequence of restricting the flexible instruments is that the industrialized countries deprive countries like Russia and China from secondary benefits and claim them for themselves. A counterargument could be that the secondary benefits in third world and Eastern European countries are smaller than in industrialized countries, but this is difficult to quantify and may easily lead to scientific disagreement and speculation (e.g. Kopp & Toman, 2000: 68).

Hypothesis 14: Uncertainty and Risk. According to hypothesis 14, the EU wanted to limit emissions trading in order to facilitate a stepwise introduction of the Kyoto Mechanisms thereby hedging against uncertainties and reducing the associated risks.

Emissions trading is surrounded by uncertainties, risks and controversies. Only a glance at the aforementioned hypotheses already reveals many of them. For instance, due to the inherent uncertainty of business-as-usual emissions, it is not clear how much hot air will be traded under the Kyoto Protocol, although most estimates vary from about 10–30% (e.g. Haites, 1997; Victor et al., 1998). Furthermore, emissions trading inherently involves price uncertainty, since the price of the permits will be determined by supply and demand on the market. The carbon price will rise, for instance, if demand increases because of economic growth or inflation (Baumol & Oates, 1988). At the time the EU made its supplementarity proposal, price estimates varied from a few dollars to more than one hundred dollar per ton of CO_2 (e.g. Jepma et al., 1998). In addition, it is not clear if and to what extent domestic action could have more positive secondary benefits relative to the cost savings that emissions trading would provide. Also JI and the CDM involve uncertainties, for instance because the emission reductions are measured from a micro-baseline which is an estimate of emissions that will never occur because of the implementation of the project. To reduce this environmental uncertainty, some authors have proposed to adjust the baseline during the project if actual circumstances deviate from the ex ante baseline assumptions (e.g. Begg et al., 1999), but this would increase economic uncertainty, because the amount of credits generated by the project may turn out to be lower than predicted if the baseline has to be adjusted downwards.

The list of uncertainties and risks goes on and on. Ybema et al. (1999) contend that a stepwise introduction of the flexible instruments would enable Annex B Parties to gain more insight into these uncertainties and risks, which are sometimes mentioned to explain the proposed ceiling on the Kyoto Mechanisms (e.g. Hourcade & Le Pesant, 1999; JIQ, 2000b; Yamin et al., 2000). The fear is that the Kyoto Mechanisms, once they are functioning, turn out to have undesirable consequences which have not been anticipated. Restricting trade would not avoid these uncertainties and risks, but it would limit their scale, thereby reducing the level and seriousness of bad choices, mistakes and unforeseen effects in setting up an international carbon trading market.

However, critics may argue that many uncertainties have already been reduced by the experience gathered in AIJ pilot projects, intra-firm emissions trading schemes (for instance by Shell and BP-Amoco) and early permit trading schemes (for instance in Denmark and the UK) (e.g. Cozijnsen, 2000; Rosenzweig et al., 2002). Furthermore, most economists agree that the Kyoto Mechanisms will lower the overall costs of reducing GHG emissions compared to domestic action (e.g. Tietenberg et al., 1999). Moreover, hot air may be traded under the emission ceilings, but the emitters will not exceed their permits (provided that compliance is assured). Finally, baselines can be standardized for (CDM) projects in order to prevent (strategic) baseline inflation. Uncertainty will never completely disappear, but it is manageable. In addition, a stepwise introduction of emissions trading does not require a ceiling on trade, but could also be pursued by starting with unrestricted trading for a limited number of (large) participants, for instance.

Hypothesis 15: Pressure on the US. According to hypothesis 15, the EU wanted to limit emissions trading in order to force the US to reduce emissions at least partly at home, since they are the biggest source of greenhouse gas emissions.

At the time the EU made its proposal on supplementarity in 1999, the US had not yet withdrawn from the Kyoto Protocol (which they did in 2001). Based on data that were then available, the US was known to be responsible for about 24% of the world's total CO_2 emissions, thereby exceeding all other large nations, such as China (12%) and Russia (8%) (OECD/IEA, 1996: 25). Hence, it was (and still is) considered important that the US substantially contributes to the international combat against global warming by reducing at least part of its domestic GHG emissions. This view draws upon several (aspects) of the aforementioned hypotheses. In particular, it was feared that the US would "buy its way out" of its historical responsibility (equity perspective) by purchasing cheap hot air from Russia and the Ukraine (hot air perspective) without initiating a necessary trajectory of curbing domestic emission growth (compliance and innovation perspective). Forcing the US to reduce emissions at home for these reasons could be an explanation of the EU proposal to place a ceiling on the use of the Kyoto Mechanisms (Hourcade & Le Pesant, 1999; Oberthür & Ott, 1999).

However, it is the question whether the EU proposal would have forced the US to reduce more of its domestic GHG emissions. There are calculations, as demonstrated in the previous section, that a restriction on trade is not binding for the US. If so, the EU proposal would induce the Americans to buy even more permits and reduce less domestically than they would have done under full trading (e.g. Bollen et al., 1999). However, this assumes relatively low marginal abatement costs for the Americans and there is still a debate in the economic literature whether this is the case (e.g. Capros et al., 2000). If this is not the case, a ceiling on trade urges the Americans to undertake (more) domestic action, which would not only make compliance in the first commitment period more expensive for them, but also endangers their ratification of the Kyoto Protocol as well as their acceptance of an ambitious target in a second commitment period. Hence, if the EU proposal is indeed intended to put pressure on the US, it could be argued that this is a risky strategy with a view to the global climate system. Another point of criticism, perhaps the most relevant one, is that the EU already "forced" the US to adopt an emission reduction commitment under the Kyoto Protocol as part of a compromise which would also allow for emissions trading. Demanding a restriction on trade can then be seen as an attempt to renegotiate the Protocol.

Hypothesis 16: Pressure on Russia. According to hypothesis 16, the EU wanted to limit emissions trading in order to prevent Russia from selling hot air and trading questionable emission reductions, since it has a weak enforcement regime.

The Russian Federation is likely to be the biggest seller of hot air, next to the Ukraine, leading to the trading of permits which are used to cover emissions that would not have been allowed in the absence of emissions trading. Furthermore, due to (presumably unintended) problems of measurement, monitoring and enforcement combined with the incentive to generate revenues (e.g. Kessler, 1998; Yamin et al., 2000), Russia might sell even more than its available hot air without taking corresponding emission reduction measures somewhere in the country. Some doubt whether the Russian government is able and willing to effectively stimulate and eventually force its companies to take the required emission reductions in order to reverse the "overselling". This fear, which combines aspects of hot air and compliance, may have led the EU to propose a quantitative ceiling on the use of the Kyoto Mechanisms (Yamin et al., 2000). A ceiling on trade might then also urge the Russians to improve their monitoring and enforcement regime.

However, critics argue that restricting trade will not improve the enforcement system of the Russian Federation, unless the EU gives them the prospect that full trading (from which the Russians are likely to gain) will be allowed after such an improvement. The EU gave no signs that it would then give up the trade restriction. Furthermore, perhaps Russia would not have accepted its stabilization target if the export of hot air would have been impossible as a compensation for its economic crisis (e.g. Bashmakov, 1999; Boom, 2000b). Finally, a ceiling on emissions trading would not only reduce Russia's incentive to sell hot air, but it would also hurt the developing countries by reducing potential CDM transfers, so that, arguably, a ceiling on trade would "overshoot" its goal.

8.4.2. Alternative Hypotheses and Limitations of the Theoretical Analysis

It is possible to formulate additional hypotheses or variants of the aforementioned 16 hypotheses. An example could be that the EU specifically wanted to limit the Kyoto Mechanisms in order to reduce the environmental impact of project-baseline uncertainty, in particular in the case of sinks under the CDM which could result in "tropical air" (JIQ, 2000b). Another example is that the EU wanted to limit these mechanisms because it would limit the possibilities of its trade rivals to lower their abatement costs (Yandle, 1998; Kopp & Toman, 2000; Wiener, 2000b).

Such alternative hypotheses or variants have not been explicitly incorporated into the research. First, they are not considered here, simply because we found them "too late" in the literature, that is, after our mailing of questionnaires to high-level EU officials in March 2000, as we will explain in the next chapter. Second, it could be argued that some possible alternative hypotheses (such as the baseline uncertainty hypothesis) are implicitly included in other hypotheses (such as the hypothesis on uncertainty and risk) which will in fact be tested empirically. Although there is no doubt that the 16 hypotheses described above reflect a large part of the literature on the supplementarity subject, the mere existence of other possible hypotheses indicates that our list of hypotheses is not complete, which imposes a limitation on our analysis.

A difficulty when formulating the hypotheses, which was done in the beginning of 2000, was that the principles, modalities, rules and guidelines for trading under the Kyoto Mechanisms had not yet been defined by the FCCC Parties. For instance, it was not clear whether Article 17 would be limited to intergovernmental trading or whether private trading would also be (or was already implicitly) included: some discussions centered on the first interpretation, for instance in the debate on market power, whereas in others the latter is assumed, for instance in the debate on permit allocation. However, in some discussions the distinction between different types of trading is less relevant, for instance with respect to the issue of negotiating power. Therefore, where possible and relevant, the starting point and the assumptions of the discussion have been indicated above when explaining the theory behind the hypotheses. Although the reader may find some hypotheses to be more obvious or likely than others, the important point is that they will be tested for their relevance in the empirical analysis of the next chapter.

8.4.3. Ex Post Clustering of the Hypotheses

For the sake of transparency and simplicity it is possible to cluster the hypotheses (ex post) into a small number of different types of conjectures, as portrayed in Table 8.1. Although there are several possibilities to form such groups, we have created four clusters of hypotheses which supposedly express the EU's:

- environmental reasons to restrict the use of the Kyoto Mechanisms (hypotheses 1, 3 and 6 on hot air, compliance and liability respectively);
- political-normative reasons to restrict the use of the Kyoto Mechanisms (hypotheses 2, 5, 11, 12 and 14 on equity, example-setting, incrementalism, allocation problems and uncertainty and risk, respectively);
- political-strategic reasons to restrict the use of the Kyoto Mechanisms (hypotheses 8, 9, 10, 15 and 16 on negotiating power regarding the flexible instruments, negotiating power regarding the FCCC, climate leadership, pressure on the US and pressure on Russia, respectively);
- technological-economic reasons to restrict the use of the Kyoto Mechanisms (hypotheses 4, 7 and 13 on technological innovation, imperfect markets and secondary benefits, respectively).

Table 8.1: Theoretical clustering of hypotheses (ex post). The EU proposed to limit the use of the Kyoto Mechanisms because of the type of reasons given below.

Type of reasons	Clustering of hypotheses
Environmental reasons	H_1, H_3, H_6
Political-normative reasons	H_2, H_5, H_{11}, H_{12}, H_{14}
Political-strategic reasons	H_8, H_9, H_{10}, H_{15}, H_{16}
Technological-economic reasons	H_4, H_7, H_{13}

Key: H_x = hypothesis x.

Here is where political science can supplement our neo-institutional economics approach. Political scientists explain that values, which are part of a nation's political culture, engage moral considerations that are conceptions of the desirable (e.g. Almond & Verba, 1965). Values do not only guide, but also constrain social action (Goodin & Klingemann, 1996: 19). Authors like van Deth & Scarbrough (1995b: 33) underline that the attitudes of political actors are partly determined by values, such as equity, but can also be influenced by other considerations, such as political-strategic ones. This proposition is reflected in our theoretical ex post clustering of hypotheses.

Two of the four clusters created are based upon value-driven hypotheses: the cluster of environmental reasons reflects the so-called "green" values on (improving) environmental integrity and the cluster of political-normative reasons reflects equity values. From our positive (not normative) perspective on social norms, it can be observed that these equity values include:

- the (historical) responsibility of industrialized countries for creating (and thus for having to solve) the problem of climate change (hypothesis 2 on equity and hypothesis 5 on example-setting),
- the desirability of incremental policy changes (and thus the undesirability of radical policy changes) (hypothesis 11 on incrementalism),
- the desirability of equal changes of the financial positions and competitive relations among comparable firms as a result of permit allocation (hypotheses 12 on allocation problems),
- the desirability to minimize environmental, economic and political uncertainties and risks (hypothesis 14 on uncertainty and risk).

The other two clusters are based upon political-strategic and technological-economic reasons to restrict the use of the Kyoto Mechanisms. These do not so much represent values, but rather include explanations for the initial EU position on supplementarity in terms of negotiating power and political pressure as well as technological and economic spillover effects of trading greenhouse gases.

However, the clustering of hypotheses in Table 8.1 does not make a distinction between the different Kyoto Mechanisms. Such a distinction would be relevant to find out why the economically superior alternative of permit trading ranks low in the political hierarchy. Therefore, an additional clustering is performed in Table 8.2 that makes a differentiation between (a) objections of EU policy makers against the Kyoto Mechanisms in general and (b) those against permit trading under IET Article 17 in particular.

Contrary to permit trading under IET Article 17, as explained in earlier chapters, JI and CDM projects do not only avoid the mobilization of hot air (and generate real reductions if they are additional), but they are also incremental as they avoid the allocation problem of explicitly (re)distributing property rights.

Table 8.2: Theoretical differentiation of hypotheses (ex post). The EU proposed to limit the use of the Kyoto Mechanisms because of the type of objections given below.

Type of objections	Clustering of hypotheses
Objections against all Kyoto Mechanisms	H_2, H_3, H_4, H_5, H_6, H_7, H_8, H_9, H_{10}, H_{13}, H_{14}, H_{15}, H_{16}
Objections against permit trading	H_1, H_{11}, H_{12}

Key: H_x = hypothesis x.

In line with these earlier findings, two clusters of hypotheses are created which supposedly express the EU's:

- Objections against all Kyoto Mechanisms (hypotheses 2, 3, 4, 5, 6, 7, 8, 9, 10, 13, 14, 15 and 16 on equity, compliance, technological innovation, example-setting, liability, imperfect markets, negotiating power regarding the flexible instruments, negotiating power regarding the FCCC, climate leadership, pressure on the US and pressure on Russia, respectively);
- Objections against permit trading (hypotheses 1, 11 and 12 on hot air, incrementalism and allocation problems, respectively).

There are "only" three hypotheses (1, 11 and 12) in the second cluster, which exclusively relates to permit trading under IET Article 17. Although the remaining hypotheses, in the first cluster, relate to objections against all Kyoto Mechanisms, it would be possible to put more hypotheses in the "permit trading cluster": some objections that can, in principle, be directed against all Kyoto Mechanisms, are mainly used, in practice, against permit trading, as can be heard in (European) policy debates. However, we have used a strict interpretation of these instruments by keeping them analytically separated.[5]

Which cluster of hypotheses (i.e. which particular type of explanation within Table 8.1 or which particular type of objections within Table 8.2) is most able to clarify why the EU proposed to quantitatively restrict the use of the Kyoto Mechanisms can only be determined by means of an empirical analysis. Such an analysis will be carried out in the next chapter. The neo-institutional economics

---

[5] An example is hypothesis 16 about putting political pressure on Russia. This hypothesis postulates that the EU wants to restrict trading to prevent that Russia sells hot air or questionable emission reductions. Hot air trading is only possible under IET Article 17, not under JI Article 6 that (should) involve(s) real emission reductions. This would suggest placing hypothesis 16 in the second cluster. However, questionable emission reductions can also be sold under JI Article 6 if its domestic enforcement regime is weak. Although we believe that the hot air argument usually dominates in policy debates (which means that the hypothesis about pressure on Russia would be primarily directed against emissions trading and not so much against project-based flexibility), we have chosen to use a strict interpretation and place hypothesis 16 in the first cluster that refers to all Kyoto Mechanisms.

literature as well as the political science literature, including the literature on path dependence (e.g. Licht, 2001), political culture (e.g. van Deth & Scarbrough, 1995a,b) and emissions trading (e.g. Bressers & Huitema, 1999), expects that values, such as equity, will play a significant role in explaining the desire of European policy makers to restrict trade.

8.5. Conclusion

There are several informal institutions, including political culture, that could reinforce a path of inefficient regulation and block or hinder the implementation of market-based climate policy (e.g. Bressers & Huitema, 1999; Licht, 2001). For some time, the economically superior arrangement of permit trading was perceived by governments as morally more suspicious than the credit-based instruments, because only the former explicitly (re)allocates "pollution rights". In this way, cultural values added to the switching costs of permit trading and contributed to the institutional lock-in.

The EU, for instance, expended years of political effort trying to reject or restrict the application of economic climate policy instruments. In the context of the Kyoto Protocol, the Europeans proposed in 1999 to make the Kyoto Mechanisms supplemental to domestic action by quantitatively limiting their use. The EU roughly wanted 50% of the Kyoto commitments to be achieved domestically and drafted rules that would not only limit demand for all Kyoto Mechanisms, but that would also exempt units resulting from JI and CDM projects from the restriction on supply (see EU Council, 1999). Although this proposal was rejected internationally in 2001, for instance because it would have reduced efficiency, an analysis of the EU supplementarity proposal is still useful to lay bare the underlying values of why some governments wanted to limit market-based climate policy, especially permit trading.

We have developed (and criticized) 16 hypotheses that could help to explain the EU proposal. An analysis of cultural values is useful, because economic interests do not explain the proposed trade restriction by the EU who is likely to be a potential buyer. By clustering the hypotheses ex post, it can be postulated that the EU wanted to restrict trading:

(a) for environmental reasons (namely to limit hot air trading, to stimulate compliance by initiating a trajectory of reducing domestic emissions and to reduce liability problems in the case of non-compliance);
(b) for political-normative reasons (namely to achieve equity by preventing that the industrialized countries "buy their way out", to show developing countries that industrialized countries are willing to reduce emissions, to facilitate

incremental change based on traditional EU environmental policy of standards, taxes and covenants, to reduce the scope for allocation problems, such as the possibility that permit allocation distorts competition, and to minimize environmental, economic and political uncertainties and risks);
(c) for political-strategic reasons (namely to wield negotiating power regarding the elaboration of the Kyoto Mechanisms in particular or regarding the FCCC negotiation process in general, to prevent that political "climate leadership" is lost to the US, to put pressure on the US, which is the largest emitter of CO_2, to reduce its emissions partly at home and to put pressure on the Russian Federation, which has a weak enforcement regime, to prevent that it sells hot air or questionable emission reductions);
(d) for technological-economic reasons (namely to stimulate technological innovation by means of domestic action, to limit the use of markets in climate policy as they may function imperfectly and to stimulate the economy because domestic action has more secondary benefits than emissions trading).

The first two clusters are based upon value-driven hypotheses: the cluster of environmental reasons reflects so-called "green" values and the cluster of political-normative reasons reflects equity values. Although attitudes are partly determined by values, they can also be influenced by other considerations, such as political-strategic ones (van Deth & Scarbrough, 1995b: 33). This is reflected in our clustering of hypotheses. Although the hypotheses could (positively) help to explain the EU proposal, it is important to note that we do not necessarily agree with the (normative) content of each hypothesis.

The official EU proposal on supplementarity was directed against unrestricted use of all Kyoto Mechanisms, not against permit trading in particular. However, it is shown that the hypotheses on hot air, incrementalism and allocating emission rights do not apply to JI and the CDM and are only relevant for permit trading under IET Article 17. Permit trading not only mobilizes the negotiated hot air, whereas projects (should) generate real reductions, but permit trading is also a non-incremental policy option that explicitly (re)allocates emission rights, whereas credit-based approaches do not require an initial distribution of such rights before trading can begin.

In the empirical analyses of the next chapter it will become clear whether and to what extent cultural objections against permit trading, including equity, are able to explain the EU proposal. The set of attitudes to be researched, on the basis of the aforementioned hypotheses, consists of the opinions of high-level officials in the EU about the European supplementarity proposal. It will also be indicated that cultural change is one of the opportunities that contribute to an institutional breakout, for instance by looking at the decision-making process of the Europeans to create permit trading in the EU from 2005 onwards.

Chapter 9

Empirical Aspects of Restricting Market-Based Climate Policy

9.1. Introduction

This chapter tests the hypotheses on restricting market-based climate policy as formulated in the previous chapter empirically by confronting them with the content of relevant EU documents, the opinions of several high-position EU officials (gathered by means of a questionnaire) and the negotiating behavior of the EU at the international climate negotiations of CoP6, while using the path dependence approach to explain the institutional breakout of the EU towards permit trading.

Neo-institutional economists recognize and analyze the role that informal institutions, like political culture, may play in the lock-in of policy arrangements. Moral resistance against "pollution trading" and the perception that they are inequitable added to the switching costs of permit trading and stimulated governments to build upon existing environmental policy by means of less efficient credit-based approaches. Although values are not directly observable, attitudes are (van Deth & Scarbrough, 1995b: 22). The set of attitudes to be researched in this chapter consists of the opinions of high-level officials in the EU, both from the environmental ministries of its Member States and from the European Commission, about the EU supplementarity proposal of 1999 to quantitatively restrict the use of the Kyoto Mechanisms (e.g. Woerdman, 2002). However, where some refer to a nation's culture as the "mother of all path dependencies", which is typically seen as an "old mother" that is resistant to change (Licht, 2001: 149, 200), we emphasize that values not only act as barriers, but also as opportunities if their content is favorable to change. Therefore, we will try to find out whether path dependence, including equity and cultural change, is able to explain why the Europeans rejected the unconstrained use of permit trading for several years, but also why they finally managed to break out by adopting a Directive that creates permit trading in the EU from 2005 onwards.

This chapter is organized as follows. Section 9.2 discusses the representativity and limitations of the empirical analysis. Section 9.3 tests the hypotheses (developed in the previous chapter) that should help to explain the EU supplementarity proposal by analyzing the content of relevant climate policy documents of the EU, the opinions on supplementarity of various high-level EU officials, gathered by means of a questionnaire, and the bargaining behavior of the EU at CoP6. Section 9.4 uses the path dependence approach to explain why EU climate institutions, which initially became locked-in, succeeded to break out in the direction of permit trading. Finally, Section 9.5 presents the conclusion.

9.2. Representativity and Limitations of the Empirical Analysis

Contrary to the EU, the US and the other so-called JUSCANZ countries (such as Canada, Australia and Japan) did not want to define supplementarity and opposed a restriction on trade. Because supplementarity was one of the major stumbling blocks that prevented the Parties from reaching consensus on the Kyoto Mechanisms at CoP6 Part I, held in The Hague in November 2000 (e.g. Churie et al., 2000), a second meeting was planned in Bonn in July 2001 under the name of CoP6 Part II. However, before this meeting took place, the Americans withdrew from the Kyoto Protocol in March 2001. Their official reason was that the developing countries were still exempted from an emission ceiling and that the Kyoto target would harm their economy (Bush, 2001), but the attempt by the EU (as well as by some developing countries) to quantitatively restrict the use of the Kyoto Mechanisms certainly did not make the US more enthusiastic to support the Kyoto Protocol.

When CoP6 Part II was actually held, the EU was willing to give up its proposal and accepted the unspecified requirement that domestic action shall be a "significant element" of Annex B countries' climate policy (CP, 2001a: 7). The EU accepted this compromise to prevent that some JUSCANZ countries or even the Russian Federation (a potential seller) would withdraw from the Kyoto Protocol, like the US had done a few months earlier. This largely unexpected US decision had changed the game and the EU, which believed in the Protocol but was skeptic to unrestricted emissions trading, rather had a market-based Protocol than no Protocol at all (e.g. Oberthür & Ott, 1999). Nevertheless, the defeat was partial because some of the environmental concerns of the EU were accommodated, for instance, by means of the adoption of restrictions on the use of sinks and the adoption of a commitment period reserve for each Annex B Party which should not drop below 90% of its assigned amount, as discussed in earlier chapters.

The official EU proposal on supplementarity was directed against unrestricted use of all Kyoto Mechanisms, not against permit trading in particular. However, it was shown in the previous chapter that the proposal does exempt credits resulting from JI and CDM projects from the restriction on supply (EU Council, 1999). In addition, some hypotheses (on hot air, incrementalism and allocating emission rights) do not apply to JI and CDM projects and are only relevant for permit trading under IET Article 17. Because a restriction on trade, as proposed by the EU, is likely to be against its own economic interests, an empirical analysis of this proposal is useful to reveal the underlying values of why some governments wanted to limit market-based climate policy, especially permit trading, even though the supplementarity proposal was rejected by the international community in 2001. Moreover, if cultural values can act as a barrier to implementing the Kyoto Mechanisms, cultural change is one of the opportunities to overcome this barrier.

The climate change literature has expressed a desire for more empirical research on the political opposition against emissions trading (e.g. Wiener, 2000a). In addition, the path dependence literature has expressed a desire for more empirical research on political culture in areas of law and economics and institutional evolution (e.g. Nooteboom, 2000: 303; Licht, 2001: 203). In this chapter, we take a modest step to meet these desires. Where the objective of the previous chapter was to provide several positive theoretical explanations for the EU proposal on supplementarity, the objective of this chapter is to test these theoretical explanations empirically. The empirical test will be performed by means of three types of analyses:

- a content analysis of official (and unofficial) documents of the EU on climate change and/or emissions trading;
- an opinion analysis of the answers regarding the central question given by high-level officials (civil servants) from the environmental ministries of the different EU Member States as well as from the European Commission;
- a bargaining analysis of the negotiating behavior of the EU at CoP6 Part I and Part II.

The approached EU civil servants are referred to as "key officials", since they are seen as experts that have contributed to the European proposal and/or have participated in preparing the decisions on the supplementarity issue made by Ministers in the context of CoP6 Part I and Part II. Their opinions have been gathered by sending them a detailed questionnaire in either March, June or September 2000 prior to CoP6 Part I. The text of the questionnaire can be found in the Appendix of this book. Considering both the official policy documents, the views of key officials and the behavior of negotiators supposedly should give a balanced insight into the question why the EU has proposed to restrict trading. The limitations of the analyses will be discussed extensively, not only in this section, but also throughout the rest of the chapter.

Next to the bargaining analysis, which is primarily based on Earth Negotiations Bulletins (ENBs), the empirical research involves a selection of EU documents as well as a selection of officials (civil servants) from EU Member States. The question then comes to mind to what extent these documents are representative of the official EU position on supplementarity. It can also be asked to what extent the opinions of the selected officials are representative of the opinions of (a) the environmental bureaucracy and (b) the environmental Ministers in the EU. Finally, one can wonder to what extent the aforementioned Bulletins are representative of the actual negotiations.

First, the content analysis of EU documents roughly covers the period from 1997 (prior to CoP3 in Kyoto) to 2000 (prior to CoP6 Part I in The Hague). This period was primarily chosen because 1997 was the year in which the Kyoto Mechanisms as well as the provision of supplementarity were laid down in the Kyoto Protocol. The period thereafter is interesting because the EU then presents documents which contain ideas about the elaboration of rules and restrictions regarding the Kyoto Mechanisms. The analysis does not consider documents after March 2000, primarily in order to obtain a valid comparison with the opinions of officials who received our questionnaire in March 2000. Two official documents on climate change and emissions trading released in 2000 were also considered to see if the EU would change its official position on supplementarity as decided upon in 1999 (EU Council, 1999). For the period 1997–2000, an attempt was made to analyze all publicly available documents. Although we cannot rule out that a document has been overlooked, the chance of this occurring is estimated to be small and the documents which have been analyzed do reflect the development and decision of the EU to propose a restriction on the use of the Kyoto Mechanisms. For the sake of completeness, a few unofficial documents have also been analyzed, such as "non-papers" and internal notes.

Second, the hypotheses described in the previous chapter have been tested by sending detailed questionnaires on the supplementarity issue in March 2000 to 41 officials from the environmental ministries of the EU Member States as well as from the European Commission.[1] Instead of asking for their personal opinions (with the exception of one particular question), we asked these officials to indicate for each hypothesis whether it has played a role in the emergence of the EU proposal on supplementarity. We imposed an upfront limitation to the empirical analysis by choosing to approach officials instead of Ministers. We assumed that

---

[1] Those who had not responded yet received the same questionnaire for the second time in June 2000. In September 2000 we sent a draft version of a research memorandum to all officials on the list, including the questionnaire for those who had not responded yet. For the latter group we stressed that we would include their answers to the memorandum no sooner than 2001 after CoP6 in order to minimize possible strategic manipulation of the data as presented in that draft. Only one questionnaire was received after the third and last mailing of September.

the Ministers would not take the time to fill out the questionnaire, not only because they would rather focus on high-priority political decision making, but also because the results of our research could have the danger of impeding political negotiations and consensus-building among the Ministers. We presumed that they would either throw the questionnaire away or delegate it to an official. In both cases, it makes more sense to approach the officials in the first place. Although we assumed that the officials would have more time and less political difficulty to fill out the questionnaire than the Ministers, we still expected a considerable (and difficult to estimate) non-response.

Although the Ministers have a higher position in the bureaucratic hierarchy than the officials, the latter are nevertheless important in the decision-making process. The Ministers decide and determine the official negotiation strategy, but the officials prepare these decisions and provide the Ministers with information and advice. Moreover, the officials approached are not "just" civil servants with a low position in the bureaucratic hierarchy or with no knowledge of the subject. Rather, we call them "key officials", not only because they are directly involved in climate change and emissions trading, but also because they are seen as experts in these fields by the European Commission. Several names and addresses of these officials were provided in February 2000 by Peter Vis, the administrator of the Climate Change Unit of the European Commission (European Union, DG-XI Environment). We extended the list by including European Commission officials participating in Working Group 1 on Flexible Mechanisms of the European Climate Change Programme and by approaching the environmental Ministries of those Member States that were not represented on the list of Peter Vis. The vast majority of the respondents stated explicitly that they wished to remain anonymous in the analysis.[2] An important exception is the response to the questionnaire by Peter Vis himself, working for and reflecting the views of the European Commission.

In the formulation of the hypotheses, and thus also in the questionnaires, the terms "emissions trading" and "Kyoto Mechanisms" are used interchangeably, following a reasoning that these flexible instruments under the Kyoto Protocol are all variants of the original emissions trading concept (e.g. Dales, 1968; Montgomery, 1972; Tietenberg, 1992). Although all objections against unrestricted trading are interesting to find out what drives "market-skeptic" governments, in particular the objections against permit trading are interesting to find why this economically superior alternative ranks low in the political hierarchy of market-based climate policy.

The 16 developed hypotheses were presented to the respondents without the extensive explanation (and critical assessment) of the theory behind them as

---

[2] The names and addresses of the respondents are known with the author.

conducted in the previous chapter. When developing the questionnaire, a trade-off had to be made in this respect. Including a discussion of the theory behind every hypothesis in the questionnaire increases the probability that different respondents interpret the hypotheses in the same way, but also increases the probability of non-response because the inquiry form would become too long and too time-consuming for busy officials to read and fill in. Consequently, not including a theoretical explanation of the theory behind every hypothesis in the questionnaire increases the potential danger of different interpretations, but lowers the risk of non-response. Because we feared a considerable level of non-response (due to constraints of time and autonomy for officials to co-operate with our research, as discussed above), we chose for the latter option not to include theoretical discussions in the questionnaire. Despite its plausible advantages and possible drawbacks, it should also be noted that the hypotheses themselves already contain a theoretical reasoning, so that the absence of comprehensive scientific discussions certainly does not mean that the questionnaire is void of any theory. Rather, the respondents had to do without a further elaboration of the theory.

Although the choice of a short questionnaire is likely to have prevented a lower level of non-response than the one we encountered, only 15 of the 41 questionnaires sent to the aforementioned European officials were returned. The others presumably did not want to answer the perhaps politically sensitive questions (which was the honest reply of one potential respondent) or they did not have the time to fill in the questionnaire. The number of officials approached is only a fraction of the environmental civil servants and, within this small number, the non-response thus amounts to 63%. On the one hand, this strongly limits the representativity of our analysis, although a survey non-response of 50% or more is no exception nowadays and these percentages are increasing internationally (e.g. de Heer, 2000). On the other hand, the officials who did answer are experts in the field of emissions trading (with a certain influence on decision making), while the 15 returned questionnaires came from almost every country in the EU as well as from the European Commission. In addition, the interviews consisted of in-depth and detailed questions. The questionnaires may therefore be considered to represent at least a relevant part of the opinions of emissions trading experts working for the environmental ministries of the 15 Member States and for the EU in Brussels.

Finally, a bargaining analysis is conducted which, next to official documents, primarily uses information from the ENBs (and occasionally draws from newspaper articles which are only brought up to refer to public statements made by government representatives or officials in the media). On the FCCC Internet site of CoP6, the ENB is described as "(…) an independent reporting service that provides daily coverage and summaries of official UN negotiations on environment and development agreements (…)" (see http://cop6.unfccc.int/enb). It is published by, but does not necessarily reflect the views of, the International

Institute for Sustainable Development (IISD) from Canada and is prepared in cooperation with the UNFCCC Secretariat. It is supported by governments and donors of the Bulletin are Ministries of international and/or environmental affairs from several industrialized countries, including the Netherlands, Canada and the US as well as the European Commission, next to UN agencies and private foundations. Although the ENB reports are not able to cover all aspects of the negotiations, such as contacts between negotiators behind closed doors (for instance, in the form of informal personal contacts or telephone calls), they do not only report on formal agenda points, official negotiating positions and negotiating outcomes, but more importantly (because the following cannot be found in official documents), they also summarize the discussions that took place during the negotiating sessions, in which, for instance, governments clarify and sometimes change their official positions. Although a summary could, in principle, leave out details that might be relevant, the ENB has built a solid reputation for its objective coverage of the climate change negotiations. The aforementioned Internet site of the FCCC states that the ENB "(…) has provided neutral, informative and objective reports on all major meetings of the UNFCCC (…)".

It follows from the remarks above that there are limitations to the empirical analysis. There is little doubt about the representativity of the document analysis and the bargaining analysis. The representativity of the analysis of questionnaires is more problematic, but still seems to be sufficient given the level of detail of the questions and answers, the level of power and knowledge of the officials and the geographical distribution of those who responded. Nevertheless, the number of officials approached is limited and the non-response is considerable, so that conclusions can only be drawn cautiously.

9.3. Empirical Analysis of the EU Proposal on Supplementarity

The 16 hypotheses that could explain the EU proposal to restrict trade are repeated and summarized in Table 9.1 for the sake of convenience. For the theory behind these hypotheses, as well as for a critical assessment, the reader is referred to the previous chapter. In this section, we will test the hypotheses empirically by analyzing the content of EU documents, the opinions of several key EU officials and the negotiating behavior of the EU at the international climate negotiations of CoP6.

9.3.1. Content Analysis of EU Documents

In this subsection, 12 EU documents are analyzed, on the basis of the aforementioned hypotheses, in chronological order starting from the beginning of 1997 to the end of 2000 (before CoP6 Part I). A first distinction is made between

Table 9.1: Summary of hypotheses.

Hypothesis number and label	Hypothesis formulation
H_1 hot air	The EU wanted to limit emissions trading in order to limit the use of "hot air" (e.g. from Russia and Ukraine), since its use would affect the environmental effectiveness of the Kyoto Protocol
H_2 equity	The EU wanted to limit emissions trading in order to stimulate domestic action with a view to equity or fairness because Annex B Parties are responsible for the majority of historical GHG emissions and should not completely "buy their way out"
H_3 compliance	The EU wanted to limit emissions trading in order to stimulate domestic action with a view to compliance because it gets more difficult for a Party to curb emissions in a second commitment period if it has implemented its Quantified Emission Limitation or Reduction Commitment of the first commitment period mainly in other countries rather than initiating a trajectory of reducing emissions at home
H_4 technological innovation	The EU wanted to limit emissions trading in order to stimulate technological innovation
H_5 example-setting	The EU wanted to limit emissions trading in order to demonstrate the willingness to reduce emissions in industrialized countries so that developing countries are stimulated to adopt commitments in the future
H_6 liability	The EU wanted to limit emissions trading in order to reduce problems in the case of non-compliance because emissions trading requires additional rules about who is responsible (buyer or seller liability) if assigned amounts, credits and/or permits have been traded which have not been backed up by real reductions
H_7 imperfect markets	The EU wanted to limit emissions trading in order to limit the use of markets in environmental policy because they are likely to function imperfectly
H_8 negotiating power flexible instruments	The EU wanted to limit emissions trading in order to wield negotiating power with respect to elaborating the rules of the Kyoto Mechanisms (Articles 17, 12 and 6) at CoP6 because at previous CoP meetings the EU has been able to exert influence on the level of country emission targets, but less on the choice of instruments

(Continued)

Table 9.1: Continued.

Hypothesis number and label	Hypothesis formulation
H_9 negotiating power FCCC	The EU wanted to limit emissions trading in order to wield negotiating power with respect to the FCCC decision-making process in general because the proposal gives the EU "something to negotiate about" which could increase the likelihood of favorable compromises at CoP6
H_{10} climate leadership	The EU wanted to limit emissions trading in order to not to lose the "climate leadership" to the US
H_{11} incrementalism	The EU wanted to limit emissions trading in order to prevent too radical policy changes (from standards/taxes to emissions trading)
H_{12} allocation problems	The EU wanted to limit emissions trading in order to reduce the scope for allocation problems associated with emissions trading, since permit allocation gives emitters an incentive to inflate baseline emissions, triggers endless debates about which distribution is fair and could induce competitive distortions
H_{13} macro-economic and secondary benefits	The EU wanted to limit emissions trading in order to stimulate the economy because domestic action (despite its higher microeconomic costs) has more positive macro-economic impacts than emissions trading (e.g. in terms of an increase in domestic employment and reduction of non-CO_2 air pollution (secondary benefits))
H_{14} uncertainty and risk	The EU wanted to limit emissions trading in order to facilitate a step-wise introduction of the Kyoto Mechanisms thereby hedging against uncertainties and reducing the associated risks
H_{15} pressure on US	The EU wanted to limit emissions trading in order to force the US to reduce emissions at least partly at home, since they are the biggest source of GHG emissions
H_{16} pressure on Russia	The EU wanted to limit emissions trading in order to prevent Russia from selling "hot air" and trading questionable emission reductions, since it has a weak enforcement regime

official and unofficial EU documents. The former category contains seven documents on climate change (and emissions trading) reflecting the official opinions, decisions, policy plans and negotiating strategies of the EU. The latter category of five documents consists of two so-called EU "non-papers" (reflecting preliminary views of the EU in preparing the FCCC negotiations), a draft reaction of the EU on proposals of other FCCC Parties, an internal discussion paper for European officials and a public letter of a Commissioner to a newspaper. A second distinction is made between direct and indirect support for a hypothesis. A reason to restrict trade corresponds directly with a hypothesis if the document relates this reason to supplementarity or to a preference for domestic action. Indirect evidence is found if the document mentions a reason to restrict trade, but uses a formulation that is somewhat different from the wording of the hypothesis or discusses it in another context or section than supplementarity and domestic action. Obviously, no support is found if the document does not mention a reason to restrict trade.

The documents are described extensively in Woerdman (2002). The content analysis of these documents is summarized in Table 9.2. It reveals direct empirical support, in particular, for hypothesis 1 on hot air (one official and two unofficial documents) and also for hypothesis 3 on compliance and hypothesis 13 on macroeconomic and secondary benefits (each in a different official document). Next to the hot air hypothesis, direct support has been found in the unofficial documents for hypothesis 11 on incrementalism. Indirect (and thus weaker) support has been found for hypothesis 14 on uncertainty and risk (one official and one unofficial document). In the official documents, indirect support was also provided for hypothesis 4 on technological innovation. Next to the uncertainty and risk hypothesis, in the unofficial documents indirect support was found for hypothesis 1 on hot air, hypothesis 3 on compliance and hypothesis 12 on allocation problems (the latter three in one and the same unofficial document). In sum, the strongest empirical support was found in EU documents for hypothesis 1 on hot air.

The document analysis revealed some interesting historical facts, which are relevant from a path dependence perspective. First, the European idea to restrict trade emerges (at least) a few months before the negotiations in Kyoto of 1997, when the EU indicated that it "(…) may want to consider allowing only a certain percentage of a Party's target to be met by JI (…)" (EU, 1997: 5). Second, when not only JI, but also international emissions trading had become part of the Kyoto Protocol, the European Commission stated in a preliminary internal note in 1998 that a ceiling on the use of the flexible instruments would not be politically acceptable for the US and Russia (EC, 1998). Third, the Commission found in 1998 that an intra-EC emissions trading scheme must conform with the rules at the international level, so that any supplementarity arrangement will also have to be respected for trading between Member States (COM, 1998b: 21). Fourth, in a nonpaper of 1998, the EU indicated that a "concrete ceiling" on the use of (not only JI but) all flexible mechanisms must be defined, possibly in relation to a number of variables, like the assigned amount, 1990 emission levels or the required effort by a Party (EU, 1998a: 1). Finally, in 1999 the EU drafted the formulas of the proposed quantitative ceiling on the use of the Kyoto Mechanisms (EU Council, 1999), as described in detail in the previous chapter. In contrast with its statement of 1998, the Commission now only wanted to restrict trade at the international level and not within the EU itself (e.g. EU Council, 1999; COM, 2000a).

Table 9.2: Content analysis of EU documents (from 1997 to 2000 in chronological order). Question: "Does the EU document provide direct support, indirect support or no support at all for hypothesis x?"

Documents		Hypothesis testing	
Number + reference	Status	Direct support	Indirect support
Document 1 (EU, 1997)	Unofficial	–	–
Document 2 (EU Council, 1997)	Official	–	–
Document 3 (COM, 1997)	Official	–	–
Document 4 (EC, 1998)	Unofficial	H_1 hot air, H_{11} incrementalism	–
Document 5 (COM, 1998b)	Official	H_{13} macro-economic and secondary benefits	H_{14} uncertainty and risk
Document 6 (EU, 1998a)	Unofficial	–	H_{14} uncertainty and risk
Document 7 (EU, 1998b)	Unofficial	–	H_1 hot air, H_3 compliance, H_{12} allocation problems
Document 8 (EU Council, 1999)	Official	H_1 hot air, H_3 compliance	H_4 technological innovation
Document 9 (COM, 1999a)	Official	–	–
Document 10 (Wallström, 1999)	Unofficial	H_1 hot air	–
Document 11 (COM, 2000a)	Official	–	–
Document 12 (COM, 2000b)	Official	–	–
Most support for hypothesis		**H_1 hot air**	**H_{14} uncertainty and risk**

Key: –, no support for any hypothesis; H_x, hypothesis x; **H_x hypothesis**, most frequently reoccurring hypothesis in the column.
Data: Total number of documents = 12.

9.3.2. Hypothesis Testing Among Key EU Officials

The hypotheses (which are summarized in Table 9.1) are tested in this subsection on the basis of detailed questionnaires on the supplementarity issue that were sent to several key EU officials involved in climate change and emissions trading from the environmental ministries of the EU Member States as well as from the European Commission. The questionnaires were sent in either March, June or

September 2000, which is not only before CoP6 Part I of November 2000 where the EU defended its proposal to restrict trade, but which is also before the US withdrew from the Kyoto Protocol in March 2001 as well as before CoP6 Part II, held in July 2001, where the EU abandoned its proposal to restrict trade (for reasons to be explained in the next section). It was already explained in the previous section that, instead of asking for their personal opinions (with the exception of one particular question), we invited these high-level EU officials to indicate for each hypothesis whether it has played a role in the emergence of the EU proposal on supplementarity. The hypotheses were presented to the respondents without a discussion of the theory behind them to lower the risk of non-response by keeping the questionnaire as short as possible. Also recall from the previous section that 41 questionnaires have been sent and 15 of them have been returned, so that conclusions can only be drawn cautiously.

The empirical analysis of the questionnaires among 14 key EU officials[3] is summarized in Table 9.3. When reading this table, it is important to keep in mind that its set-up is directly related to the interview questions. The hypotheses are mentioned in the first column. As can be seen in the Appendix (where the text of the questionnaire can be found), the respondents were asked to indicate for each hypothesis whether it has been a major reason, a side issue or no reason at all for the EU to propose a limit on emissions trading. This is reflected in the next three columns. After they had done this for each hypothesis, the respondents were also asked to draw a conclusion (to "double-check" the consistency of their answers) by indicating which hypothesis they think, finally, is the principal reason for the EU to propose a ceiling on trade. This is reflected in the final column. The highest number of respondents in each column is printed in bold type characters. Each number of respondents (that supported a particular hypothesis) equal to or higher than 8, which is one more than half of the total of 14 officials that responded, is underlined. To see where the European Commission stands, we have marked the answers of Peter Vis reflecting the views of the Commission (included in the indicated number of respondents) by using the abbreviation "*com*".

The analysis of questionnaires shows the highest level of support for hypothesis 2 on equity (13 respondents). These insiders believe that limiting the use of the Kyoto Mechanisms is fair because Annex B Parties are responsible for the majority of historical greenhouse gas (GHG) emissions and should not completely "buy their way out". Equity is also mentioned as the principal reason (7 respondents) when the respondents are asked to draw a conclusion about which hypothesis best explains the EU position on supplementarity. This result is not surprising from the perspective that equity plays an important (if not sometimes dominant) role in politics. The prevalence of equity is surprising,

---

[3] One of the 15 respondents did not fill in the hypothesis questions.

Table 9.3: Analysis of questionnaires among EU officials (March/June/September 2000). Question: "Is hypothesis x a major reason, a side issue or no reason at all for the EU to propose a limit on emissions trading?"

Hypotheses	Level of support from respondents			Conclusion by respondents:
	Major reason	Side issue	No reason at all	Principal reason
H_1 hot air	11 respondents *com*	3 respondents	—	3 respondents[a]
H_2 equity	**13 respondents** *com*	1 respondent	—	**7 respondents**[b]
H_3 compliance	2 respondents *com*	**11 respondents**	1 respondent	2 respondents *com*
H_4 technological innovation	4 respondents *com*	8 respondents	2 respondents	—
H_5 example-setting	5 respondents *com*	7 respondents	2 respondents	—
H_6 liability	1 respondent	4 respondents	9 respondents *com*	—
H_7 imperfect markets	—	3 respondents	11 respondents *com*	—
H_8 negotiating power flex-mex	1 respondent	5 respondents	8 respondents *com*	—
H_9 negotiating power FCCC	2 respondents	3 respondents	9 respondents *com*	—
H_{10} climate leadership	1 respondent	3 respondents	10 respondents *com*	—
H_{11} incrementalism	—	1 respondent	**13 respondents** *com*	—
H_{12} allocation problems	—	3 respondents	11 respondents *com*	—

(*Continued*)

Table 9.3: Continued.

Hypotheses	Level of support from respondents			Conclusion by respondents:
	Major reason	Side issue	No reason at all	Principal reason
H_{13} secondary benefits	–	3 respondents	11 respondents *com*	–
H_{14} uncertainty and risk	2 respondents	5 respondents *com*	7 respondents	–
H_{15} pressure on US	6 respondents *com*	7 respondents	1 respondent	1 respondent
H_{16} pressure on Russia	6 respondents *com*	6 respondents	2 respondents	–

Key: –, no respondent mentioned this level of support for the hypothesis; H_x, hypothesis *x*; *n* **respondents**, highest number of respondents in the column; *n* respondents, number of respondents equal to or higher than 8 (more than half of total number); *com*, answer by Peter Vis reflecting the views of the European Commission (included in indicated number of respondents).
Data: Total number of respondents = 14 (one of the 15 respondents did not fill in the hypothesis questions); according to one respondent, there is no principal reason.
[a] Two respondents mentioned both H_1 and H_2.
[b] One respondent mentioned both H_2 and H_4.

however, from the perspective that hot air (and not equity) is the most frequently recurring reason mentioned by the EU itself in its official and unofficial documents and that hot air is generally seen as the primary motivation behind the EU proposal (e.g. Baron et al., 1999). We should not exaggerate the result, though, because hot air is still mentioned as the second major reason for the EU to propose a cap on trade (11 respondents) and also appears in second place when asked to draw a conclusion (3 respondents). Furthermore, nobody claimed that equity and hot air are "no reason at all" for the EU to propose a concrete ceiling on the flexible instruments.

Compliance with a view to the second commitment period is the most frequently mentioned side issue when asked to explain the rationale behind the EU proposal (11 respondents). Although only 2 respondents call it a major reason, Peter Vis even states when reflecting the views of the European Commission that compliance (and not equity as pointed out by the majority of respondents) should be seen as the principal explanation for the EU position on supplementarity. Although this dissimilarity is a clear indication that the EU is no unitary actor, there is agreement among most of the respondents of the Member States as well as between them and Peter Vis of the Commission, that liability, market imperfection, climate leadership, incrementalism, allocation problems and secondary benefits are no reason at all or a side issue for the EU to propose a restriction on trade. Whereas a call for incrementalism or the problem of allocating emission rights have hardly played a role in the EU proposal to restrict international trade in emissions, they are still likely to be significant barriers to set up a cross-border permit trading scheme, for example, within the EU where they are major and continuing issues of debate (e.g. COM, 2000a, 2001a; EP, 2002).

From an economic perspective, it is interesting that hypothesis 13 on the macro-economic and secondary benefits of domestic action is rejected as an explanation by almost all interviewees (11 respondents), which could indicate that not much is expected from these effects in the first place. Also Peter Vis, reflecting the views of the Commission, finds it no reason at all, which conflicts with earlier Commission statements (COM, 1998b) that secondary benefits are in fact a reason for the EU to propose a ceiling on trade (as we have found in the previous subsection).

From a political perspective, it is not clear to what extent the EU has proposed a ceiling on emissions trading because of some fear for market mechanisms. It seems that such a fear was limited, but also that the uncertainties which surround these mechanisms did play a role. On the one hand, almost all EU officials, including Peter Vis reflecting the views of the European Commission, agree that a limit on trade is not proposed, among other things, to prevent too radical policy changes (13 respondents), imperfect markets (11 respondents) or allocation problems (11 respondents). On the other hand, uncertainty and risk are said to have played their part in the idea of limiting the use of the Kyoto Mechanisms, albeit a minor one, as 2 respondents call it a major reason and 5 respondents as well as Peter Vis from the European Commission label it as a side issue.

Respondents had rather different perceptions about the role the EU proposal was intended to play in wielding negotiating power in the context of CoP6. Negotiating power with respect to the flexible instruments or the FCCC process in general is no reason at all for the EU to propose a ceiling on trade according to both Peter Vis reflecting the views of the Commission as well as 8 and 9 respondents, respectively, but this can be doubted for several reasons.

First, negotiating power with respect to the flexible mechanisms is a major reason according to one respondent and a side issue according to 5 respondents, while negotiating power with respect to the FCCC process in general is still a major reason according to 2 respondents and a side issue according to 3 respondents. Second, previous EU proposals, for instance, its proposal dating from 1997 to make industrialized countries reduce 15% of their GHGs, were also intended and used as bargaining chips, which is also recognized by the EU itself (COM, 1997: 18). Third, some respondents acknowledged that the EU proposal dating from 1999 to restrict trade was at least partly developed to put pressure on the US and Russia. Fourth, the EU did not have a strong bargaining position to start with, making it more likely that it was in need of ways to improve its position, because JUSCANZ countries formed a broader "Umbrella Group" that also included the Russian Federation and the Ukraine to explore the possibilities to use Article 4 of the Kyoto Protocol and circumvent the supplementarity restrictions under Articles 6, 12 and 17 by means of transferring emissions under a "bubble" (Oberthür & Ott, 1999: 149). Fifth, it will be shown in a next subsection that the EU proposal to limit trade was in fact used (intended or not) as a bargaining chip during CoP6 Part I and Part II.

Interestingly, there is a difference when the respondents are asked about negotiating power in general terms (hypotheses 8 and 9) and in specific terms (hypotheses 15 and 16). In general terms, most respondents find that the EU proposal was not intended to wield negotiating power at CoP6, but in specific terms most respondents (including Peter Vis reflecting the views of the European Commission) agree that pressure on the US and on Russia are major reasons (each 6 respondents) or side issues (7 and 6 respondents, respectively) in explaining why the EU put forward the proposal to limit the use of the Kyoto Mechanisms. One respondent even concludes that pressure on the US is the principal explanation for the EU position on supplementarity.

Besides equity, hot air, compliance and pressure on the US and Russia, there is also substantial support for hypothesis 4 on technological innovation through domestic action and hypothesis 5 on example-setting for developing countries. Peter Vis, reflecting the views of the European Commission, as well as 4 and 5 respondents, respectively, label them as major reasons for the EU to propose a quantitative ceiling, whereas still 8 and 7 respondents, respectively, call them side issues. The above analysis of the questionnaires thus seems to reveal that the EU has not one, but several reasons to propose a ceiling on emissions trading. And although hot air and compliance are frequently mentioned explanations, equity is seen by EU officials as the principal reason behind the EU position on supplementarity.

The prominence of equity, but also of hot air and political pressure reasons behind the EU proposal to restrict trade, is emphasized once more in Table 9.4 where a ranking of the hypotheses is performed based on the number of respondents in the "major reason" and "side issue" columns of Table 9.3. An ordering is made based on the scores in the former column and when two or more hypotheses have equal scores, the scores in the latter column decide which hypothesis ranks higher. The advantage of this ranking is its simplicity. The disadvantages are (a) that it does not cluster the hypotheses on the basis of their content (as performed in the previous chapter), (b) that, although the rank ordering itself is systematic, it constructs the so-called "final" rank ordering more or less on an ad hoc basis (that, nevertheless, takes into account the "distances" between the various scores) and (c) that it does not weigh the different types of scores (by explicitly incorporating the "no reason at all" column and/or the "principal reason" column). We will tackle these disadvantages in another ranking hereafter.

On a theoretical level, for the sake of transparency, we have clustered the hypotheses (ex post) in the previous chapter into a small number of different types of conjectures. This appears to nuance the conclusion provided by the EU officials themselves that equity is the principal reason behind the EU proposal. Although there are several possibilities to form such groups, we have created four clusters of hypotheses which supposedly express the EUs:

- environmental reasons to restrict emissions trading (hypotheses 1, 3 and 6 on hot air, compliance and liability, respectively);
- political-normative reasons to restrict emissions trading (hypotheses 2, 5, 11, 12 and 14 on equity, example-setting, incrementalism, allocation problems and uncertainty and risk, respectively);
- political-strategic reasons to restrict emissions trading (hypotheses 8, 9, 10, 15 and 16 on negotiating power regarding the flexible instruments, negotiating power regarding the FCCC, climate leadership, pressure on the US and pressure on Russia, respectively);
- technological-economic reasons to restrict emissions trading (hypotheses 4, 7 and 13 on technological innovation, imperfect markets and secondary benefits, respectively).

On an empirical level, the scores we have calculated for the clusters in Table 9.5 are based on the number of respondents for each hypothesis in Table 9.3. The scores are weighted in the calculations by multiplying the number of

Table 9.4: Simple ranking of hypotheses based on empirical scores.

"Final" rank	Rank	Hypothesis	Score
1	1	H_2 equity	$13 + 1$
	2	H_1 hot air	$11 + 3$
2	3	H_{15} pressure on US	$6 + 7$
	4	H_{16} pressure on Russia	$6 + 6$
3	5	H_5 example-setting	$5 + 7$
	6	H_4 technological innovation	$4 + 8$
4	7	H_3 compliance	$2 + 11$
5	8	H_{14} uncertainty and risk	$2 + 5$
	9	H_9 negotiating power FCCC	$2 + 3$
	10	H_8 negotiating power flex-mex	$1 + 5$
	11	H_6 liability	$1 + 4$
	12	H_{10} climate leadership	$1 + 3$
6	13	H_7 imperfect markets	$0 + 3$
	14	H_{12} allocation problems	$0 + 3$
	15	H_{13} secondary benefits	$0 + 3$
	16	H_{11} incrementalism	$0 + 1$

Key: H_x, hypothesis x.
Data: The scores are based on the number of respondents in the "major reason" and "side issue" columns for each hypothesis in Table 9.3. The rank ordering is based on the scores in the "major reason" column of Table 9.3 and when two or more hypotheses have equal scores, the scores in the "side issue" column of Table 9.3 decide which hypothesis ranks higher. The "final" rank ordering is constructed more or less on an ad hoc basis by taking into account the "distances" between the various scores (for a more systematic and sophisticated analysis see next tables).

respondents with a factor 2 when the hypothesis is mentioned as a "major reason", 1 as a "side issue" and 0 as "no reason at all". The sensitivity analysis also contains the mention of the "principal reason", which is weighted with a factor 3. The final rank ordering of the clusters is based on the average weighted score for each cluster calculated as the total weighted score divided by the number of hypotheses in that cluster. As an example, we show the calculations of the average weighted score of the cluster of political-normative hypotheses (the scores of the other clusters are calculated in the same way):

(1) *Calculating scores of clustered hypotheses (example)*:
Major reason: $H_2 + H_5 + H_{11} + H_{12} + H_{14} = 13 + 5 + 0 + 0 + 2 = 20$

Side issue: $H_2 + H_5 + H_{11} + H_{12} + H_{14} = 1 + 7 + 1 + 3 + 5 = 17$
No reason: $H_2 + H_5 + H_{11} + H_{12} + H_{14} = 0 + 2 + 13 + 11 + 7 = 33$
Total weighted score: $(20 \times 2) + (17 \times 1) + (33 \times 0) = 57$
Average weighted score: $57/5 = 11.4$

(2) Sensitivity analysis (example):
Principal reason: $H_2 + H_5 + H_{11} + H_{12} + H_{14} = 7 + 0 + 0 + 0 + 0 = 7$
Total weighted score: $(7 \times 3) + (20 \times 2) + (17 \times 1) + (33 \times 0) = 78$
Average weighted score: $78/5 = 15.6$

The empirical analysis based on the ex post clustering of hypotheses assigns the first rank to environmental reasons (such as hot air) and the second rank to political-normative reasons (such as equity) behind the EU proposal on supplementarity (see Table 9.5). Political-strategic reasons (such as negotiating power) also play a role, albeit a smaller one, whereas technological-economic reasons (such as secondary benefits) are the least important. The sensitivity analysis does not change these ranks. However, it was already shown above that the EU officials give equity the first (and not the second) rank when they are asked to rank the individual (and not the clustered) hypotheses themselves because the majority of them sees equity as the major and principal reason (and hot air as the second reason) behind the EU proposal to restrict emissions trading (see Tables 9.3 and 9.4).[4]

Hence, confronting the testing of individual hypotheses with the ex post clustering of hypotheses reveals that political-normative reasons, such as equity, are an important factor next to environmental reasons, such as hot air, to explain why the EU has proposed to restrict the use of the Kyoto Mechanisms. This also confirms, as neo-institutional economists would expect, that values are important in climate change politics. In this and the previous subsection we have demonstrated empirically that the so-called "green" values, reflected in the cluster of environmental reasons, as well as equity values, reflected in the cluster of political-normative reasons, play a significant role in explaining the desire of European policy makers to restrict trade.

However, the clustering of hypotheses in Table 9.5 does not make a distinction between the different Kyoto Mechanisms. Such a distinction, as argued in the previous chapter, would be relevant to find out why the economically superior alternative of permit trading ranks low in the political hierarchy. Therefore, an additional clustering and empirical test is performed in Table 9.6 that makes

---

[4] The final rank ordering of the clusters would change if it would have been based on the total weighted score for each cluster. Political-normative reasons would become the primary explanatory factor (total weighted score of 57), also in the sensitivity analysis, and political-strategic reasons would even end up higher than environmental reasons (total weighted scores of 56 and 46, respectively), although not in the sensitivity analysis. However, calculating average scores is more desirable as it takes away any correlation between the number of hypotheses in a cluster and the total score of that cluster.

Table 9.5: Theoretical and empirical clustering of hypotheses (ex post). The EU proposed to limit the use of the Kyoto Mechanisms because of the type of reasons given below.

Type of reasons	Clustering of hypotheses	Total weighted score	Average weighted score	Rank
Environmental reasons	H_1, H_3, H_6	46 (61)	15.3 (20.3)	1
Political-normative reasons	H_2, H_5, H_{11}, H_{12}, H_{14}	57 (78)	11.4 (15.6)	2
Political-strategic reasons	H_8, H_9, H_{10}, H_{15}, H_{16}	56 (59)	11.2 (11.8)	3
Technological-economic reasons	H_4, H_7, H_{13}	22 (22)	7.3 (7.3)	4

Key: H_x, hypothesis x; (xx), sensitivity analysis.
Data: The scores are based on the number of respondents for each hypothesis in Table 9.3. The scores are weighted in the calculations with a factor 2 for major reason, 1 for side issue and 0 for no reason at all (see Table 9.3). The sensitivity analysis includes a factor 3 for principal reason in the weighted scores. The average weighted score is calculated as the total weighted score divided by the number of hypotheses in that clustering. The rank ordering is based on the average weighted scores.

a differentiation between (a) objections of EU policy makers against the Kyoto Mechanisms in general and (b) those against permit trading under IET Article 17 in particular.

Contrary to permit trading under IET Article 17, as explained in other chapters, JI and CDM projects do not only avoid the mobilization of hot air (and generate real reductions if they are additional), but they are also incremental as they avoid the explicit (re)allocation of property rights. In line with these earlier findings, the previous chapter provided two clusters of hypotheses which supposedly express the EUs objections against permit trading (hypotheses 1, 11 and 12 on hot air, incrementalism and allocating emission rights, respectively) and its objections against all Kyoto Mechanisms (all remaining hypotheses, including those on equity and negotiating power). The empirical analysis based on this particular ex post differentiation of hypotheses (which is, obviously, different from the one presented earlier) uses the same calculation method as portrayed and explained above. It appears that the first rank is assigned to objections against all Kyoto Mechanisms (such as equity) and the second rank to objections against permit trading (such as hot air) to explain the EU proposal on supplementarity (see Table 9.6). The sensitivity analysis does not change these ranks. This means that the Europeans primarily developed their proposal to restrict the use of all Kyoto Mechanisms, rather than just one of them.

Table 9.6: Theoretical and empirical differentiation of hypotheses (ex post). The EU proposed to limit the use of the Kyoto Mechanisms because of the type of objections given below.

Type of objections	Clustering of hypotheses	Total weighted score	Average weighted score	Rank
Objections against all Kyoto Mechanisms	H_2, H_3, H_4, H_5, H_6, H_7, H_8, H_9, H_{10}, H_{13}, H_{14}, H_{15}, H_{16}	152 (182)	11.7 (14)	1
Objections against permit trading	H_1, H_{11}, H_{12}	29 (38)	9.7 (12.7)	2

Key: H_x, hypothesis x; (xx), sensitivity analysis.
Data: The scores are based on the number of respondents for each hypothesis in Table 9.3. The scores are weighted in the calculations with a factor 2 for major reason, 1 for side issue and 0 for no reason at all (see Table 9.3). The sensitivity analysis includes a factor 3 for principal reason in the weighted scores. The average weighted score is calculated as the total weighted score divided by the number of hypotheses in that clustering. The rank ordering is based on the average weighted scores.

This empirical result also suggests that an analysis of the EU proposal on supplementarity, although it can reveal cultural objections against unrestricted use of the Kyoto Mechanisms, is not suitable to explain why the economically superior alternative of permit trading ranks low in the political hierarchy. However, two observations must be made to nuance this view. First, the difference between the average weighted scores for the different types of objections is not large (a score of 11.7 for all mechanisms compared to a score of 9.7 for permit trading), which indicates that objections against permit trading, albeit not dominant, still played a considerable role in the EU negotiating position on limiting trade. Second, on a theoretical level, we have been able to identify specific objections against permit trading, not against any of the credit-based approaches in particular. On an empirical level, one of these arguments against permit trading that led the EU to propose a limited use of the Kyoto Mechanisms, namely the hot air hypotheses, turned out to be highly relevant, ranking first in the content analysis of EU documents and second in the opinion analysis of key EU officials. The other two hypotheses associated with permit trading, however, namely those on incrementalism and allocation problems, although mentioned in EU documents, hardly gained support among these officials (although direct support has been found in the EU documents for the former hypothesis and indirect support for the latter).

The empirical relevance of the hot air hypothesis in the survey among high-level officials makes it reasonable to conclude that objections against permit trading are not a dominant, but still a significant factor to explain the initial wish of the EU to quantitatively restrict the use of all Kyoto Mechanisms. The empirical relevance of equity values and environmental values underlines that political culture, followed by political-strategic considerations, is the primary explanation of the EU proposal on supplementarity.

9.3.3. Analysis of Questions on Supplementarity Among Key EU Officials

In the questionnaire, the EU officials were not only asked to judge the explanatory power of the 16 hypotheses, but they were also invited to answer a set of additional questions on supplementarity. Although conclusions can only be drawn cautiously due to the high level of non-response, a surprising result is that the interviewed key officials throughout the EU are strongly divided on the issue whether a quantitative ceiling should be placed on the use of the Kyoto Mechanisms.

To avoid that the officials would simply duplicate the official EU position, this was the only question in the questionnaire in which we asked them for their personal opinion about the EU proposal. In a personal capacity, almost half of the interviewees (7 respondents) indicated that supplementarity should be achieved "quantitatively" through a ceiling on emissions trading, similar to the EU proposal. However, the other albeit smaller half (6 respondents or 40%) pointed out — more in line with the opinion of the JUSCANZ countries — that supplementarity should be advanced "qualitatively" through internationally agreed requirements with respect to domestic climate policy (like the requirement to demonstrate adequate efforts to control emissions, for instance, via tighter energy-efficiency norms). One respondent preferred to combine a quantitative ceiling with qualitative requirements.[5] Interestingly, Peter Vis reflecting the views of the European Commission did not answer this particular and highly relevant question.

The withdrawal of the US from the Protocol in March 2001 was an external shock that had reduced the bargaining power of the EU because the other JUSCANZ countries could make a credible threat to make a similar move if the EU continued to insist on a quantitative ceiling. The fact that about half of the interviewed officials personally do not support the official European proposal for a concrete ceiling points to internal differences within the EU, which could further undermine its bargaining power. This also means that the JUSCANZ countries had

---

[5] One of the interviewees remarked that qualitative requirements are acceptable as long as they attain the desired effects (presumably on domestic action). Another respondent noted that qualitative requirements are preferable because a quantitative ceiling has an uncertain effect on policies and measures as well as on energy efficiency. Finally, one respondent found that those who do little in policies and measures could be named (and shamed).

some opportunity to try to exploit such underlying differences within the European environmental bureaucracy. Because the officials prepare the policy and negotiating positions of their Ministers, the JUSCANZ countries — in the end — may not have had much difficulty in convincing relevant EU Ministers of the arguments against a ceiling (despite their unanimous decision, if only or partly for reasons of bargaining power, to propose a ceiling on trade in the negotiations), because half of their officials was already convinced — even before the withdrawal of the US.

However, it is difficult to judge to what extent the EU decision makers from the different Member States have supported or opposed the official EU proposal to limit emissions trading internally before, during and after CoP6. On the one hand, the doubt among European policy makers about the official EU proposal seemed to be reinforced by the fact that almost half of the interviewees (6 respondents) — even including Peter Vis reflecting the views of the European Commission — agreed, and the other albeit larger half did not agree (9 respondents), that a ceiling on trade unnecessarily limits potential cost savings as well as potential revenues for developing countries and countries with economies in transition. On the other hand, most of them (12 respondents), including Peter Vis reflecting the views of the European Commission, strongly disagreed with the idea that a quantitative ceiling on trading would be undesirable because (among other things) it would raise the costs of reducing emissions, rejecting the reasoning that the trade restriction would make it more difficult for Parties to adopt ambitious targets in a second commitment period or make it harder for new countries to join the set of Annex B Parties. Rather the opposite, several respondents claimed. Three respondents clarified — in different wordings — that by limiting the use of the Kyoto Mechanisms, one also limits the purchase of hot air, which implies more domestic action and thus a lower (emissions) starting point for the next negotiations, so that more ambitious targets can be set in a second commitment period.

Although 2 respondents explicitly stated that they did not want to make predictions about the amount of hot air to be traded during the first commitment period because of its inherent uncertainty, the other 13 respondents were convinced that hot air will constitute a major part (more than 20%) of emissions trading. This could be interpreted as a rather pessimistic estimate compared to more optimistic figures that can be found in some model surveys in which hot air projections vary between practically zero to almost 25% of the Kyoto commitments (Haites, 1998). Nevertheless, the pessimistic estimate of the respondents is in line with the hot air concern expressed in several EU documents (which have been analyzed in a previous subsection). However, most EU officials (10 respondents), including Peter Vis reflecting the views of the European Commission, also agreed that (not so much the economic inclusion of emissions trading, but rather) the distributional inclusion of hot air in the Kyoto Protocol was necessary for the US to initially accept an internationally agreed emission reduction commitment (while 3 respondents disagreed and 2 did not answer the question).

In a sequential bargaining model, Dupont (1994) underlines the importance of (creating) perceptions. An actor can obtain negotiating power if it successfully convinces the other actors, or makes them believe, that it has fixed domestic constraints which reduce its international bargaining space. Similarly, the fact that most of the consulted key EU officials view hot air as an important environmental problem, but at the same time realize that this hot air has been a political and economic precondition for the US to accept a significant emission reduction target is good news for the bargaining position of the US and the other JUSCANZ countries. It would probably be more difficult for them to "preserve" the already negotiated hot air if the EU had no understanding for such concerns. The European political and economic perceptions thus also make it more difficult for the EU to convincingly plead for an environmental limit on the use of hot air, for instance, by means of a quantitative restriction on all Kyoto Mechanisms.

The prominent position of equity as discovered in the hypotheses analysis of the previous subsection is reconfirmed in a separate question to which — presumably the same — 13 EU officials answered (see Table 9.3) that a ceiling on emissions trading promotes international equity or responsibility because it will force Annex B Parties to "clean up their own mess" by reducing emissions domestically. Only 2 respondents disagreed with this line of reasoning and one of them expressed the neoclassical economic view that equity would be achieved through equalization of marginal costs.

Some economists believed that the EU proposal to restrict trade could only be put forward because too few economists and too many lawyers and engineers were involved. However, this conjecture was unanimously rejected by 12 respondents (whereas the other 3 respondents said that they did not know). One respondent even remarked that many economists were, in fact, involved. Nevertheless, another respondent added that political interests (rather than economic motivations) have mainly shaped the EU proposal.

In 2001, the EU indicated to be willing to make compromises (Hanks et al., 2001a). The above analysis explains this on the basis of external and internal factors. First, there was external pressure on the EU to leave its supplementarity proposal because the US had already withdrawn from the Protocol and there was a perceived danger that others might follow. Second, we have demonstrated that there was also some internal pressure within the EU, at least on the level of the bureaucracy, given that almost half of a number of high-level EU officials that responded to our questionnaire disagrees with the EU proposal to restrict trade. Moreover, although these key civil servants prepare but do not make the final political decisions, it appeared that they see hot air as a major environmental problem, but at the same time recognize that the US would probably not have accepted (and will not ratify) the Kyoto Protocol without it. Next to hot air, the separate analysis in this subsection (not directly linked to the hypothesis testing) reconfirms the importance of equity in explaining the content of the EU proposal on supplementarity.

However, it should be kept in mind that high-level officials are not politicians. Although these officials are likely to have had an influence on their Ministers, because they prepare their decisions, as well as on the negotiations, because Ministers and high-level officials met several times during CoP6 to continue negotiations in closed meetings (e.g. Hanks et al., 2001b: 2), future empirical research should establish to what extent key officials in the EU actually had such an influence.

9.3.4. Bargaining Behavior of the EU at CoP6

An additional albeit less systematic test for the hypotheses on negotiating power is the observed bargaining behavior of the EU at the international climate negotiations. The EU did not enter the negotiations of CoP6 with a strong negotiating position. Apart from the internal pressures revealed above (40% of the key EU officials was against the proposal), we also found that most EU officials, although they believed that hot air will constitute a major part (more than 20%) of emissions trading, at the same time agreed that hot air was necessary for the US to initially accept (and subsequently ratify) an emission reduction commitment under the Kyoto Protocol. Moreover, the EU also faced external pressures as the Americans withdrew from the Kyoto Protocol, so that other countries could now make a credible threat to withdraw as well if the EU would continue to insist on restricting the use of flexible instruments.

When no agreement on the mechanisms was reached at CoP6 Part I in The Hague in November 2000, largely as a result of the supplementarity issue, the negotiations were resumed at CoP6 Part II held in Bonn in July 2001. The EU signaled to be "ready for compromises" already at the start of these negotiations (Hanks et al., 2001a: 3), although they were eager to restrict hot air trading and keep sinks out of the CDM due to their environmental uncertainties (e.g. Chow, 2001: 1). The aforementioned internal and external pressures that emerge from our empirical research help to explain this attitude change. The Europeans were also more willing to make a compromise for political-strategic reasons, namely to show the world that it does not need the US to obtain international (environmental) co-operation. Finally, the EU gave up its proposal for a quantitative ceiling and accepted the unspecified requirement that domestic action shall be a "significant element" of Annex B countries' climate policy (CP, 2001a: 7).

Although "(…) many of the deals struck in Bonn fell in the direction the US has long favored" (Kopp, 2001: 1), the EU gave up its proposal for a quantitative restriction on trade in return for some favors, such as restrictions on the use of sinks (like reforestation projects) and the establishment of a commitment period reserve for each Annex B Party, which implies that a country may sell 10% of its hot air without any repercussions, but more hot air can only be traded if and to the extent that its reviewed emissions are lower than 90% of its assigned amount. This compromise not only underlines the weak bargaining position of the EU, but also suggests that the EU proposal to limit trade was in fact used as a bargaining chip during CoP6. This fact reduces the credibility of those officials in our research who have claimed that the EU proposal was not intended to exert bargaining power, also because some of these respondents acknowledged that the EU proposal was partly (or even principally) developed to put pressure on the US and Russia. The observed bargaining behavior of the EU provides some additional support for the hypotheses on negotiating power, in particular hypothesis 8 regarding the Kyoto Mechanisms.

9.4. The Institutional Breakout of EU Climate Policy

We have found that informal institutions, like political culture, play a role in the lock-in of extant environmental policy arrangements. Moral resistance against hot air trading and the perception that "pollution rights" are inequitable added to the switching costs of permit trading and stimulated governments to broaden traditional environmental policy by means of incremental albeit less efficient credit-based approaches. Primarily for reasons driven by equity and "green" values, the EU expended years of political effort trying to reject or limit emissions trading. However, the Europeans took a surprising next step at the beginning of the new millennium. Against the background of Denmark and the UK developing domestic emissions trading schemes of their own, the EU managed to break out by adopting a Directive in 2003 that creates permit trading in the EU from 2005 onwards.

When reading this rather amazing piece of history, which exhibits both impediments and incentives for evolution to efficiency, the question comes to mind why the EU made such a U-turn in climate policy. Those who write about this remarkable attitude change usually present a list of (more or less relevant) ad hoc explanations without providing an overarching theoretical framework (e.g. Christiansen & Wettestad, 2003; Convery et al., 2003). In this section, however, we take an institutional economics perspective by answering the question on the basis of the path dependence approach (e.g. Woerdman, 2004b).

9.4.1. A Path-Dependent History of Market-Based Climate Policy in the EU

Damro & Méndez (2003: 71) believe that the adoption of emissions trading by the Europeans is "best explained as a process of policy transfer" from the US to the EU. We do not fully agree. At most it can be *described* as such, but other approaches and concepts are still necessary to explain what happened. Moreover, although Damro and Méndez acknowledge the role of sunk costs and learning, for instance, they fail to analyze switching costs, scale advantages, drivers of cultural change and possible institutional lock-in effects. In addition, we disagree with their characterization of emissions trading as a "marginal change" in climate policy, which would be "nothing more than the introduction of an instrument" because it leaves the policy goals untouched. Here, they forget to make a distinction between permit trading and credit-based flexible instruments, the latter of which is more incremental (and less efficient) than the former. This distinction is, in fact, necessary to understand the subtleties in the history and implementation of emissions trading in America, as many authors underline (e.g. Tietenberg et al., 1999), and more recently also of that in Europe. The path dependence approach provides an explanation and sheds new light on this piece of history.

Despite the superiority of permit trading, seen from the viewpoint of neoclassical economics, certain Member States were inclined to opt for credit trading, either or not based on existing voluntary agreements. This fits the observation by Bressers & Huitema (1999: 180) that new economic instruments are often based on existing legal instruments. The incrementalism literature, however, that could be expected to provide some explanation, is predominantly normative and empirical. It explains incremental policy change by referring to factors such as conflicting interests, the power of large companies and incomplete information, but it does not offer a systematic positive theory and, although the concept of incrementalism is often used by scientists and policy makers, it has not produced a cumulating line of research (e.g. Weiss & Woodhouse, 1992). The path dependence approach, on the contrary, is more promising in this respect because it does offer a systematic positive theory to explain incremental (as well as radical) policy change. Incremental change by building upon existing policy has the advantage of making use of its sunk costs, learning effects and increases in institutional scale, thereby avoiding the perceived costs of switching to a completely new policy paradigm.

This is exactly what credit trading does: it builds upon existing environmental policy and avoids the perceived switching costs of permit trading. And this is exactly what initially happened in various Member States that already had some climate policy to build upon (for a country overview see IEA, 2002). In particular, the Netherlands and Belgium, but to some extent also Germany, Sweden and the UK, that had already introduced energy-efficiency standards under voluntary agreements for the energy-intensive industry, were tempted to make the existing framework more flexible by adding credit trading to it. Climate policy in the Netherlands is a clear case in point. Apart from the Social-Economic Council that, in its capacity as advisor to the Dutch Government, pleaded in favor of permit trading, there were various Ministries as well as the so-called Vogtländer advisory committee which initially pleaded in favor of credit trading by building upon existing standards for those sectors of industry that were energy-intensive and competing internationally (for a policy overview see Woerdman et al., 2002). The factors mentioned earlier, such as making use of the sunk costs and learning effects of extant policy, help to explain their position.

The aforementioned EU Member States found themselves in a (as it later appeared to be temporary) situation of third degree path dependence in which a superior alternative is known but not chosen. The situation was different for other Member States, in particular those in the South of Europe, such as Portugal, Spain and Greece. These countries hardly had any existing climate policy, let alone a well-established tradition of environmental policy instruments, to build upon. Also on an overarching European level (as against individual Member State level), there was virtually no existing climate policy. The path-dependent history is illustrative: the European carbon tax as proposed in the early 1990s failed to be adopted in the Council of Ministers, which provided some institutional void, in terms of policy instruments, making the European institutions themselves less vulnerable to third degree path dependence. When permit trading became the cornerstone of the Directive, European climate policy, after years of uncertainty, can now be said to be en route to an institutional breakout. This still does not explain, however, how the attitudes of policy makers in the EU have changed. The path dependence approach is capable of providing an answer, though, by focusing on the conditions of an institutional breakout.

Attitudes (which are observable) have changed, but this does not mean that values (which are unobservable) have changed as well or that market-based instruments have gained in acceptance as a result of cultural change. It is important to realize, as we have seen before, that there are elements other than values in attitudes (van Deth & Scarbrough, 1995a,b). We hypothesize that the attitudes of policy makers in the EU have changed as a result of path-dependent internal pressures and external "shocks" (that were difficult if not impossible to influence), which has contributed to a process of cultural change (and not just the other way around). There is significant evidence, which runs along the lines of the path dependence approach, that confirms the hypothesis.

First, the problem-solving capacity of actual and planned policies and measures, in the European regulatory tradition of standards, taxes and voluntary agreements, came under pressure. In the policy community, the perception took hold not only that the effectiveness of the existing policy framework was decreasing (e.g. COM, 2000e: 2–4), but also that emissions trading (next to efficiency) would enhance effectiveness (e.g. COM, 2000a: 4). In the Netherlands, for instance, the GHG emissions had risen by about 10% in 2000 relative to 1990 emissions, whereas it had pledged to stabilize emissions (COM, 2000e).

Second, existing environmental policy has sunk costs, but the perceived switching costs of permit trading were steadily decreasing. The idea became more widespread that Europe would miss the opportunity of saving costs if no use would be made of trading (e.g. COM, 2000e: 3). The Commission performed the role of policy pioneer and argued in favor of starting an experimental EU-wide carbon trading scheme among large emitters by 2005 (e.g. Drexhage, 2001). The Commission later drafted a Directive to establish such a scheme, whereas some EU Member States, notably Denmark and the UK, had already started to develop domestic emissions trading schemes, as we have seen before. This is an indication that cultural barriers towards the introduction of markets in climate policy were breaking down in some entrepreneurial policy arenas and in some countries. New interests (as opposed to the vested interests), such as emission market brokers, also pushed for the acceptance of permit trading.

Third, interlinked with the aforementioned processes, the availability, quality and dissemination of information on permit trading among policy makers improved over time. To obtain what it perceived to be meaningful emission targets from countries like the US and the Russian Federation, the EU accepted the inclusion of emissions trading in the Kyoto Protocol of 1997. Because, from then on, EU policy makers had the perception that emissions trading was now a permanent part of the policy "landscape", they started to invest more time and rigor in studying this market-based option with which they had been largely unfamiliar (e.g. COM, 1999a: 14–16). The Commission itself later recognized that the Kyoto Protocol had put emissions trading on the political agenda of the EU (COM, 2000a: 7). Here, commissioners Zapfel and Vainio (2002: 5–12) distinguish three phases to which no specific time periods are attached: in the first phase emissions trading was "widely unknown and misunderstood", in the second phase there was an "increasing understanding of the participants" and in the third phase the EU adopted "a proposal for a Directive on EU-wide trading in GHG permits", as they write.

Fourth, an external political shock occurred. Although in particular the Americans, but also countries like Canada and Japan, had bargained hard, and with success, to introduce emissions trading in the Kyoto Protocol, in 2001 the US rather unexpectedly withdrew from the Protocol. This meant that the EU and the rest of the world were left with an agreement full of flexibility instruments that initially were a pre-condition for the US to accept the emission reduction target which they now rejected. This fait accompli was exogenous to the extent that the earlier EU supplementarity proposal to quantitatively restrict emissions trading had no influence on the US decision that followed shortly after this particular proposal was made by the Europeans.

To prevent that countries like Canada and Japan would follow the US example, which could now make a credible threat to do so, the EU had to give up its resistance against full trading under emission ceilings for private entities. The EU wanted to keep those countries on board not only for environmental reasons, but also for political-strategic reasons, namely to show that it still regards itself as a climate leader which does not need the US to make international climate policy succeed (e.g. Hanks et al., 2001a: 14). The Russian Federation, however, became the next stumbling block for the EU: in 2003 it threatened not to ratify the Kyoto Protocol. Various Members of the European Parliament stated that this external threat accelerated the internal co-decision procedure on EU-wide emissions trading (e.g. Houlder, 2003). An early agreement should stimulate Russia to ratify by signaling that the EU takes climate policy and market instruments seriously and that the Russians, although the Americans had left, can still gain from trading emissions with the Europeans. It should also stimulate the US to come back to the international climate change table.

The aforementioned (exogenous) path-dependent developments of, first, forceful US target acceptance conditions and, then, the sudden unilateral US withdrawal and the resulting threat power of other market-oriented countries pleading in favor of trading (or against the Kyoto Protocol), as well as the increasing sense of a necessity to reduce compliance costs in climate policy, have shaped the perception among an increasing number of EU politicians and civil servants that unrestricted use of emissions trading among private entities (albeit in their view, to some extent, undesirable) is de facto unavoidable. The unrelenting attempts of the Commission to get permit trading accepted, mainly by means of performing studies, but also by means of lobbying, were factors of internal pressure in the EU. Consequently, the attitudes of policy makers have changed, which, in its turn, triggered a path-dependent process of cultural change, as a result of internal pressures and external "shocks" mainly caused by (exogenous) international political developments that were difficult if not impossible to influence.

In the Northian sense of informal constraints, this provided a window of opportunity for permit trading. This "window" was even enlarged by a path-dependent shift in formal constraints: whereas the carbon tax was a financial matter that required unanimity in the Council of (Financial) Ministers, emissions trading was an environmental issue that "only" required a qualified majority in the co-decision procedure between the Council of (Environmental) Ministers and the European Parliament (Christiansen & Wettestad, 2003; Convery et al., 2003). Internal and external pressures, both formally and informally, triggered an attitude change in the EU which, in its turn, caused cultural values to slowly change towards, what Bernstein (2002) calls, "liberal environmentalism".

9.4.2. A Path-Dependent Future of Market-Based Climate Policy in the EU?

Thanks to decreasing set-up costs, information improvements, a deteriorating problem-solving capacity of extant policy as well as external shocks and policy entrepreneurs, the EU has developed a Directive that enables CO_2 permit trading for large emitters to start in 2005. Also outside the EU, various countries intend to build national tradeable emission rights systems, such as Switzerland, Norway, Japan and Canada, which could eventually be linked to the EU scheme provided that they mutually recognize their transferable units. The decision of the EU to use permit trading across Europe is a remarkable institutional breakout. Some therefore conclude that the permit-versus-credit discussion is now politically out of date because the EU Directive defines "allowances" (not credits) in Article 3, authorizing the holder to emit 1 tonne of carbon dioxide equivalent during a specified period.

However, elements of credit trading can still be brought into this permit trading regime through the backdoor, for instance, based on Annex III of the Directive that requires quantities of allowances to be consistent with the (technological) potential of activities to reduce emissions. Some companies and policy makers have tried to steer the national allocation plans in the direction of credit trading, by linking the height of ceilings for individual companies, within the ceiling for an industry as a whole, to the size of their production (e.g. EZ, 2002). This sort of linkage, which is advocated on fairness grounds by many (energy-intensive) companies, and even by some scientists (e.g. Groenenberg & Blok, 2002), is not fully efficient, as explained before. On the level of the individual firm, it is then signaled that production growth implies free emission space. Economists know, however, that no such thing exists as a "free lunch". The price of the extra emission space should make clear that an expansion of carbon-intensive production can lead to destroying economic value because it would necessitate relatively expensive, additional emission reduction measures elsewhere in the economy. Moreover, credit trading can still be used for installations not covered under the emissions trading Directive. Hazardous or municipal waste installations are exempted, for instance, as well as the transport sector or those parts of the chemical industry that fall below the 20 MW threshold.

Would it be a problem if individual firms obtain flexibility without being subject to emission caps or if credit trading is created for installations not covered under the Directive? Some contend that it is not problematic to start with credit trading, assuming that such a scheme can later be transformed into a more efficient and effective permit trading system (e.g. Tietenberg & Victor, 1994). On the basis of the path dependence approach, we have explained, though, that this comes at a risk. The political choice of credit trading, a sub-optimal type of emissions trading, can result in an institutional lock-in from which it may be difficult to escape in the future. Four factors can then be identified that contribute to a possible institutional lock-in of credit trading.

First, credit trading profits from the learning effects associated with building on existing environmental policy. Learning effects lower the average costs of running the established system. Second, policy makers will be more persuaded to opt for credit trading if there is a predominant perception that the problem-solving capacity of the existing environmental laws is growing or stable. If the effort of policy makers is directed to "satisficing" rather than "optimizing", they are less receptive to theoretically superior alternatives such as permit trading. Third, credit trading can profit from network or co-ordination benefits by building on extant policy. The differential administrative costs decline as the institutional scale increases, which can be done by expanding an existing environmental instrument to cover extra target groups, such as more segments of industry, or by adding an element such as credit trading. Fourth, credit trading builds on the sunk costs of existing environmental policy. These start-up costs that have already been incurred play no role in the decision to continue current environmental policy without emission ceilings, whether or not modified to take account of credit trading. Although permit trading reduces running costs, it involves relatively high political transaction costs because it implies crossing over to a new institutional arrangement. Resistance by vested interests contributes to these switching costs. Contrary to permit trading, the industry does not have to purchase extra emission rights if companies seek to expand their production under a credit trading regime.

From the perspective of path dependence, there is a risk that starting with credit trading for some installations, firms or sectors triggers a self-reinforcing process from which it may be difficult to escape. Although there are opportunities for an institutional breakout, EU Member States should at least acknowledge and consider this risk when constructing their national allocation plans because they might find themselves stuck with a differentiated, partly sub-optimal emissions trading system in a few years time, that may then be difficult if not impossible to change. In fact, any government that is involved in designing a domestic emissions trading scheme, as well as company representatives and scientists that want to contribute to the permit-versus-credit discussion, should take this risk into account.

9.5. Conclusion

Our empirical analysis of the EU supplementarity proposal of 1999 to quantitatively restrict the use of the Kyoto Mechanisms demonstrates that informal institutions, including equity values, have contributed to the institutional lock-in of inefficient environmental regulation and that cultural change has helped to force a breakout. Following a desire in the literature for more empirical research on the political opposition against emissions trading (e.g. Wiener, 2000a), we have tested the hypotheses on the EU proposal, developed in the previous chapter, by analyzing (a) the content of 12 EU documents, (b) the opinions of high-level EU officials, gathered by means of a questionnaire and (c) the bargaining behavior of the EU at CoP6. Although the EU proposal to restrict trade was rejected internationally in 2001 because it would have reduced efficiency, an empirical analysis is still useful to see why some governments wanted to limit market-based climate policy, especially permit trading.

Although the non-response was high (only 15 out of 41 key officials responded), the questionnaires that have been received not only contain new and detailed information, but also came from almost every country in the EU as well as from the European Commission. In line with our conjectures, the vast majority of these EU officials mentioned equity as the primary motivation behind the EU proposal to prevent that industrialized countries "buy their way out". This shows that equity can be the enemy of efficiency, contrary to the neoclassical idea that emissions trading is neutral to equity. When the hypotheses are clustered, the opinion analysis assigns the first rank to environmental reasons (such as hot air) and the second rank to political-normative reasons (such as equity) to explain the EU proposal. Political-strategic reasons (such as negotiating power) also play a role, albeit a smaller one, whereas technological-economic reasons (such as secondary benefits) are the least important. This shows that a political culture dominated by equity values and "green" values is the primary explanation of the EU proposal to restrict market-based climate policy. Although the EU documents primarily mentioned hot air, equity turned out to be at least as important for key EU officials.

Analyzing their opinions based on another clustering of the hypotheses suggests that the EU proposal was made to restrict all Kyoto Mechanisms rather than to restrict permit trading alone. This seems to imply that we cannot use these data to explain why permit trading may rank low in the political hierarchy of market-based instruments. However, the differences between the scores in the empirical analysis are small (11.7 for all mechanisms and 9.7 for permit trading). Furthermore, one of the arguments against permit trading, namely hot air (which cannot be mobilized by means of JI), ranked first in the content analysis of EU documents and second in the opinion analysis among high-level EU officials.

In addition, the EU wanted to exempt JI and CDM projects from a limit on supply (EU Council, 1999). Therefore, objections against permit trading are not a dominant, but still a significant factor to explain the EU proposal.

The EU did not enter the negotiations of CoP6 with a strong bargaining position. Our empirical analysis revealed some internal political divisions as 40% of the key EU officials from different Member States declared (in a personal capacity) to disagree with the official EU proposal decided upon by their Ministers to restrict trade. Moreover, most EU officials in our survey recognized that hot air was necessary for the US to initially accept (and subsequently ratify) an emission reduction commitment. After the US withdrew from the Kyoto Protocol, the divided EU feared that others might withdraw as well and soon gave up its proposal. In return for some favors, like restrictions on the use of sinks, the EU accepted the unspecified requirement that domestic action shall be a "significant element" of climate policy in Annex B countries (CP, 2001a: 7). Another interesting finding is that there is a difference when the EU officials are asked about negotiating power in general terms and in specified terms. In general terms most officials state that the EU proposal on supplementarity was not intended to wield negotiating power, but in specified terms several of them acknowledge that the proposal was intended to put pressure on the US and the Russian Federation. Intended or not, the EU proposal was in fact used as a bargaining chip at CoP6 to obtain other favors.

Although the Europeans opposed to emissions trading for several years, they managed to break out in 2003 by adopting a Directive that creates permit trading in the EU from 2005 onwards. Where most economists stop by saying that market-based instruments have gained in acceptance as a result of cultural change, we take a neo-institutional approach to demonstrate that the attitudes of policy makers have also changed as a result of path-dependent internal pressures and external "shocks", which has contributed to this process of cultural change (and not just the other way around). The European Commission exerted internal pressure by adopting a pioneering role and the perception took hold that the problem-solving capacity of existing policy was decreasing. While switching costs decreased as cultural resistance against "pollution rights" crumbled and information on permit trading improved, an external "shock" occurred in the form of the withdrawal of the US from the Kyoto Protocol, which gave other countries favorable to unrestricted private trading credible threat power. EU politicians acknowledged that the threat of the Russians not to ratify the Protocol then accelerated the co-decision procedure on EU-wide emissions trading because an early agreement should signal that the EU takes climate policy and market instruments seriously and that the Russians, although the Americans had left, can still gain from trading emissions with the Europeans.

Although the permit trading Directive has been adopted, a full-scale institutional breakout in the EU is not guaranteed. Some firms and policy makers still try to steer the national allocation of emission rights in the inefficient direction of credit trading, by linking the height of ceilings for individual companies, within the ceiling for an industry as a whole, to the size of their production. Moreover, credit trading can still be used for installations not covered by the Directive. The path dependence approach, however, emphasizes that starting with (elements of) credit trading comes at a risk: it can result in an institutional lock-in and reinforce a path from which it may be difficult to escape. Governments, firms and scientists should at least acknowledge this risk.

PART V
CONCLUSION

Chapter 10

Conclusion

10.1. Developments in Environmental Economics

The existing literature on emissions trading (a) hardly considers the political transaction costs to set up market-based climate policy, (b) insufficiently recognizes that climate institutions (like technologies) exhibit patterns of path dependence that might lead to lock-in situations and (c) typically focuses on formal institutions usually without applying law and economics perspectives and without considering, also empirically, the impact of informal institutions, like political culture. Our book tries to fill these gaps in environmental economics, although much remains to be researched.

The objective of this book is to analyze the institutional barriers to implementing market-based climate policy, as well as to provide some opportunities to overcome them. The IPCC considers such an analysis as a priority area for research (e.g. Banuri et al., 2001: 71). Our approach is that of institutional economics, with special emphasis on political transaction costs and path dependence. Instead of rejecting the neoclassical approach, we use it where fruitful and show when and why it is necessary to employ a new or neo-institutionalist approach. The result is that we consider equity next to efficiency, that we study the evolution and possible lock-in of both formal and informal climate institutions and that we pay attention to the politics and law of economic instruments for climate policy, including some new empirical analyses.

If the Kyoto Protocol would enter into force, the world's largest market-oriented institution in the realm of climate policy becomes reality. It imposes absolute emission ceilings on industrialized countries and establishes three market-based instruments, the so-called Kyoto Mechanisms, namely International Emissions Trading (IET), Joint Implementation (JI) and the Clean Development Mechanism (CDM). However, some governments support the Kyoto Protocol, like the Member States of the EU, while others reject it, like the US. In addition, some economists support the Kyoto Protocol as an important first step that took years of negotiations, while others reject it as it would do "too little, too fast". Therefore, the theories and concepts used in this book concern market-based climate policy in general, but we have presented applications and examples in the context of the Kyoto Mechanisms where relevant.

North (1990, 1991) defines institutions as the humanly devised constraints that structure political, economic and social interaction and makes a distinction between formal constraints, including laws and property rights, and informal constraints, including cultures and customs. These constraints usually evolve incrementally throughout history, he argues, and determine the costs of transacting. New institutional economics studies these transaction costs, based on neoclassical assumptions of rationality and cost minimization. Neo-institutional economics takes a step further by studying political transaction costs, path dependence and informal institutions, recognizing that boundedly rational actors may bring about inefficient outcomes. Formal institutions can then be analyzed by using law and economics, ranging from neoclassical to new and neo-institutional approaches.

This is reflected in the structure of this book. Part I presents the general institutional economics framework by discussing the design and implementation issues of market-based climate policy and develops an institutional path dependence approach that is able to explain the lock-in of sub-optimal environmental policy instruments, for instance, by looking at the political transaction costs of establishing more efficient ones. Part II considers the new institutional economics by studying the impact of institutional design and operation of market-based climate policy on environmental effectiveness and compares the (political) transaction costs of different types of market-based climate policy instruments. Part III uses law and economics approaches, both neoclassical and neo-institutional ones, to specify the formal constraints of market-based climate policy by formulating the conditions, in terms of efficiency and equity, under which international differences in the domestic allocation of emission rights distort competition and lead to actionable subsidies under WTO rules or the state aid under EC law, including an empirical analysis in the UK and Denmark. Part IV applies insights from neo-institutional economics to specify the informal constraints of market-based climate policy by developing 16 hypotheses that could help explain the EU proposal to quantitatively restrict the Kyoto Mechanisms, including equity as a cultural barrier, and tests them empirically, for instance, by analyzing the opinions of several high-position EU officials. Part V contains the conclusion.

10.2. The Institutional Economics of Market-Based Climate Policy

When designing market-based climate policy, a choice must be made between various types of instruments. Credit-based instruments, including credit trading as well as project-based trading, are inefficient and their effectiveness is uncertain.

Credit trading means that a firm can create credits voluntarily by reducing its emissions below the emission level required by existing regulation, like energy-efficiency standards. The environmental scarcity is not reflected in a price for each unit of emissions: when the economy grows, the supply of credits increases as well because polluters do not have an emission ceiling. Project-based flexibility, like a JI or CDM project, means that an investor receives credits for achieving emission reductions at a (usually foreign) host, measured from a baseline that estimates emissions if the project had not taken place. Such projects have relatively high transaction costs because transactions require pre-approval. Permit trading, however, is superior according to neoclassical economics. Under permit trading, also called allowance trading or cap-and-trade, polluters receive emission rights under an emission ceiling. This is both efficient and effective: when the economy grows, the demand for emission rights increases, but the supply of such rights remains constant because of the emission ceiling. Without pre-approval, transaction costs in the market are relatively low.

However, when implementing market-based climate policy, politicians are tempted to make existing environmental policy more flexible by adding credit trading to it. An example is the EU where several Member States developed such plans. Also the international community initially started with (experimental) emission reduction projects and avoided the trading of emission allowances. Apparently, the economic hierarchy and the political hierarchy of market-based climate policy instruments do not necessarily coincide. Nevertheless, history has shown that they may converge. The international community now accepts permit trading under the Kyoto Protocol, next to the project-based mechanisms, and the EU will start its own permit trading scheme in 2005.

To explain why a sub-optimal design like credit trading may "lock-in" institutionally and when an institutional "breakout" in the direction of permit trading might occur, we have developed an institutional path dependence approach. The concept of political transaction costs is placed in a historical and evolutionary setting, not only by distinguishing sunk costs from switching costs, but also by considering positive feedbacks, self-reinforcing mechanisms and lock-in effects. This explains why policy-making often leads to incremental changes and makes clear when a switch to new institutions might occur. Building upon Arthur (1989) and North (1990), we acknowledge that an evolution over time to the most efficient alternative not necessarily occurs. An institutional lock-in refers to the dominance of a sub-optimal institutional arrangement, such as a (set of) inefficient policy instrument(s), in the presence of a superior institutional arrangement. An institutional breakout then means that the superior institutional arrangement is, in fact, adopted and implemented.

The conditions for an institutional lock-in are the existence of a superior alternative, incomplete information, a problem-solving capacity of existing policy which is perceived to be increasing or stable, as well as large set-up costs. Such a lock-in is strengthened when the superiority of the superior alternative is contested, for instance, due to theoretical ambiguity or uncertainty. The conditions for an institutional breakout mirror those of a lock-in. The chances for the superior alternative improve when information quality is enhanced and when set-up costs decrease against the background of a deteriorating problem-solving capacity of extant policy. External (political) shocks and additional (economic) incentives can also provide pressures for regulatory change.

The political transaction costs that have to be made to create a more efficient institutional arrangement are an important element in explaining an institutional lock-in. Examples of such set-up costs that the government incurs are the costs of gathering and processing information, the costs of developing the required legal framework, the costs of (re)allocating property rights, and the costs of dealing with lobbying efforts and cultural resistance. Set-up costs can be subdivided into sunk costs (of the existing arrangement) and switching costs (of a new arrangement). The former are not relevant for the decision whether or not to continue and extend the existing arrangement because they were made in the past ("bygones are forever bygones", according to economic theory), but switching costs are relevant when establishing a new one because they still have to be made. The switching costs that arise from these formal and informal institutions are extensively analyzed in our book.

In environmental policy, Bressers & Huitema (1999: 180) observe that "(…) new 'economic' instruments are often based on existing legal instruments (…)". We believe that institutional path dependence and lock-in are an important, but often overlooked part of the explanation. Moreover, where Stavins (2002: 15) argues that market-based instruments "have moved center stage", also in Europe, we have emphasized that some market-based instruments initially moved more to the center than others as a result of path dependence.

10.3. The New Institutional Economics of Market-Based Climate Policy

Just like Liebowitz & Margolis (2000) clarified the survival of the QWERTY-keyboard by contesting the superiority of its alternatives, it helps to explain the inclination of policy makers to add sub-optimal designs to the existing environmental policy framework by nuancing the effectiveness advantages of permit trading as well as the presumed ineffectiveness of credit-based approaches.

First, hot air trading was considered to be the most important effectiveness problem of permit trading. A country has hot air if its business-as-usual emissions remain below its official emission ceiling, like the Russian Federation under the Kyoto Protocol. Emission rights can then be traded and used to cover emissions that might have remained unused without emissions trading. The emission ceilings are still respected when hot air is traded, so that effectiveness is achieved in its formal interpretation. However, hot air trading does disturb effectiveness in an ethical interpretation because it can make overall emissions higher with than without emissions trading. Without hot air trading, the actual emissions of all emission sources together could have been lower than the overall target. Permit trading was considered environmentally "fit" in some (formal), but not all (ethical) respects. This relative fitness problem helps to explain the institutional lock-in. The uncertainty in science and the absence of consensus in politics about whether hot air trading is an environmental problem or not made permit trading look suspicious instead of superior.

Second, policy makers were confronted with a growing literature about the institutional opportunities to improve the effectiveness of credit-based approaches. JI and CDM projects, for instance, face the baseline problem of estimating future emissions at the project site in the absence of the project. Because the baseline is a counterfactual (that can be set too high), effectiveness is uncertain, but baseline standardization, if adopted, ensures that project partners have less possibilities to claim more credits by inflating baseline emissions. This appeared to be politically acceptable because it also reduces transaction costs as it will not be necessary anymore to construct a baseline for each individual project. Project-based flexibility was also perceived to be more institutionally "fit" than contended in the neoclassical economic hierarchy of market-based climate policy when it became clear that project baselines could be enhanced and that baselines could be derived from existing environmental regulation (if present). These issues made governments more reluctant to switch to permit trading.

In addition, relatively high political transaction costs are involved to set up this superior alternative. Permit trading largely replaces existing environmental policy by explicitly (re)allocating property rights, whereas credit-based approaches have lower set-up costs because they can use existing environmental policy (as the baseline) from which to calculate the tradeable emission reductions. Politicians were well aware of this (e.g. Oberthür & Ott, 1999: 204), which gave rise to another relative fitness problem: permit trading was considered superior in some (market transaction cost), but not in all (political transaction cost) respects. This nuances the transaction cost advantages of permit trading and helps to explain why some politicians are tempted to opt for less efficient credit markets. Ambiguity about the superiority of permit trading, and uncertainty about its (high) level of set-up costs, contributed to the institutional lock-in of sub-optimal designs.

This was strengthened, once more, by the existence of institutional opportunities to lower the market transaction costs for credit-based instruments. For instance, credit trading does not necessarily require a pre-approval of each transaction and compliance can be checked at the end of the year. Moreover, the transaction costs of project-based trading can be lowered by standardizing (baseline) procedures, which was done under the Kyoto Protocol in the so-called "fast" institutional track created for JI projects: if the host Party has a reliable emission registration system, it may verify reductions as being additional.

North (1990: 51) suggests that political transaction costs are generally higher (and more difficult to measure) than market transaction costs. If this was also the perception among politicians, for instance, in the context of the Kyoto Protocol, then political transaction costs have played a (major) role in governmental decision making, which helps to explain why they are tempted to add sub-optimal designs to extant policy instead of readily accepting permit trading. There are various formal and informal institutional barriers, including equity, that contribute to these political transaction costs.

10.4. The Institutional Law and Economics of Market-Based Climate Policy

Permit allocation is the largest impediment to implementing permit trading (e.g. Ellerman, 1998). Neoclassical economists find this hard to understand because permit allocation does not affect efficiency which should make permit trading "neutral" to any equity consideration (e.g. Ciorba et al., 2001: 8). However, permit allocation leads to inadmissible subsidization under WTO law or EC law if the concept of competitive distortion is interpreted in equity instead of efficiency terms. Politicians feared that competition may be distorted when some governments would grandfather their permits, while others would auction them. This helps to explain why permit trading may not rank high in the political hierarchy of market-based climate policy.

In a neoclassical interpretation, (efficient) competition is not distorted because not only auctioning, but also grandfathering entails costs for firms (e.g. Nentjes et al., 1995). Emission rights allocated for free have opportunity costs when they are used to cover the emissions of the permit owner. A firm with gratis emission rights has to include these costs (equal to the permit price) in the product price if it does not want to go bankrupt in the longer term. This means that it cannot ask lower product prices than its competitor with auctioned rights abroad, so that there is no need to harmonize permit allocation procedures. However, in a neo-institutional interpretation, (fair) competition is distorted because grandfathered permits are a capital gift, which implies that a company with gratis permits has more financial resources than a comparable foreign firm with auctioned allowances. Although the lump sum subsidy of grandfathering leaves efficiency unaffected, it does imply a financial advantage that affects equity. The permit allocation itself leads to unequal changes of the competitive relations among competing firms. The "level playing field" is said to be distorted, so that it is desirable to harmonize permit allocation procedures. These neoclassical and neo-institutional perspectives are crucial to understand the potential legal barrier of grandfathering under WTO and EC rules.

Grandfathering is not an actionable subsidy under the WTO Agreement on Subsidies and Countervailing Measures (ASCM) in the efficiency interpretation of competitive distortions. There is no benefit because of its opportunity costs and the auctioned industry abroad is not injured as grandfathering does not affect efficiency. However, grandfathering could be seen as an actionable subsidy in the equity interpretation of competitive distortions because grandfathered firms have a stronger financial position than their auctioned competitors abroad. This means that grandfathered firms have a benefit, which could be seen as to affect the interests of the domestic industry in the country that uses auctioning. Moreover, grandfathering could imply that revenue otherwise due is foregone because the government would otherwise have collected revenues in the alternative of auctioning. Although there was a perceived legal ambiguity about whether grandfathering constitutes an actionable subsidy, even within the WTO itself (e.g. Vaughan, 1999), politicians never agreed upon an equity-driven international harmonization of allocation methods under the WTO regime, for instance, because they perceived it as unnecessary from an efficiency point of view or as undesirable from a state sovereignty point of view.

Subsidization is also regulated in the EU. Grandfathering is not state aid under EC Article 87(1) from a neoclassical perspective because it does not distort (efficient) competition. Trade is not affected because grandfathering does not affect efficiency. Moreover, grandfathered firms are not favored because they have to include the opportunity cost of their permits in the product price. This means that grandfathered firms are not advantaged (in the sense of having lower costs). From a neo-institutional perspective, however, grandfathering could be seen as state aid because it distorts (fair) competition. Grandfathered permits are a capital gift, which gives the firm more financial resources than a comparable firm abroad with auctioned permits. The state favors specific firms by giving them a financial advantage over their auctioned competitors in another Member State. This lump sum subsidy affects trade, not in efficiency terms, but in equity terms by unequally altering the competitive relations (the level playing field) among competing firms. Although there is not a genuine transfer of resources from the government, the so-called "state origin" criterion, which is one of the criteria to determine state aid, is also satisfied if the State will receive less revenues as in the case of grandfathering, which amounts to giving the (hypothetical) auction revenue to the polluters (e.g. Welch, 1983: 168). In the equity view, grandfathered firms are advantaged, due to the mere process of permit allocation, so that it is desirable to harmonize permit allocation procedures.

We have used this law and economics framework to analyze the state aid cases of permit trading in Denmark and the UK empirically. It appears that the European Commission indeed considered grandfathering as state aid by using the state origin criterion: the State foregoes revenue which could derive from auctioning the valuable permits (COM, 2000d, 2001d). Nevertheless, the Commission exempted the aid by using environmental, economic, legal as well as political arguments: the grandfathering was allowed, among other things, by following EC Article 87(3)(c) that exempts state aid if it helps to develop certain activities or areas and by stating a political desire to gain experience with and prepare for emissions trading. Although the Commission mentioned neither the impact of opportunity costs nor the desire for a level playing field, grandfathering was interpreted as a wealth transfer which could affect the equal treatment of firms. This set a political (albeit not legal) precedent to interpret grandfathering in the EU in terms of equity.

The latter conjecture was largely supported by the Directive, adopted in 2003, that creates permit trading in the EU from 2005 onwards with grandfathering as the harmonized rule. The choice for permit trading means that efficiency has guided the general economic design of the Directive, but the choice for harmonization means that equity explains the specific legal form of the Directive. This harmonization would not have been necessary from an efficiency point of view because grandfathered permits have opportunity costs, so that competition is not distorted when other governments auction their permits.

Although grandfathering is financially more attractive and acceptable to firms than auctioning, it is unclear whether grandfathering is also readily acceptable to governments because it could constitute an actionable subsidy under WTO law or state aid under EC law if it is seen as a distortion of fair competition. Permit allocation dissimilarities were perceived to be legally "fit" in some (efficiency), but not necessarily in all (equity) respects. This legal ambiguity added to the switching costs of permit trading, which made some contribution to the institutional lock-in of incrementally building credit-based instruments upon extant environmental policy. The allocation of emission rights remains implicit under these sub-optimal arrangements, whereas permit trading requires an explicit choice between grandfathering and auctioning.

The equity view made no headway in the WTO context, but it was in fact used in the EU. Although the European policy makers managed to break out by choosing the efficient path of permit trading, equity still explains not only their initial perception that differences in permit allocation methods between countries might distort competition and could lead to state aid (e.g. COM, 2001a: 11), but also their subsequent legal choice to harmonize grandfathering.

10.5. The Neo-Institutional Economics of Market-Based Climate Policy

Equity also plays a role in the informal constraints to implementing market-based climate policy. There are several informal institutions, including political culture, that could reinforce a path of inefficient regulation and block or hinder the implementation of market-based climate policy (e.g. Licht, 2001). For some time, the economically superior arrangement of permit trading was perceived by governments as morally more suspicious than the credit-based instruments because only the former explicitly allocates (what was seen as) "pollution rights". In this way, cultural values added to the switching costs of permit trading and contributed to the institutional lock-in.

The EU, for instance, expended years of political effort trying to reject or restrict the application of market-based climate policy. In the context of the Kyoto Protocol, the Europeans proposed in 1999 to make the Kyoto Mechanisms supplemental to domestic action by quantitatively limiting their use. The EU roughly wanted 50% of the Kyoto commitments to be achieved domestically. Although this proposal was rejected internationally in 2001 because it would have reduced efficiency, an analysis of the EU supplementarity proposal is still useful to reveal why some governments wanted to limit market-based climate policy, especially permit trading. We have developed (and criticized) 16 hypotheses that could help to explain the EU proposal. An analysis of cultural values is useful because economic interests do not explain the proposed trade restriction by the EU who is likely to be a potential buyer.

By clustering the hypotheses ex post, it can be postulated that the EU wanted to restrict trading: (a) for environmental reasons (for instance, to limit hot air trading or to stimulate compliance by initiating a trajectory of reducing domestic emissions); (b) for political-normative reasons (for instance, to achieve equity by preventing that the industrialized countries "buy their way out", to facilitate incremental change based on traditional EU environmental policy or to reduce the scope for allocation problems, such as the possibility that permit allocation distorts competition); (c) for political-strategic reasons (for instance, to wield negotiating power regarding the elaboration of the Kyoto Mechanisms, to put pressure on the US or to put pressure on the Russian Federation); and (d) for technological-economic reasons (for instance, to stimulate technological innovation by means of domestic action or to stimulate the economy because domestic action has more secondary benefits than emissions trading). The cluster of environmental reasons reflects so-called "green" values and the cluster of political-normative reasons reflects equity values. Although attitudes are partly determined by values, they can also be influenced by other considerations, such as political-strategic ones (van Deth & Scarbrough, 1995a,b: 33).

The EU proposal on supplementarity was officially directed against unrestricted use of all Kyoto Mechanisms, not against permit trading in particular. However, the hypotheses on hot air, incrementalism and allocating emission rights do not apply to JI and the CDM and are only relevant for permit trading under IET Article 17. Permit trading mobilizes the negotiated hot air, whereas projects generate real reductions (if the baseline is correct), and permit trading is a non-incremental policy option that explicitly (re)allocates emission rights, whereas credit-based approaches do not require an initial distribution of such rights before trading can begin.

Our empirical analysis of the EU position on supplementarity demonstrates that informal institutions, including equity values, have contributed to the institutional lock-in of inefficient environmental regulation and that cultural change has helped to force a breakout. Following a desire in the literature for more empirical research on the political opposition against emissions trading (e.g. Wiener, 2000a), we have tested the aforementioned hypotheses on the EU proposal by analyzing (a) the content of 12 EU documents, (b) the opinions of high-level EU officials, gathered by means of a questionnaire and (c) the bargaining behavior of the EU at CoP6.

Although the non-response was high (only 15 out of 41 key officials responded), the questionnaires which have been received not only contain new and detailed information, but also came from almost every country in the EU as well as from the European Commission. The vast majority of these EU officials mentioned equity as the primary motivation behind the EU proposal to prevent that industrialized countries "buy their way out". When the hypotheses are clustered ex post, the opinion analysis assigns the first rank to environmental reasons (such as hot air) and the second rank to political-normative reasons (such as equity) to explain the EU proposal. Political-strategic reasons (such as negotiating power) also play a role, albeit a smaller one, whereas technological-economic reasons (such as secondary benefits) are the least important. This shows that a political culture dominated by equity values and "green" values is the primary explanation of the EU proposal to restrict market-based climate policy. Although the EU documents primarily mentioned hot air, equity turned out to be at least as important in the eyes of key EU officials.

Analyzing their opinions based on another clustering of the hypotheses suggests that the EU proposal was made to restrict all Kyoto Mechanisms rather than to restrict permit trading alone, in line with official EU statements. This seems to imply that we cannot use these data for our purpose to explain why permit trading may rank low in the political hierarchy of market-based instruments. However, the differences between the scores in the empirical analysis are small (11.7 for all mechanisms and 9.7 for permit trading). Furthermore, one of the arguments against permit trading, namely, hot air (which cannot be mobilized by means of JI), ranked first in the content analysis of EU documents and second in the opinion analysis among high-level EU officials. In addition, the EU wanted to exempt JI and CDM projects from a limit on supply (EU Council, 1999). Therefore, objections against permit trading are not a dominant, but still a significant factor to explain the EU proposal.

The EU did not enter the negotiations of CoP6 with a strong bargaining position. Our empirical analysis revealed some internal political divisions as 40% of the key EU officials from different Member States declared (in a personal capacity) to disagree with the official EU proposal decided upon by their Ministers to restrict trade. Moreover, most EU officials in our survey recognized that hot air was necessary for the US to initially accept (and subsequently ratify) an emission reduction commitment. After the US withdrew from the Kyoto Protocol, the divided EU feared that others might withdraw as well and soon gave up its proposal. Another interesting finding is that there is a difference when the EU officials are asked about negotiating power in general terms and in specified terms. In general terms, most officials state that the EU proposal on supplementarity was not intended to wield negotiating power, but in specified terms, several of them acknowledge that the proposal was intended to put pressure on the US and the Russian Federation. Intended or not, the EU proposal was in fact used as a bargaining chip at CoP6 to obtain other favors, like restrictions on the use of sinks.

Although the Europeans opposed to emissions trading for several years, they managed to break out in 2003 by adopting a Directive that creates permit trading in the EU from 2005 onwards. Where most economists stop by saying that market-based instruments have gained in acceptance as a result of cultural change, we take a neo-institutional approach to demonstrate that the attitudes of policy makers have also changed as a result of path-dependent internal pressures and external "shocks", which has contributed to this process of cultural change (and not just the other way around). The European Commission exerted internal pressure by adopting a pioneering role and the perception took hold that the problem-solving capacity of existing policy was decreasing. While switching costs decreased as cultural resistance against "pollution rights" crumbled and information on permit trading improved, an external "shock" occurred in the form of the withdrawal of the US from the Kyoto Protocol, which gave other countries favorable to unrestricted private trading credible threat power. EU politicians said that the threat of the Russians not to ratify the Protocol accelerated the co-decision procedure on EU-wide emissions trading because an early agreement should signal that the EU takes climate policy and market instruments seriously and that the Russians, although the Americans had left, can still gain from trading emissions with the Europeans.

10.6. Some Policy Implications

We have seen that market-based climate policy can be designed on the basis of permit trading or credit-based approaches. Permit trading is superior according to neoclassical economics. Credit trading and project-based flexible instruments are inefficient and their effectiveness is uncertain. The environmental scarcity is not reflected in a price for each unit of emissions: when the economy grows, the supply of credits increases as well because polluters do not have an emission ceiling. Under permit trading, polluters do have an emission ceiling. This design option is both efficient and effective: when the economy grows, the demand for emission rights increases, but the supply of such rights remains constant because of the emission ceiling. The external costs of climate change are then internalized.

However, institutional economics learns that credit trading is politically attractive, not only for firms because they do not have to purchase new emission rights when they expand, but also for policy makers because the set-up costs (or political transaction costs) of permit trading are relatively high. Permit trading comes to replace existing environmental policy, while credit trading builds incrementally on extant policy by using it (as a baseline) to calculate the transferable emission reductions.

The sub-optimal credit-based approaches then profit from the learning effects associated with building on existing environmental policy. Learning effects lower the average costs of running the established system. Moreover, policy makers may expand the existing arrangement with credit trading because they are unacquainted, or are not sufficiently well acquainted, with permit trading. In addition, credit trading builds on the sunk costs of existing environmental policy. These start-up costs that have already been incurred play no role in the decision to continue current environmental policy without emission ceilings, whether or not modified to take account of credit trading. Although permit trading reduces running costs, it involves relatively high start-up costs because it implies crossing over to a new legal arrangement. Opposition by vested interests as well as legal problems and cultural resistance, for instance, driven by equity considerations, contribute to these switching costs.

Building credit trading upon extant environmental policy, ineffective and inefficient as it may be, is referred to as an institutional lock-in. Policy makers may get stuck in the existing regulatory arrangements. An institutional breakout towards efficiency is still possible, though, for instance when the information on permit trading improves, when the costs of switching to a permit trading system decline or when the problem-solving capacity of traditional environmental regulation deteriorates.

However, even when politicians adopt permit trading, a full-scale institutional breakout is not guaranteed. Some firms and policy makers may still try to steer the national allocation of emission rights in the inefficient direction of credit trading, for instance, by linking the height of ceilings for individual companies, within the ceiling for an industry as a whole, to the size of their production. Moreover, credit trading can still be used for installations, firms or sectors not covered by the permit trading scheme.

Some argue that it is not problematic to start with (elements of) credit trading, assuming that it can later evolve into a more efficient permit trading system. The path dependence approach, however, emphasizes that this comes at a risk: starting with a sub-optimal type of emissions trading can result in an institutional lock-in and reinforce a path from which it may be difficult to escape. Governments, firms and scientists should at least acknowledge this risk.

Appendix (Questionnaire)

Dear Mr. / Mrs. X,

The following document is a questionnaire on the supplementarity of emissions trading.

Filling out this questionnaire should take approximately 10 minutes.

We are specifically interested in *your* knowledge. But if you do not have the time to fill out the questionnaire, please delegate it to a competent colleague in order to avoid non-response.

Thank you very much.

Yours sincerely,

Edwin Woerdman

University of Groningen

The Netherlands

Appendix

The EU Proposal on Supplementarity: a Questionnaire

March 2000

Edwin Woerdman
Department of Economics and Public Finance (ECOF), Faculty of Law, University of Groningen, PO Box 716, 9700 AS Groningen, The Netherlands. Tel.: +31-50-363-5261; fax: +31-50-363-7101, e.woerdman@rechten.rug.nl

Introduction: Purpose of this Questionnaire

The EU proposal on supplementarity roughly implies that 50% of the Kyoto commitments should be achieved domestically via a ceiling on emissions trading (see EU Council Conclusions, May 1999).

The purpose of this questionnaire is to obtain opinions from both experts and "insiders" on the question why the EU has proposed a ceiling on GHG emissions trading.

The questionnaire is also interesting for you. Firstly, you will of course be among the first ones to receive the preliminary results and the final article, which will contain the views of several of your colleagues. Secondly, the questionnaire itself is worthwhile, since it provides an overview which summarizes many — if not all — reasons which have been put forward to limit emissions trading. We ask you to judge these reasons for their explanatory power of the EU position by answering 12 questions.

The answers obtained from this questionnaire will be incorporated in a policy-oriented article on the EU proposal by Andries Nentjes, Ger Klaassen and Edwin Woerdman (forthcoming).

Andries Nentjes is a professor in economics and Edwin Woerdman is a political scientist, both working at the University of Groningen (RuG/ECOF) in Groningen, the Netherlands.
Ger Klaassen is an economist working at the International Institute for Applied Systems Analysis (IIASA) in Laxenburg, Austria.

We kindly request you to fill out this questionnaire (which should take approximately 10 min) and return it to the above-mentioned address as soon as possible. We thank you very much for your effort.

Yours sincerely,

Edwin Woerdman

QUESTIONNAIRE: Possible Explanations of the EU Proposal on Supplementarity

Question 1 In recent literature (for example Baron et al., 1999; Gusbin et al., 1999; Hourcade & Le Pesant, 1999) several possible explanations can be found why the EU has proposed a ceiling on emissions trading. Please indicate for each of these "hypotheses" not so much what you think about them personally, but rather whether it has played a role in the shaping of the EU proposal on supplementarity.

(Please indicate for each hypothesis whether it has been a *major reason*, a *side issue* or *no reason at all* for the EU to propose a limit on emissions trading).

Hypothesis 1

The EU wants to limit emissions trading in order to limit the use of "hot air" (e.g. from Russia and Ukraine), since its use would affect the environmental effectiveness of the Kyoto Protocol.

( ) *major reason*, ( ) *side issue*, ( ) *no reason at all* for the EU to propose a limit on emissions trading.

Hypothesis 2

The EU wants to limit emissions trading in order to stimulate domestic action with a view to equity or fairness, because Annex B Parties are responsible for the majority of historical GHG emissions and should not completely "buy their way out".

( ) *major reason*, ( ) *side issue*, ( ) *no reason at all* for the EU to propose a limit on emissions trading.

Hypothesis 3

The EU wants to limit emissions trading in order to stimulate domestic action with a view to compliance, because it gets more difficult for a Party to curb emissions in a second commitment period if it has implemented its QELRC of the first commitment period mainly in other countries rather than initiating a trajectory of reducing emissions at home.

( ) *major reason*, ( ) *side issue*, ( ) *no reason at all* for the EU to propose a limit on emissions trading.

Hypothesis 4

The EU wants to limit emissions trading in order to stimulate technological innovation.

( ) *major reason*, ( ) *side issue*, ( ) *no reason at all* for the EU to propose a limit on emissions trading.

284   Appendix

Hypothesis 5

The EU wants to limit emissions trading in order to demonstrate the willingness to reduce emissions in industrialized countries so that developing countries are stimulated to adopt commitments in the future.

( ) *major reason*, ( ) *side issue*, ( ) *no reason at all* for the EU to propose a limit on emissions trading.

Hypothesis 6

The EU wants to limit emissions trading in order to reduce problems in the case of non-compliance, because emissions trading requires additional rules about who is responsible (buyer or seller liability) if assigned amounts, credits and/or permits have been traded which have not been backed up by real reductions.

( ) *major reason*, ( ) *side issue*, ( ) *no reason at all* for the EU to propose a limit on emissions trading.

Hypothesis 7

The EU wants to limit emissions trading in order to limit the use of markets in environmental policy because they are likely to function imperfectly.

( ) *major reason*, ( ) *side issue*, ( ) *no reason at all* for the EU to propose a limit on emissions trading.

Hypothesis 8

The EU wants to limit emissions trading in order to wield negotiating power with respect to elaborating the rules of the Kyoto Mechanisms (Articles 17, 12 and 6) at CoP6, because at previous CoP meetings the EU has been able to exert influence on the level of country emission targets, but less on the choice of instruments.

( ) *major reason*, ( ) *side issue*, ( ) *no reason at all* for the EU to propose a limit on emissions trading.

Hypothesis 9

The EU wants to limit emissions trading in order to wield negotiating power with respect to the FCCC decision-making process in general, because the proposal gives the EU "something to negotiate about" which could increase the likelihood of favorable compromises at CoP6.

( ) *major reason*, ( ) *side issue*, ( ) *no reason at all* for the EU to propose a limit on emissions trading.

Hypothesis 10

The EU wants to limit emissions trading in order to not to loose the "climate leadership" to the USA.

( ) *major reason*, ( ) *side issue*, ( ) *no reason at all* for the EU to propose a limit on emissions trading.

Hypothesis 11

The EU wants to limit emissions trading in order to prevent too radical policy changes (from standards/taxes to emissions trading).

( ) *major reason*, ( ) *side issue*, ( ) *no reason at all* for the EU to propose a limit on emissions trading.

Hypothesis 12

The EU wants to limit emissions trading in order to reduce the scope for allocation problems associated with emissions trading, since permit allocation gives emitters an incentive to inflate baseline emissions, triggers endless debates about which distribution is fair and could induce competitive distortions.

( ) *major reason*, ( ) *side issue*, ( ) *no reason at all* for the EU to propose a limit on emissions trading.

Hypothesis 13

The EU wants to limit emissions trading in order to stimulate the economy, because domestic action (despite its higher micro-economic costs) has more positive macro-economic impacts than emissions trading (e.g. in terms of an increase in domestic employment and reduction of non-CO_2 air pollution (secondary benefits)).

( ) *major reason*, ( ) *side issue*, ( ) *no reason at all* for the EU to propose a limit on emissions trading.

Hypothesis 14

The EU wants to limit emissions trading in order to facilitate a step-wise introduction of the Kyoto Mechanisms thereby hedging against uncertainties and reducing the associated risks.

( ) *major reason*, ( ) *side issue*, ( ) *no reason at all* for the EU to propose a limit on emissions trading.

Hypothesis 15

The EU wants to limit emissions trading in order to force the USA to reduce emissions at least partly at home, since they are the biggest source of greenhouse gas emissions.

( ) *major reason*, ( ) *side issue*, ( ) *no reason at all* for the EU to propose a limit on emissions trading.

Hypothesis 16

The EU wants to limit emissions trading in order to prevent Russia from selling "hot air" and trading questionable emission reductions, since it has a weak enforcement regime.

( ) *major reason*, ( ) *side issue*, ( ) *no reason at all* for the EU to propose a limit on emissions trading.

Remarks (such as alternative explanations of the EU proposal to limit emissions trading):

Question 2 Consider the 16 above-mentioned hypotheses once more. If you have denominated more than one hypothesis as a "major reason" for the EU to propose a limit on emissions trading, please indicate which of these "major reasons" is the principal one.

What hypothesis do you think is the *principal* reason for the EU to propose a ceiling on emissions trading?

(Please indicate the number of the hypothesis (H_x) on the dotted line).

Principal reason: hypothesis number………

Remarks:

Question 3 The projections of "hot air" seem to differ widely. Do you think that "hot air" will constitute a major or only a minor part of emissions trading?

( ) minor (less than 20%)
( ) major (more than 20%)

Remarks:

Question 4 Some observers argue that (not so much the economic inclusion of emissions trading, but rather) the distributional inclusion of "hot air" in the Kyoto Protocol was necessary for the USA to accept an internationally agreed emission stabilization or reduction commitment. They contend that without the "hot air" the overall cap on Annex I emissions would have been lower. Do you agree?

( ) yes
( ) no

Remarks:

Question 5 Do you think, personally, that supplementarity should be achieved "quantitatively" through a ceiling on emissions trading (e.g. the EU proposal) or "qualitatively" through internationally agreed requirements with respect to domestic climate policy (e.g. the requirement to demonstrate adequate efforts to control emissions, for instance via tighter energy efficiency norms)?

( ) quantitative ceiling
( ) qualitative requirements

Remarks:

Question 6 Some observers argue that a ceiling on emissions trading unnecessarily limits potential cost savings as well as potential revenues for developing countries and countries with economies in transition. Do you agree?

( ) yes
( ) no

Remarks:

Question 7 Some observers argue that a ceiling on emissions trading promotes international equity or responsibility because it will force Annex B Parties to "clean up their own mess" by reducing emissions domestically. Do you agree?

( ) yes
( ) no

Remarks:

288 Appendix

Question 8 Some observers argue that the Annex B emission cap does not change from emissions trading, so that a ceiling on trade should only apply to the CDM. Do you agree?

( ) yes
( ) no

Remarks:

Question 9 Some observers argue that mainly lawyers and engineers have shaped the EU proposal to limit emissions trading. Hence, they contend that this EU proposal could only be put forward because too few economists have been involved. Do you agree?

( ) yes
( ) no

Remarks:

Question 10 Some observers argue that the nuclear energy lobby has been effective in influencing the EU position on emissions trading: stimulating domestic action would increase the attractiveness of nuclear energy which does not contribute to GHG emissions. Do you agree?

( ) yes
( ) no

Remarks:

Question 11 In different contexts, some authors have proposed to set a minimum carbon price in order to stimulate technological innovation and reduce uncertainty (cf. Hourcade & Le Pesant, 1999; Rolfe et al., 1999). Do you think this a good idea?

( ) yes
( ) no

Remarks:

Question 12 Some observers argue that a quantitative ceiling on trading is not desirable, because (among other things) it raises the costs of reducing emissions, thereby making it more difficult for Parties to adopt ambitious targets in a second commitment period and making it harder for new countries to join the set of Annex B Parties. Do you agree with this line of reasoning?

( ) yes
( ) no

Remarks:

Name and Confidential Treatment

Note on confidential treatment   We will *not* use or mention your name or affiliation in our forthcoming article if you desire anonymity. Nevertheless, understandably, we would prefer to be able to refer to your name and your expertise in our research. Can we mention your name and affiliation in our article when discussing the answers to this questionnaire?

(   ) yes, you may mention my name and affiliation
(   ) no, I desire anonymity

Remarks:

Please fill out your name (confidential): ....

___

Thank you once more both for your effort and for your valuable contribution to our research.

You will receive the results of this questionnaire as soon as possible.

___

References

AGBM. (1997). *Implementation of the Berlin Mandate: Proposals from Parties*, FCCC/AGBM/1997/MISC.1 (including Add.1), 19 February 1997. Ad Hoc Group on the Berlin Mandate (AGBM), Bonn.

Aidt, T. S., & Dutta, J. (2001). *Transitional Politics: Emerging Incentive-Based Instruments in Environmental Regulation*, Nota di Lavoro 78.2001. Fondazione Eni Enrico Mattei (FEEM), Milan.

Aldy, J. E., Barrett, S., & Stavins, R. N. (2003). *Thirteen Plus One: A Comparison of Global Climate Policy Architectures*, Nota di Lavoro 64.2003. Fondazione Eni Enrico Mattei (FEEM), Milan.

Alexander, G. (2001). Institutions, path dependence and democratic consolidation. *Journal of Theoretical Politics*, **13**, 3, 249–270.

Allen, D. (2000). Transaction costs. In: B. Bouckaert, & G. de Geest (Eds), *Encyclopedia of Law and Economics* (pp. 893–926). Edward Elgar, Cheltenham.

Almond, G. A., & Verba, S. (1965). *The Civic Culture: Political Attitudes and Democracy in Five Nations*. Little Brown, Boston.

Anderson, J. W. (1997). *Climate Change, Clinton and Kyoto: The Negotiations over Global Warming*. Resources for the Future (RFF), Washington, November 1997.

Anderson, D., Roland, K., Schreiner, P., & Skjelvik, J. M. (1999). Designing a domestic GHG emissions trading system. In: C. J. Jepma, & W. P. van der Gaast (Eds), *On the Compatibility of Flexible Instruments* (pp. 109–124). Kluwer, Dordrecht.

Arrow, K. J. (1969). *The Organization of Economic Activity: Issues Pertinent to the Choice of Market Versus Non-Market Allocation*, The Analysis and Evaluation of Public Expenditures: The PBB-System, Joint Economic Committee 91/1 (1). Government Printing Office, Washington, DC.

Arthur, W. B. (1989). Competing technologies, increasing returns, and lock-in by historical small events. *Economic Journal*, **99**, 116–131.

Arthur, W. B. (1994). *Increasing Returns and Path Dependence in the Economy*. University of Michigan Press, Michigan.

Aslam, M. A. (1999). The clean development mechanism: unravelling the "mystery". In: C. J. Jepma, & W. P. van der Gaast (Eds), *On the Compatibility of Flexible Instruments* (pp. 33–45). Kluwer, Dordrecht.

Baker, D., & Barrett, J. (1999). *Cleaning Up the Kyoto Protocol: Emission Permit Trading Would Let Developing Nations Reap Profits from Green Policies*, EPI Issue Brief 131. Economic Policy Institute (EPI), Washington, DC.

Banuri, T., et al. (1996). Equity and social considerations. In: J. P. Bruce, H. Lee, & E. F. Haites (Eds), *Climate Change 1995: Economic and Social Dimensions of Climate Change*

(Contribution of Working Group III to the Second Assessment Report of the IPCC, pp. 79–124). Cambridge University Press, Cambridge.

Banuri, T., et al. (2001). *Climate Change 2001: Mitigation (Technical Summary)*, Report of Working Group III of the Intergovernmental Panel on Climate Change (IPCC).

BAPA. (1998). *Buenos Aires Plan of Action (BAPA)*, FCCC/CP/1998/16/Add.1, Preliminary Version of CoP4 Decisions and Resolutions (subject to final editing), 27 November.

Barde, J. P. (1995). Environmental policy and policy instruments. In: H. Folmer, H. Landis, H. Gabel, & H. Opschoor (Eds), *Principals of Environmental and Resource Economics* (pp. 201–227). Edward Elgar, Aldershot.

Barker, T., Kram, T., Oberthür, S., & Voogt, M. (2000). *The Role of Domestic Greenhouse Gas Mitigation Options*, Draft Paper Presented at the European Forum on Integrated Environmental Assessment Second Climate Workshop, 18–19 April 2000, Amsterdam, The Netherlands.

Barker, T., Kram, T., Oberthür, S., & Voogt, M. (2001). The role of EU internal policies in implementing greenhouse gas mitigation options to achieve Kyoto targets. *International Environmental Agreements: Politics, Law and Economics*, **1**, 243–265.

Baron, R. (1997). *Economic/Fiscal Instruments: Competitiveness Issues Related to Carbon/Energy Taxation*, Annex I Expert Group on the United Nations Framework Convention on Climate Change, Working Paper No. 14, OCDE/GD(97)190. Organization for Economic Co-operation and Development (OECD), Paris.

Baron, R. (1999a). *An Assessment of Liability Rules for International GHG Emissions Trading*. International Energy Agency (IEA), Paris, October 1999.

Baron, R. (1999b). The Kyoto mechanisms: how much flexibility do they provide? In: J. Pershing (Ed.), *Emissions Trading and the Clean Development Mechanism: Resource Transfers, Project Costs and Investment Incentives*, Bonn CoP5 October–November 1999. International Energy Agency (IEA), Paris.

Baron, R., Bosi, M., Lanza, A., & Pershing, J. (1999). *A Preliminary Analysis of the EU Proposals on the Kyoto Mechanisms*, Draft 28 May/8 June. International Energy Agency (IEA), Paris.

Bashmakov, I. (1999). Strengthening the economy through climate change policies: the case of the Russian Federation. In: C. J. Jepma, & W. P. van der Gaast (Eds), *On the Compatibility of Flexible Instruments* (pp. 17–30). Kluwer, Dordrecht.

Baumert, K. A., Bhandari, R., & Kete, N. (1999). *What Might a Developing Country Climate Commitment Look Like?*, Climate Notes, May 1999. World Resources Institute, Washington.

Baumol, W. J., & Oates, W. E. (1988). *The Theory of Environmental Policy*. Cambridge University Press, Cambridge, 2nd ed.

Begg, K., & Parkinson, S. (2001). JI and CDM: lessons from pilot project assessment. *Energy & Environment*, **12**, 5/6, 475–486.

Begg, K., Parkinson, S., Jackson, T., Morthorst, P.-E., & Bailey, P. (1999). *Overall Issues for Accounting for the Emissions Reductions of JI Projects*, Paper Presented at the CDM Workshop on Baseline for CDM, 25–26 February, Tokyo, Japan.

Bernstein, S. (2002). International institutions and the framing of domestic policies: the Kyoto Protocol and Canada's response to climate change. *Policy Sciences*, **35**, 203–236.

Black-Arbelaez, T., Nondek, L., Mintzer, I., Moorcroft, D., & Kalas, P. J. (2000). *The World Bank/Donor Supported Program of National CDM/JI Strategy Studies: The NSS Program*. World Bank National Strategy Studies (NSS), Washington, DC.

Bohm, P. (1997). *Joint Implementation as Emission Quota Trade: An Experiment Among Four Nordic Countries*, TemaNord 1997:4. Nordic Council of Ministers, Copenhagen.

Bohm, P. (1999). *International Greenhouse Gas Emission Trading — With Special Reference to the Kyoto Protocol*, TemaNord 1999:506. Department of Economics, Stockholm.
Bolin, B. (1993). A joint scientific and political process for a convention on climate change. In: G. Sjöstedt, U. Svedin, & B. Hägergäll-Aniansson (Eds), *International Environmental Negotiations: Process, Issues and Contexts* (pp. 155–163). Sage Publications, Newbury Park.
Bollen, J. C., Gielen, A. M., & Timmer, H. R. (1999). Clubs, ceilings and CDM: macro-economics of compliance with the Kyoto Protocol. *Energy Journal*, 177–206, Special Issue, The costs of the Kyoto Protocol: a multi-model evaluation.
Boom, J. T. (2000a). *International Emissions Trading Under the Kyoto Protocol: Credit Trading*, Economic Discussion Paper No. 7/2000. University of Southern Denmark, Odense.
Boom, J. T. (2000b). *The Effect of Emission Trade on International Environmental Agreements*, Working Paper 00-1. Department of Economics, The Aarhus School of Business.
Boom, J. T., & Nentjes, A. (2000). *Level of International Emissions Trading: Should Governments Trade, or Should Firms?*, Economic Discussion Paper No. 4/2000. University of Southern Denmark, Odense.
Bovenberg, A. L. (1993). Policy instruments for curbing CO_2 emissions: the case of the Netherlands. *Environmental and Resource Economics*, **3**, 233–244.
Bovenberg, L., & Cnossen, S. (1995). Public economics and the environment in an imperfect world: an introductory summary. In: L. Bovenberg, & S. Cnossen (Eds), *Public Economics and the Environment in an Imperfect World*. Kluwer, Dordrecht.
Boyd, E., Hanks, J., Schipper, L., Sell, M., Spence, C., & Voinov, J. (2001). Summary of the seventh conference of the parties to the framework convention on climate change: 29 October–10 November 2001, CoP-7 Final. *Earth Negotiations Bulletin*, **12**, 189, 1–16.
Bressers, H. Th. A., & Huitema, D. (1999). Economic instruments for environmental protection: can we trust the "magic carpet"? *International Political Science Review*, **20**, 2, 175–196.
Brockmann, K., Stronzik, M., & Bergmann, H. (1999). *Emissionsrechtehandel — eine neue Perspektive für die Deutsche Klimapolitik nach Kioto*. Physica-Verlag, Heidelberg.
Burtraw, D., Palmer, K., Bharvirkar, R., & Paul, A. (2001). *The Effect of Allowance Allocation on the Cost of Carbon Emission Trading*, RFF Discussion Paper 01-30. Resources for the Future (RFF), Washington.
Bush, G. W. (2001). *Text of a Letter from the President to Senators Hagel, Helms, Craig and Roberts*. US Newswire/White House Press Office, Washington, 13 March.
Bush, G. W. (2002). *President Announces Clear Skies & Global Climate Change Initiatives*, Office of the Press Secretary, February 14, 2002, News Release by the White House, www.whitehouse.gov/news/releases/2002/02.
Butzengeiger, S., Betz, R., & Bode, S. (2001). *Making GHG Emissions Trading Work: Crucial Issues in Designing National and International Emissions Trading Systems*, HWWA Discussion Paper 154. Hamburg Institute of International Economics (HWWA), Hamburg.
Calabresi, G., & Melamed, A. D. (1972). Property rules, liability rules, and inalienability: one view of the cathedral. *Harvard Law Review*, **85**, 6, 1089–1128.
Capros, P., Mantzos, L., Vainio, M., & Zapfel, P. (2000). *Economic Efficiency of Cross-Sectoral Emission Trading in CO_2 in the European Union*, Paper Presented at the Conference on Instruments for Climate Policy: Limited Versus Unlimited Flexibility?, 19–20 October. University of Gent (Belgium), Gent.

Carpenter, C., et al. (1998). Report of the fourth conference of the parties to the UN framework convention on climate change: 2–13 November 1998. *Earth Negotiations Bulletin*, **12**, 97, 1–14.

Chomitz, K. M. (1999). *Baselines for Greenhouse Gas Reductions: Problems, Precedents, Solutions*, Paper Presented at the CDM Workshop on Baseline for CDM, 25–26 February, Tokyo, Japan.

Chow, K. K. (2001). Chairman's comments on COP-6 negotiations. *Global Greenhouse Emissions Trader*, **9**, 1.

Christiansen, A. C., & Wettestad, J. (2003). The EU as a frontrunner on greenhouse gas emissions trading: how did it happen and will the EU succeed? *Climate Policy*, **3**, 3–18.

Churie, A., Hanks, J., Schipper, L., Sell, M., Spence, C., & Voinov, J. (2000). Summary of the sixth conference of the parties to the framework convention on climate change: 13–25 November 2000, CoP-6 Final. *Earth Negotiations Bulletin*, **12**, 163, 1–19.

Cini, M., & McGowan, L. (1998). *Competition Policy in the European Union*. MacMillan, London.

Ciorba, U., Lanza, A., & Pauli, F. (2001). *Kyoto Protocol and Emission Trading: Does the US Make a Difference?* Nota di Lavoro 90.2001. Fondazione Eni Enrico Mattei (FEEM), Milan.

Coase, R. H. (1960). The problem of social cost. *Journal of Law and Economics*, **3**, 1–44.

Cogen, J., Rosenzweig, R., & Varilek, M. (2003). *Overview of Emerging Markets for Greenhouse Gas Commodities*, Greenhouse Gas Market 2003: Emerging but Fragmented. International Emissions Trading Association (IETA), Geneva.

Cole, D. H. (1999). Clearing the air: four propositions about property rights and environmental protection. *Duke Environmental Law & Policy Forum*, **10**, 103, 103–130.

Cole, D. H. (2000). New forms of private property: property rights in environmental goods. In: B. Bouckaert, & G. de Geest (Eds), *Encyclopedia of Law and Economics* (pp. 274–307). Edward Elgar, Cheltenham.

COM. (1997). *Climate Change — The EU Approach for Kyoto*, Communication from the Commission, 1 October. European Commission, Brussels.

COM. (1998a). *XXVIIth Report on Competition Policy 1997*. Office for Official Publications of the European Communities, Luxembourg.

COM. (1998b). *Climate Change — Towards an EU Post-Kyoto Strategy*, Communication from the Commission to the Council and the European Parliament, Document COM(1998)353. European Commission, Brussels.

COM. (1999a). *Preparing for Implementation of the Kyoto Protocol*, Commission Communication to the Council and the Parliament, Document COM(1999)230, 19 May. European Commission, Brussels.

COM. (1999b). *XXVIIIth Report on Competition Policy 1998*. Office for Official Publications of the European Communities, Luxembourg.

COM. (2000a). *Green Paper on Greenhouse Gas Emissions Trading Within the European Union*, Green Paper Presented by the Commission, 8 March. European Commission, Brussels.

COM. (2000b). *XXIXth Report on Competition Policy 1999*, Document SEC(2000)720 Final, 5 May. European Commission, Brussels/Luxembourg.

COM. (2000c). *Eighth Survey on State Aid in the European Union*, Document COM(2000)205 Final, 11 April. European Commission, Brussels/Luxembourg.

COM. (2000d). *State Aid No. N 653/99 — CO_2 Quotas (Statsstøttesag nr. N 653/99 — CO_2 kvoter)*, Letter by Mario Monti to the Danish Government, English Version (Draft), 12 April. European Commission, Brussels/Luxembourg.

COM. (2000e). *EU Policies and Measures to Reduce Greenhouse Gas Emissions: Towards a European Climate Change Programme*, Communication from the Commission to the Council and the European Parliament, Document COM(2000)88, 8 March. European Commission, Brussels.

COM. (2001a). *Proposal for a Directive of the European Parliament and of the Council Establishing a Framework for Greenhouse Gas Emissions Trading Within the European Community and Amending Council Directive 96/61/EC (Presented by the Commission)*, Document COM(2001)581, 23 October. European Commission, Brussels/Luxembourg.

COM. (2001b). *Proposal for a Directive of the European Parliament and of the Council Establishing a Framework for Greenhouse Gas Emissions Trading Within the European Community*, Document COM(2001)xxx, Version for Interservice Consultation, 31 May. European Commission, Brussels.

COM. (2001c). *XXXth Report on Competition Policy 2000*, Document SEC(2001)694 final, 7 May. European Commission, Brussels/Luxembourg.

COM. (2001d). *State Aid No. N 416/2001 — United Kingdom Emission Trading Scheme*, Letter by Mario Monti to the UK Government, C(2001)3739 Final, 28 April. European Commission, Brussels/Luxembourg.

COM. (2003). *The EU Emissions Trading Scheme: How to Develop a National Allocation Plan*, Non-paper, 2nd Meeting of Working Group 3 of the Monitoring Mechanism Committee, 1 April 2003. European Commission, Brussels/Luxembourg.

COM. (2004). *Communication from the Commission on Guidance to Assist Member States in the Implementation of the Criteria Listed in Annex III to Directive 2003/87/EC Establishing a Scheme for Greenhouse Gas Emission Allowance Trading Within the Community and Amending Council Directive 96/61/EC, and on the Circumstances Under Which Force Majeure is Demonstrated*, Document COM(2003)830 Final, 7 January 2004. European Commission, Brussels/Luxembourg.

Commissie Vogtländer. (2002). *Handelen voor een beter klimaat: Haalbaarheid van een nationaal systeem voor CO_2-emissiehandel*, Samenvatting eindadvies Commissie CO_2-handel, Januari 2002. KPMG Milieu, De Meern, in Dutch.

Conrad, K., & Kohn, R. E. (1996). The US market for SO_2 permits: policy implications of the low price and trading volume. *Energy Policy*, **24**, 12, 1051–1059.

Convery, F. J., Redmond, L., Louise Dunne, L., & Ryan, L. B. (2003). *Assessing the European Union Emissions Trading Directive*, Paper Presented at the 12th Annual Conference of the European Association of Environmental and Resource Economists (EAERE), June 28–30, Bilbao, Spain.

Cools, S. (2003). Belgium bases allocation on voluntary agreements. *Carbon Market Europe*, p. 2, June 20.

Cooper, G., & Nicholls, M. (2000). Trading around the corner. *Environmental Finance*, Supplement, XII–XIV, October 2000.

CoP1. (1995). *Activities Implemented Jointly Under the Pilot Phase*, FCCC/CP/1995/7/Add.1, 6 June, Decision 5/CP.1 of 7 April, CoP1 (28 March–7 April), Berlin, Germany.

Cordato, R. (1994). Efficiency. In: P. J. Boettke (Ed.), *The Elgar Companion to Austrian Economics* (pp. 131–136). Edward Elgar, Cheltenham.

Cowan, R., & Gunby, P. (1996). Sprayed to death: path dependence, lock-in and pest control strategies. *The Economic Journal*, **106**, 521–542.

Cowan, R., & Hultén, S. (1996). Escaping lock-in: the case of the electric vehicle. *Technological Forecasting and Social Change*, **53**, 61–79.

Cozijnsen, J. (2000). International developments in emissions and reductions trade. *Change: Research and Policy Newsletter on Global Change from the Netherlands*, **51**, 1–5.

CP. (2001a). *Implementation of the Buenos Aires Plan of Action*, Decision 5/CP.6 ('Bonn Agreement'), Document FCCC/CP/2001/L.7, 24 July 2001. CoP6 Part II, Bonn.

CP. (2001b). *Report of the Conference of the Parties on its Seventh Session*, Document FCCC/CP/2001/13/Add. 1–4 ('Marrakesh Accords'), 21 January 2002. CoP7, Marrakesh.

Crane, A. T., Holmes, K. J., & Friedman, R. M. (1998). *Designs for Domestic Carbon Emissions Trading*. The H. John Heinz III Center for Science, Economics and the Environment, Washington, DC.

Criqui, P., Mima, S., & Viguier, L. (1999). Marginal abatement costs of CO_2 emission reductions, geographical flexibility and concrete ceilings: an assessment using the POLES model. *Energy Policy*, **27**, 10, 585–601.

Cullet, Ph., & Kameri-Mbote, A. P. (1998). Joint implementation and forestry projects: conceptual and operational fallacies. *International Affairs*, **74**, 2, 393–408.

Dales, J. H. (1968). *Pollution, Property and Prices: An Essay in Policy-Making and Economics*. Toronto University Press, Toronto.

Damro, C., & Méndez, P. L. (2003). Emissions trading at Kyoto: from EU resistance to union innovation. *Environmental Politics*, **12**, 2, 71–94.

David, P. A. (1985). Clio and the economics of QWERTY. *American Economic Review*, **75**, 2, 332–336, AEA Papers and Proceedings.

DEFRA. (2002). *A Summary Guide to the UK Emissions Trading Scheme*. Global Atmosphere Division/Department for Environment, Food & Rural Affairs (DEFRA), London.

de Heer, W. (2000). Een symposium over non-respons by enquêtes. *Facta: Sociaal-Wetenschappelijk Magazine*, **8**, 5, 14–16.

Demsetz, H. (1967). Toward a theory of property rights. *American Economic Review*, **57**, 347–359.

Denne, T. (2000). *Sharing the Benefits: Mechanisms to Ensure the Capture of Clean Development Mechanism Project Surpluses*. Center for Clean Air Policy (CCAP), Washington.

de Savornin Lohman, L. (1994). Economic incentives in environmental policy: why are they white ravens? In: H. Opschoor, & K. Turner (Eds), *Economic Incentives and Environmental Policies: Principles and Practice* (pp. 55–67). Kluwer, Dordrecht.

Devlin, R. A., & Grafton, R. Q. (1998). *Economic Rights and Environmental Wrongs: Property Rights for the Common Good*. Edward Elgar, Cheltenham.

Dietz, F. J., & Vollebergh, H. R. J. (1999). Explaining instrument choice in environmental policies. In: J. C. J. M. van den Bergh (Ed.), *Handbook of Environmental and Resource Economics* (pp. 339–351). Edward Elgar, Cheltenham.

Dijkstra, B. R. (1998). *The Political Economy of Instrument Choice in Environmental Policy*, Dissertation. University of Groningen, Groningen.

Dijkstra, B. R. (1999). *The Political Economy of Environmental Policy: A Public Choice Approach to Market Instruments*. Edward Elgar, Cheltenham.

Dixit, A. K. (1996). *The Making of Economic Policy: A Transaction-Cost Politics Perspective*. MIT Press, Cambridge.

Drexhage, J. (2001). *Proposal for a Directive of the European Parliament and of the Council: Establishing a Framework for Greenhouse Gas Emissions Trading Within the European Community — An Analysis of Some Salient Elements*, Note for the Domestic Emissions Trading Working Group. International Institute for Sustainable Development (IISD).

Dudek, D. J., & Wiener, J. B. (1996). *Joint Implementation and Transaction Costs Under the Climate Change Convention*, Restricted Discussion Document ENV/EPOC/GEEI(96)1. Organization for Economic Co-operation and Development (OECD), Paris.

Dupont, C. (1994). Domestic politics and international negotiations: a sequential bargaining model. In: P. Allan, & C. Schmidt (Eds), *Game Theory and International Relations: Preferences, Information and Empirical Evidence*. Edward Elgar, Aldershot.

Dutschke, M., & Michaelowa, A. (1999). Creation and sharing of credits through the clean development mechanism under the Kyoto Protocol. In: C. J. Jepma, & W. P. van der Gaast (Eds), *On the Compatibility of Flexible Instruments* (pp. 47–64). Kluwer, Dordrecht.

EC. (1998). *Emissions Trading, Joint Implementation and Climate Policy*, European Commission, Internal Note (Draft, 27 March). European Commission, Brussels.

ECCP. (2001). *European Climate Change Programme — Report June 2001*. European Commission, Brussels/Luxembourg.

Eckersley, R. (1993). Free market environmentalism: friend or foe? *Environmental Politics*, **2**, 1, 1–19.

EcoSecurities. (2000). *Financing and Financing Mechanisms for Joint Implementation (JI) Projects in the Electricity Sector*. EcoSecurities Ltd (Environmental Finance Solutions), Oxford, November 2000.

Ellerman, A. D. (1998). *Obstacles to Global CO_2 Trading: A Familiar Problem*. Massachusetts Institute of Technology (MIT), Cambridge, MA.

Ellerman, A. D. (2000). Supplementarity: an invitation to monopsony? *Energy Journal*, **21**, 4, 1–23.

Ellerman, A. D., Schmalensee, R., Joskow, P. L., Montero, J. P., & Bailey, E. M. (1997). *Emissions Trading Under the U.S. Acid Rain Program: Evaluation of Compliance Costs and Allowance Market Performance*. Massachusetts Institute of Technology (MIT), Cambridge, MA.

Ellis, J. (1999a). *Experience with Emission Baselines Under the AIJ Pilot Phase*, OECD Information Paper 78456. Organization for Economic Co-operation and Development (OECD), Paris.

Ellis, J. (1999b). *Experience with Emission Baselines Under the AIJ Pilot Phase*, OECD Information Paper ENV/EPOC(99)23/Final. Organization for Economic Co-operation and Development (OECD), Paris.

Ellis, J., & Bosi, M. (1999). *Options for Project Emission Baselines*, OECD and IEA Information Paper, October 1999. Organization for Economic Co-operation and Development (OECD)/International Energy Agency (IEA), Paris.

ENB. (1999). Technical workshop on mechanisms under Articles 6, 12 and 17 of the Kyoto Protocol: 9–15 April 1999. *Earth Negotiations Bulletin*, **12**, 98, 1–27.

ENB. (2000). 'The Kyoto Protocol and the WTO', special report on selected side events at UNFCCC COP-6 (21 November). *Earth Negotiations Bulletin*, **7**, 1.

Endres, A. (1999). Assessing the different instruments in climate change mitigation from the perspective of economics. In: J. Hacker, & A. Pelchen (Eds), *Goals and Economic Instruments for the Achievement of Global Warming Mitigation in Europe* (pp. 409–425). Kluwer, Dordrecht.

EP. (2002). *Draft Report on the Proposal for a European Parliament and Council Directive on Establishing a Scheme for Greenhouse Gas Emission Allowance Trading within the Community and Amending Council Directive 96/61/EC*, Provisional Document 2001/0245(COD), PR/385503EN.doc, Committee on the Environment, Public Health and Consumer Policy (Rapporteur Jorge Moreira da Silva), 8 April. European Parliament, Brussels.

Estache, A., & Martimort, D. (1999). *Politics, Transaction Costs and the Design of Regulatory Institutions*, Policy Research Working Paper 2073. World Bank, Washington.

EU. (1997). *EU Comments on Framework Compilation of Proposals from Parties for the Elements of a Protocol or Another Legal Instrument, First Draft*. European Union (EU), Brussels.

EU. (1998a). *Non-paper on Joint Implementation*, Preliminary Views of the EU and Its Member States Plus Switzerland, Poland, Czech Republic, Hungary, Slovakia, Croatia, Latvia, Slovenia and Bulgaria, 11 June 1998. Subsidiary Body for Scientific and Technological Advice (SBSTA), Bonn.

EU. (1998b). *Non-paper on Principles, Modalities, Rules and Guidelines for an International Emissions Trading Regime*, Non-paper (Preliminary Views) Submitted by the United Kingdom on Behalf of the EU and Its Member States Plus Switzerland, Poland, Czech Republic, Hungary, Slovakia, Croatia, Latvia, Slovenia and Bulgaria, 12 June 1998, FCCC/SB/1998/MISC.1/Add.3/Rev.1. Subsidiary Body for Scientific and Technological Advice (SBSTA), Bonn.

EU Council. (1997). *EU Conclusions on Climate Change as Agreed by the 1990th Meeting of the EU Council of Ministers (Environment)*, 3 March 1997, http://www.ji.org/usiji/events.1997.

EU Council. (1999). *Council Conclusions on a Community Strategy on Climate Change*, Draft 12 May 1999, Document 8226/99 (Limited). European Union (EU), Brussels.

EU Council. (2002). *Amended Proposal for a Directive of the European Parliament and of the Council Establishing a Framework for Greenhouse Gas Emission Allowance Trading within the Community and Amending Council Directive 96/61/EC — Political Agreement*, Document 2001/0245 (COD), 11 December. Council of the European Union, Brussels.

Eyckmans, J., & Cornillie, J. (2000). *Efficiency and Equity in the EU Bubble Agreement*, Paper Presented at the Conference on Instruments for Climate Policy: Limited versus Unlimited Flexibility? 19–20 October. University of Gent (Belgium), Gent.

EZ. (2002). *Allocation of CO_2 Emission Allowances: Distribution of Emission Allowances in a European Emissions Trading Scheme*, The Netherlands Ministry of Economic Affairs/Ministerie van Economische Zaken (EZ). EZ/KPMG/ECOFYS, The Hague, 8 October.

Faure, M. (1998). Harmonisation of environmental law and market integration: harmonising for the wrong reasons? *European Environmental Law Review*, June, 169–175.

Faure, M. (2001). Regulatory competition versus harmonization in EU environmental law. In: D. C. Esty, & D. Geradin (Eds), *Regulatory Competition and Economic Integration — Comparative Perspectives* (pp. 263–286). Oxford University Press, Oxford.

Faure, M., Gupta, J., & Nentjes, A. (2003). Key instrumental and institutional design issues in climate change policy. In: M. Faure, J. Gupta, & A. Nentjes (Eds), *Climate Change and the Kyoto Protocol: The Role of Institutions and Instruments to Control Global Change* (pp. 3–24). Edward Elgar, Cheltenham.

Fermann, G. (1997). The requirement of political legitimacy: burden sharing criteria and competing conceptions of responsibility. In: G. Fermann (Ed.), *International Politics of Climate Change: Key Issues and Critical Actors* (pp. 179–192). Scandinavian University Press, Oslo.

Fichtner, W., Göbelt, M., Wietschel, M., & Rentz, O. (1999). *Suitable Project Types for Flexible Instruments*, Presentation at the Workshop on Project Types for Flexible Instruments — The Situation After Buenos Aires, 28–29 January. Institute for Industrial Production (IIP)/University of Karlsruhe (TH), Karlsruhe.

Fichtner, W., Graehl, S., & Rentz, O. (2003). The impact of private investor's transaction costs on the cost effectiveness of project-based Kyoto Mechanisms. *Climate Policy*, **3**, 249–259.

Field, A. J. (2000). New economic history and law and economics. In: B. Bouckaert, & G. de Geest (Eds), *Encyclopedia of Law and Economics* (pp. 728–749). Edward Elgar, Cheltenham.

Fisher, R. A. (1958). *The Genetical Theory of Natural Selection*. Dover Publications, New York.

Fisher, B. S., et al. (1996). An economic assessment of policy instruments for combatting climate change. In: J. P. Bruce, H. Lee, & E. F. Haites (Eds), *Climate Change 1995: Economic and Social Dimensions of Climate Change* (Contribution of Working Group III to the Second Assessment Report of the IPCC, pp. 397–439). University Press, Cambridge.

Folketinget. (1999). *Bill on CO_2 Quotas for Electricity Production*, Act No. 376 of 2 June 1999, Unauthorised Translation of 30 June 1999. Folketinget (Danish Parliament), Denmark.

Foxon, T. J. (2002). *Technological and Institutional Lock-In as a Barrier to Sustainable Innovation*, Working Paper, November 2002. Imperial College Centre for Energy Policy and Technology (ICCEPT), London.

Furubotn, E. G., & Richter, R. (1997). *Institutions and Economic Theory: The Contribution of the New Institutional Economics*. The University of Michigan Press, Ann Arbor.

Gerlagh, R., & Hofkes, M. W. (2002). *Escaping Lock-In: The Scope for a Transition Towards Sustainable Economic Growth?*, Nota di Lavoro 12.2002. Fondazione Eni Enrico Mattei (FEEM), Milan.

GGET. (2001). Editorial. *Global Greenhouse Emissions Trader*, **10**, 1, May 2001.

Ghosh, P. (1999). The Asian Development Bank and capacity building for the CDM. *Joint Implementation Quarterly*, **5**, 1, 3–4.

Goldberg, D. M., Porter, S., LaCasta, N., & Hillman, E. (1998). *Responsibility for Non-compliance Under the Kyoto Protocol's Mechanisms for Cooperative Implementation, Draft*. Center for International Environmental Law/Euronatura, Washington.

Golub, J. (1998). Global competition and EU environmental policy: introduction and overview. In: J. Golub (Ed.), *Global Competition and EU Environmental Policy* (pp. 1–33). Routledge, London.

Goodin, R. E., & Klingemann, H. D. (1996). Political science: the discipline. In: R. E. Goodin, & H. D. Klingemann (Eds), *A New Handbook of Political Science* (pp. 3–49). Oxford University Press, Oxford.

Gosseries, A. P. (1999). The legal architecture of joint implementation: what do we learn from the pilot phase? *N.Y.U. Environmental Law Journal*, **7**, 51–117.

Goulder, L. H., Parry, I. W. H., Williams, R. C. III, & Burtraw, D. (1999). The cost-effectiveness of alternative instruments for environmental protection in a second-best setting. *Journal of Public Economics*, **72**, 3, 329–360.

Grafton, R. Q., & Devlin, R. A. (1996). Paying for pollution: permits and charges. *Scandinavian Journal of Economics*, **98**, 2, 275–288.

Grimeaud, D. J. E. (2001). *An Overview of the Policy and Legal Aspects of the International Climate Change Regime (Part I)*, METRO Transnational Legal Research Institute. Maastricht University, Maastricht.

Groenenberg, H., & Blok, K. (2002). Benchmark-based emission allocation in a cap-and-trade system. *Climate Policy*, **2**, 105–109.

Groenewegen, J. (1996). Transaction cost economics and beyond: why and how? In: J. Groenewegen (Ed.), *Transaction Cost Economics and Beyond* (pp. 1–10). Kluwer, Dordrecht.

Groenewegen, J., & Vromen, J. (1997). Theory of the firm revisited: new and neo-institutional perspectives. In: L. Magnusson, & J. Ottosson (Eds), *Evolutionary Economics and Path Dependence* (pp. 33–56). Edward Elgar, Cheltenham.

Grubb, M. (2003). Russian Kyoto delay not to affect the EU ETS. *Carbon Market Europe*, p. 2, October 3.

Gupta, J., Ringius, L., & Ott, H. (2000). *EU Leadership: Between Ambition and Reality*, Paper Presented at the Second EFIEA Climate Policy Workshop 'From Kyoto to The Hague — European Perspectives on Making the Kyoto Protocol Work', Draft, 14 April 2000.

Gusbin, D., Klaassen, G., & Kouvaritakis, N. (1999). Costs of a ceiling on Kyoto flexibility. *Energy Policy*, **27**, 14, 833–844.

Haddad, B. M., & Palmisano, J. (2001). Market Darwinism versus market creationism: adaptability and fairness in the design of greenhouse gas trading mechanisms. *International Environmental Agreements: Politics, Law and Economics*, **1**, 427–446.

Hahn, R. W., & Hester, G. L. (1989). Marketable permits: lessons for theory and practice. *Ecology Law Quarterly*, **16**, 361–406.

Hahn, R. W., & Stavins, R. N. (1999). *What has Kyoto Wrought? The Real Architecture of International Tradable Permit Markets*, Discussion Paper 99-30. Resources for the Future (RFF), Washington.

Haites, E. F. (1997). *Briefing Document on Emissions Trading for Greenhouse Gases*, Prepared for the Economic Instruments Committee of the National Round Table on the Environment and Economy. Margaree Consultants, Toronto.

Haites, E. F. (1998). *Estimate of the Potential Market for Cooperative Mechanisms 2010*. Margaree Consultants, Toronto, September 11.

Haites, E. F. (2000). *Institutional Features of the Kyoto Mechanisms and the COP-6 Decisions*, Paper Presented at the Quantifying Kyoto Workshop, August 30–31. Royal Institute of International Affairs (RIIA), London.

Haites, E., Skjelvik, J., Harrison, D., Ward, M., Zarganis, N., & Yamaguchi, K. (2000). Emissions reduction policies and measures in Annex I countries. In: R. J. Kopp, & J. B. Thatcher (Eds), *The Weathervane Guide to Climate Policy: An RFF Reader* (pp. 83–101). Resources for the Future (RFF), Washington.

Hamilton, C. (1998). *The Evolution of the Global Market for Greenhouse Gas Emissions Allowances*, The Australia Institute, Background Paper No. 16, Lyneham, Australia.

Hamwey, R., & Baranzini, A. (1999). Sizing the global GHG offset market. *Energy Policy*, **27**, 123–127.

Hanks, J., Schipper, L., Sell, M., Spence, C., & Voinov, J. (2001a). Summary of the resumed sixth session of the conference of the parties to the framework convention on climate change: 16–27 July 2001, CoP-6.bis Final. *Earth Negotiations Bulletin*, **12**, 176, 1–15.

Hanks, J., Schipper, L., Sell, M., Spence, C., & Voinov, J. (2001b). UNFCCC COP-6 Part II highlights Friday, 20 July 2001, CoP-6.bis #6. *Earth Negotiations Bulletin*, **12**, 171, 1–2.

Hardin, R. (1982). *Collective Action*. Johns Hopkins University Press, Baltimore, MD.

Hargrave, T. (1998). *US Carbon Emissions Trading: Description of an Upstream Approach*. Center for Clean Air Policy (CCAP), Washington.

Hargrave, T. (1999). *Identifying the Proper Incidence of Regulation in a European Union Greenhouse Gas Emissions Allowance Trading System*, Draft June 1999. Center for Clean Air Policy (CCAP), Washington.

Hargrave, T. (2000). *An Upstream/Downstream Hybrid Approach to Greenhouse Gas Emissions Trading*. Center for Clean Air Policy (CCAP), Washington.

Hargrave, T., Helme, N., & Puhl, I. (1999a). *Options for Simplifying Baseline Setting for Joint Implementation and Clean Development Mechanism Projects*, Paper Presented at the CDM Workshop on Baseline for CDM, 25–26 February, Tokyo, Japan.

Hargrave, T., Helme, N., Denne, T., Kerr, S., & Lefevere, J. (1999b). *Design of a Practical Approach to Greenhouse Gas Emissions Trading Combined with Policies and Measures in the EC*. Center for Clean Air Policy (CCAP), Washington.

Harrison, D., & Radov, D. B. (2002). *Evaluation of Alternative Initial Allocation Mechanisms in a European Union Greenhouse Gas Emissions Allowance Trading Scheme*. National Economic Research Associates (NERA).

Hathaway, O. A. (2003). *Path Dependence in the Law: The Course and Pattern of Legal Change in a Common Law System*, Working Paper 270. Yale Law School.

Heister, J., Michaelis, P., & Mohr, E. (1992). *The Use of Tradable Emission Permits for Limiting CO_2 Emissions*, European Economy: The Economics of Limiting CO_2 Emissions, Special Edition No. 1. Commission of the European Communities, Brussels/Luxembourg, pp. 27–61.

Heller, T. (1998). The path to EU climate change policy. In: J. Golub (Ed.), *Global Competition and EU Environmental Policy* (pp. 108–141). Routledge, London.

Heller, T. C. (1999). Additionality, transactional barriers and the political economy of climate change. In: C. J. Jepma, & W. P. van der Gaast (Eds), *On the Compatibility of Flexible Instruments* (pp. 77–89). Kluwer, Dordrecht.

Helpman, E., & Krugman, P. R. (1989). *Trade Policy and Market Structure*. MIT Press, London.

Heyes, A., & Dijkstra, B. (1999). *The Political Economy of the Environment*, Draft Paper, 17 October 1999. University of London/Royal Holloway College (Department of Economics), Egham.

Hildebrand, D. (1998). *The Role of Economic Analysis in the EC Competition Rules*. Kluwer Law International, The Hague.

Hinchy, M., Fisher, B. S., & Graham, B. (1998). *Emissions Trading in Australia*, ABARE Research Report 98.1. Australian Bureau of Agricultural and Resource Economics (ABARE), Canberra.

Hoel, M. (1997). Coordination of environmental policy for transboundary environmental problems? *Journal of Public Economics*, **66**, 199–224.

Holtsmark, B. J., & Alfsen, K. H. (1998). *Coordination of Flexible Instruments in Climate Policy*, CICERO Report 1998:4, Center for International Climate and Environmental Research (CICERO). University of Oslo, Oslo.

Houlder, V. (2003). EU paves way for emissions trading. *Financial Times*, June 26.

Hourcade, J. C., & Le Pesant, T. (1999). Negotiating targets, negotiating flex-mex: the economic background of a US–EU controversy. In: C. Carraro (Ed.), *Efficiency and Equity of Climate Change Policy, Draft Version* (pp. 1–14). Kluwer/Fondazione Eni Enrico Mattei (FEEM), Dordrecht/Milan.

Hourcade, J. C., et al. (1996). Estimating the costs of mitigating greenhouse gases. In: J. P. Bruce, H. Lee, & E. F. Haites (Eds), *Climate Change 1995: Economic and Social Dimensions of Climate Change* (Contribution of Working Group III to the Second Assessment Report of the IPCC, pp. 263–296). University Press, Cambridge.

Hurrell, A., & Kingsbury, B. (1992). The international politics of the environment: an introduction. In: A. Hurrell, & B. Kingsbury (Eds), *The International Politics of the Environment* (pp. 1–50). Clarendon Press, Oxford.

IEA. (1999). *CO_2 Emissions from Fuel Combustion 1971–1997*. International Energy Agency (IEA), Paris.

IEA. (2002). *Dealing with Climate Change: Policies and Measures in IEA Member Countries (2002 Edition)*. International Energy Agency (IEA), Paris.

Ingham, A. (1992). The market for sulphur dioxide permits in the USA and UK. *Environmental Politics*, **1**, 1, 98–122.

Jackson, T. (1995). Joint implementation and cost-effectiveness under the framework convention on climate change. *Energy Policy*, **23**, 2, 117–138.

Jäger, J., & O'Riordan, T. (1996). History of climate change science and politics. In: T. O'Riordan, & J. Jäger (Eds), *Politics of Climate Change: A European Perspective* (pp. 1–31). Routledge, London.

Jans, J. H. (1995). *European Environmental Law*. Kluwer Law International/Wolters-Noordhoff, Den Haag/Groningen.

Janssen, J. (2000). *Will Joint Implementation Survive International Emissions Trading? Distinguishing the Kyoto Mechanisms*, FEEM Working Papers, Nota di Lavoro 60.00. Fondazione Eni Enrico Mattei (FEEM), Milan.

Jensen, J., & Rasmussen, T. N. (1998). *Allocation of CO_2 Emission Permits: A General Equilibrium Analysis of Policy Instruments*. Danish Ministry of Business and Industry, Copenhagen, Denmark.

Jepma, C. J. (1999a). Editor's note on standardization of baselines. *Joint Implementation Quarterly*, **5**, 1, 1 p.

Jepma, C. J. (1999b). *Determining a Baseline for Project Co-operation Under the Kyoto Protocol: A General Overview*, Paper Presented at the CDM Workshop on Baseline for CDM, 25–26 February, Tokyo, Japan.

Jepma, C. J. (2002). Hurdles of EU emissions trading (editor's note). *Joint Implementation Quarterly*, **8**, 1, 1 p.

Jepma, C. J., & Munasinghe, M. (1998). *Climate Change Policy: Facts, Issues and Analyses*. Cambridge University Press, Cambridge.

Jepma, C. J., & van der Gaast, W. P. (1999). 'Flexible instruments' carbon credits after Kyoto. In: C. J. Jepma, & W. P. van der Gaast (Eds), *On the Compatibility of Flexible Instruments* (pp. 3–15). Kluwer, Dordrecht.

Jepma, C. J., van der Gaast, W. P., & Woerdman, E. (1998). *The Compatibility of Flexible Instruments Under the Kyoto Protocol*, Dutch National Research Programme on Global Air Pollution and Climate Change (NRP), NRP Report No. 410 200 026. Joint Implementation Network (JIN)/NRP, Groningen/Bilthoven.

Jepma, C. J., van der Gaast, W. P., Schrijver, N., & Giesberger, D. (1999). *The Role of the Private Sector in Joint Implementation with Central and Eastern European Countries: An Interpretation of Article 6 of the Kyoto Protocol.* Joint Implementation Network (JIN), Groningen.

JIQ. (1996). First results of Nordic JI study project. *Joint Implementation Quarterly,* **2**, 4, 4–5.

JIQ. (1997). Nordic study on 10 projects in Eastern Europe. *Joint Implementation Quarterly,* **3**, 4, 10–11.

JIQ. (1999). World Bank Executive Board approves Prototype Carbon Fund. *Joint Implementation Quarterly,* **5**, 3, 5 p.

JIQ. (2000a). GHG abatement in the CDM via forestry: a case study and modernising heat supply through AIJ. *Joint Implementation Quarterly,* **6**, 1, 8–9.

JIQ. (2000b). *Joint Implementation Quarterly,* Special Issue for the Lyon Preparatory Conference for CoP6 (11–15 September), August 2000. Joint Implementation Network (JIN), Paterswolde.

JIQ. (2001a). The national strategy study: lessons for CDM and JI policy design. *Joint Implementation Quarterly,* **7**, 1, 6–7.

JIQ. (2001b). Early lessons from Prototype Carbon Fund. *Joint Implementation Quarterly,* **7**, 1, 4 p.

Jones, T. (1993). *Operational Criteria for Joint Implementation,* OCDE/GD(93)88. Organization for Economic Co-operation and Development (OECD), Paris.

Joskow, P. L., Schmalensee, R., & Bailey, E. M. (1998). The market for sulfur dioxide emissions. *The American Economic Review,* **88**, 4, 669–685.

Karani, P. (1997). *Constraints for Activities Implemented Jointly (AIJ) Technology Transfer in Africa.* World Bank, Washington.

Katz, M., & Shapiro, C. (1985). Network externalities, competition and compatibility. *American Economic Review,* **75**, 3, 424–440.

Kemp, R. P. M. (1995). *Environmental Policy and Technical Change: A Comparison of the Technological Impact of Policy Instruments,* Dissertation. University of Groningen, Groningen.

Kerr, S. (1998). *Enforcing Compliance: The Allocation of Liability in International GHG Emissions Trading and the Clean Development Mechanism,* Resources For the Future (RFF) Climate Issue Brief 15 (Internet Edition).

Kerr, S. (1999). *Allocation of Greenhouse Gas Reduction Responsibilities Among and Within the Countries of the European Union,* Scoping Paper No. 5, Study for the European Commission DG XI, Designing Options for Implementing an Emissions Trading Regime for Greenhouse Gases in the EC. Center for Clean Air Policy (CCAP), Washington.

Kerr, S., & Maré, D. (1997). *Transaction Costs and Tradeable Permit Markets: The United States Lead Phasedown,* Paper Prepared for the EAERE Conference, 26–28 June, Tilburg, The Netherlands.

Kessler, E. (1998). Russia's protected areas in transition: the impacts of perestroika, economic reform and the move towards democracy. *Ambio: A Journal of the Human Environment,* **3**, 1, Internet Edition.

Kim, J. A. (2000). *Potential Limits Imposed by the Multilateral Trading System in Implementing Flexibility Mechanisms,* Paper Presented at the Conference on Instruments for Climate Policy: Limited versus Unlimited Flexibility? 19–20 October, Centre for

Environmental Economics and Environmental Management (CEEM). University of Gent (Belgium), Gent.

Klaassen, G., & Nentjes, A. (1997). Sulfur trading under the 1990 CAAA in the US: an assessment of the first experiences. *Journal of Institutional and Theoretical Economics*, **153**, 2, 384–410.

Klaassen, G., & Nentjes, A. (2002). *On the Quality of Compliance Mechanisms in the Kyoto Protocol, Draft Version*. University of Groningen/International Institute for Applied Systems Analysis (IIASA), Groningen/Laxenburg.

Klein, P. G. (2000). New institutional economics. In: B. Bouckaert, & G. de Geest (Eds), *Encyclopedia of Law and Economics* (pp. 456–489). Edward Elgar, Cheltenham.

Kok, M. T. J., & Verweij, W. (1999). *The Kyoto Protocol: Implications for Research*, NRP Report No. 410 200 018 (1999). Dutch National Research Programme on Global Air Pollution and Climate Change (NRP), Bilthoven.

Kopp, R. J. (2001). An analysis of the Bonn Agreement. *Weathervane*, **134**, 1–3.

Kopp, R. J., & Toman, M. A. (2000). International emissions trading: a primer. In: R. J. Kopp, & J. B. Thatcher (Eds), *The Weathervane Guide to Climate Policy: An RFF Reader* (pp. 65–70). Resources for the Future (RFF), Washington.

Koutstaal, P. R. (1997). *Economic Policy and Climate Change: Tradable Permits for Reducing Carbon Emissions*. Edward Elgar, Cheltenham.

Koutstaal, P., & Nentjes, A. (1995). Tradable carbon permits in Europe: feasibility and comparison with taxes. *Journal of Common Market Studies*, **33**, 2, 219–233.

Kram, T., & Hill, D. (1996). A multinational model for CO_2 reduction: defining boundaries of future CO_2 emissions in nine countries. *Energy Policy*, **24**, 1, 39–51.

Krutilla, K. (1999). Environmental policy and transactions costs. In: J. C. J. M. van den Bergh (Ed.), *Handbook of Environmental and Resource Economics* (pp. 249–264). Edward Elgar, Cheltenham.

Kuik, O., & Gupta, J. (1996). Perspectives on Africa and the global debate on joint implementation. In: R. S. Maya, & J. Gupta (Eds), *Joint Implementation: Carbon Colonies or Business Opportunities? Weighing the Odds in an Information Vacuum*. Southern Centre for Energy & Environment, Harare.

Kyoto Protocol. (1997). *Kyoto Protocol to the United Nations Framework Convention on Climate Change*, FCCC/CP/1997/L.7/Add.1, 10 December 1997, CoP3. FCCC, Kyoto.

Langlois, R. (1994). The "new" institutional economics. In: P. J. Boettke (Ed.), *The Elgar Companion to Austrian Economics* (pp. 535–540). Edward Elgar, Cheltenham.

Lefevere, J., & Yamin, F. (1999). *EC Trade and Competition Law Issues Raised by the Design of an EC Emissions Trading System*, Scoping Paper No. 1, Study for the European Commission DG XI, Designing Options for Implementing an Emissions Trading Regime for Greenhouse Gases in the EC. Foundation for International Environmental Law and Development (FIELD), London.

Licht, A. N. (2001). The mother of all path dependencies: toward a cross-cultural theory of corporate governance systems. *Delaware Journal of Corporate Law*, **26**, 147–205.

Liebowitz, S. J., & Margolis, S. E. (1995). Path dependence, lock-in and history. *The Journal of Law, Economics and Organization*, **11**, 1, 205–226.

Liebowitz, S. J., & Margolis, S. E. (2000). Path dependence. In: B. Bouckaert, & G. de Geest (Eds), *Encyclopedia of Law and Economics* (pp. 981–998). Edward Elgar, Cheltenham.

Lindblom, C. E. (1959). The science of "muddling through". *Public Administration Review*, **39**, 79–88.
Liski, M. (1999). *Thin Versus Thick CO_2 Markets*, Working Paper. Helsinki School of Economics, Helsinki.
Luhmann, H.-J., Beuerman, C., Fischedick, M., & Ott, H. (1995). *Making Joint Implementation Operational: Solutions for Some Technical and Operational Problems of JI in the Fossil Fuel Power Sector*, Wuppertal Paper No. 31. Wuppertal Institute, Wuppertal.
Lyon, R. M. (1982). Auctions and alternative procedures for allocating pollution rights. *Land Economics*, **58**, 1, 16–32.
Mackaay, E. (2000). History of law and economics. In: B. Bouckaert, & G. de Geest (Eds), *Encyclopedia of Law and Economics* (pp. 65–94). Edward Elgar, Cheltenham.
Magnusson, L., & Ottosson, J. (1996). Transaction costs and institutional change. In: J. Groenewegen (Ed.), *Transaction Cost Economics and Beyond* (pp. 351–364). Kluwer, Dordrecht.
Magnusson, L., & Ottosson, J. (1997). Introduction. In: L. Magnusson, & J. Ottosson (Eds), *Evolutionary Economics and Path Dependence* (pp. 1–9). Edward Elgar, Cheltenham.
Mahoney, J. (2000). Path dependence in historical sociology. *Theory and Society*, **29**, 507–548.
Malueg, D. A. (1990). Welfare consequences of emission credit trading programs. *Journal of Environmental Economics and Management*, **18**, 66–77.
Mariñoso, B. G. (2001). Technological incompatibility, endogenous switching costs and lock-in. *The Journal of Industrial Economics*, **3**, 281–295.
Masten, S. E. (1996). Empirical research in transaction cost economics: challenges, progress, directions. In: J. Groenewegen (Ed.), *Transaction Cost Economics and Beyond* (pp. 43–64). Kluwer, Dordrecht.
Matsuo, N. (1999). *Baseline as the Critical Issue of CDM: Possible Pathway to Standardization*, Paper Presented at the CDM Workshop on Baseline for CDM, 25–26 February, Tokyo, Japan.
Mauch, S. P., von Stokar, T., & North, N. (1999). Meeting the Kyoto commitments using JI, CDM and IET: opportunities, risks and constraints at the practical implementation level. In: C. J. Jepma, & W. P. van der Gaast (Eds), *On the Compatibility of Flexible Instruments* (pp. 143–150). Kluwer, Dordrecht.
McNutt, P. A. (1996). *The Economics of Public Choice*. Edward Elgar, Cheltenham.
Medema, S. G., Mercuro, N., & Samuels, W. J. (2000). Institutional law and economics. In: B. Bouckaert, & G. de Geest (Eds), *Encyclopedia of Law and Economics* (pp. 418–450). Edward Elgar, Cheltenham.
Metz, B., Berk, M., Kok, M. T. J., van Minnen, J. G., de Moor, A., & Faber, A. (2001). How can the European Union contribute to a COP-6 agreement? An overview for policy makers. *International Environmental Agreements: Politics, Law and Economics*, **1**, 167–185.
Michaelowa, A. (1995). *Incentive Aspects of Activities Implemented Jointly*, Paper Presented at the UNEP Workshop on Activities Implemented Jointly in the European Region, June 21–23.
Michaelowa, A. (1998a). Joint implementation — the baseline issue: economic and political aspects. *Global Environmental Change*, **8**, 1, 81–92.
Michaelowa, A. (1998b). *Actors in International Greenhouse Gas Emissions Trading and the Starting Point of Trading*, Working Paper, St.-Cloud, France.
Michaelowa, A. (1999). *Clean Development Mechanism and Joint Implementation — Which Instrument is Likely to have a Higher Impact?*, Paper Presented at the IIP/TH Workshop on

Project Types for Flexible Instruments — The Situation After Buenos Aires, January 28–29. Institute for Industrial Production (IIP)/University of Karlsruhe (TH), Karlsruhe.

Michaelowa, A., & Dutschke, M. (1998). Interest groups and efficient design of the clean development mechanism under the Kyoto Protocol. *International Journal of Sustainable Development*, **1**, 1, 24–42.

Michaelowa, A., & Koch, T. (1999a). *Critical Issues in Current Climate Policy*, HWWA-Report No. 194. Institute for Economic Research (HWWA), Hamburg.

Michaelowa, A., & Koch, T. (1999b). *An International Registration and Tracking System for Greenhouse Gas Emission Trade: Elements, Possibilities, Problems and Issues for Further Discussion*, Working Paper, St.-Cloud, France.

Michaelowa, A., & Stronzik, M. (2002). *Transaction Costs of the Kyoto Mechanisms*, HWWA Discussion Paper. Hamburg Institute of International Economics (HWWA), Hamburg.

Michaelowa, A., Michaelowa, K., & Vaughan, S. (1998). Joint implementation and trade policy. *Aussenwirtschaft*, **53**, 4, 573–589.

Michaelowa, A., Dutschke, M., & Stronzik, M. (1999). *Convergence Criteria for Participation in the Flexible Mechanisms of the Kyoto Protocol*, HWWA Discussion Paper No. 82. Hamburg Institute of International Economics (HWWA), Hamburg.

Montgomery, W. D. (1972). Markets in licences and efficient pollution control programs. *Journal of Economic Theory*, **5**, 3, 395–418.

Morlot, J. C. (1998). *Ensuring Compliance with a Global Climate Change Agreement*, ENV/EPOC(98)5/REV1. Organization for Economic Co-operation and Development (OECD), Paris.

Mullins, F., & Baron, R. (1997). *International GHG Emission Trading, Policies and Measures for Common Action*, Working Paper 9, March 1997. Organization for Economic Co-operation and Development (OECD)/International Energy Agency (IEA), Paris.

Nash, J. R. (2000). Too much market? Conflict between tradable pollution allowances and the "polluter pays" principle. *Harvard Environmental Law Review*, **465**, 1–59.

Nelson, R. R., & Sampat, B. N. (2001). Making sense of institutions as a factor shaping economic performance. *Journal of Economic Behavior & Organization*, **44**, 31–54.

Nentjes, A. (1998). Two views of emissions trading. *Change: Research and Policy Newsletter on Global Change from the Netherlands*, **44**, 4–7.

Nentjes, A. (2000). *Eigen verantwoordelijkheid en beslissingsruimte voor de energie-intensieve sectoren in het internationale klimaatbeleid: convenanten en handelsystemen*, Paper Presented at the RMNO Workshop, 13 October, The Hague, The Netherlands, in Dutch.

Nentjes, A. (2001). *Emissions Trading in the Netherlands*, Draft Paper Presented at the CATEP Meeting, Paris, 14 May 2001.

Nentjes, A., & Dijkstra, B. R. (1994). The political economy of instrument choice in environmental policy. In: M. Faure, J. Vervaele, & A. Weale (Eds), *Environmental Standards in the European Union in an Interdisciplinary Framework*. Maklu, Antwerpen.

Nentjes, A., & Klaassen, G. (2004). On the quality of compliance mechanisms in the Kyoto Protocol. *Energy Policy*, **32**, 531–544.

Nentjes, A., & Woerdman, E. (2000). *The EU Proposal on Supplementarity in International Climate Change Negotiations: Assessment and Alternative*, ECOF Research Memorandum 2000 (28). University of Groningen, Groningen.

Nentjes, A., Koutstaal, P., & Klaassen, G. (1995). *Tradeable Carbon Permits: Feasibility, Experiences, Bottlenecks*, Dutch National Research Programme on Global Air Pollution and Climate Change (NRP), NRP Report No. 410 100 114. RuG/NRP, Groningen/Bilthoven.

Nentjes, A., Boom, J. T., Dijkstra, B. R., Koster, M., Woerdman, E., & Zhang, Z. X. (2002). *National and International Emissions Trading for Greenhouse Gases*, Dutch National Research Programme on Global Air Pollution and Climate Change (NRP), NRP Report No. 410 200 093. NRP, Bilthoven.

Neumayer, E. (2000). In defence of historical accountability for greenhouse gas emissions. *Ecological Economics*, 33, 185–192.

Nicolaides, P., & Bilal, S. (1999). State aid rules: do they promote efficiency? In: S. Bilal, & P. Nicolaides (Eds), *Understanding State Aid Policy in the European Community: Perspectives on Rules and Practice*, European Institute of Public Administration (EIPA). Kluwer Law International, Maastricht.

Nilsson, C., & Huhtala, A. (2000). *Is CO_2 Trading Always Beneficial? A CGE-Model Analysis on Secondary Environmental Benefits*. National Institute of Economic Research, Stockholm.

Nooteboom, B. (2000). *Learning and Innovation in Organizations and Economies*. Oxford University Press, Oxford.

North, D. C. (1990). *Institutions, Institutional Change and Economic Performance*. Cambridge University Press, Cambridge.

North, D. C. (1991). Institutions. *Journal of Economic Perspectives*, 5, 1, 97–112.

North, D. C. (1997). Transaction costs through time. In: C. Menard (Ed.), *Transaction Cost Economics: Recent Developments* (pp. 149–160). Edward Elgar, Cheltenham.

North, D. C., & Thomas, R. P. (1973). *The Rise of the Western World: A New Economic History*. Cambridge University Press, Cambridge.

Oberthür, S., & Ott, H. E. (1999). *The Kyoto Protocol: International Climate Policy for the 21st Century*. Springer, Berlin.

O'Connor, M., Faucheux, S., & van den Hove, S. (1997). *EU Climate Policy: Research Support for Kyoto and Beyond*, Policy/Research Interface Workshop Series: A Synthesis, DG XII RTD/DG XI/C3ED/Université de Versailles.

OECD/IEA. (1996). *Climate Change Policy Initiatives*, 1995/96 Update, Volume II, Selected Non-IEA Countries. Organization for Economic Co-operation and Development (OECD)/ International Energy Agency (IEA), Paris.

OJ. (1994). Community guidelines on state aid for environmental protection. *Official Journal of the European Communities C 72*, 37, 3–9, 10 March 1994.

OJ. (2001). Community guidelines on state aid for environmental protection. *Official Journal of the European Communities C 37*, 03, 3–16, 3 February 2001.

OJ. (2003). Directive 2003/87/EC of the European Parliament and of the Council of 13 October 2003 establishing a scheme for greenhouse gas emission allowance trading within the community and amending council directive 96/61/EC. *Official Journal of the European Union L 275*, 32–46, 25 October 2003.

Olson, M. (1965). *The Logic of Collective Action: Public Goods and the Theory of Groups*. Harvard University Press, Cambridge.

O'Riordan, T., & Jäger, J. (1996). Social institutions and climate change. In: T. O'Riordan, & J. Jäger (Eds), *Politics of Climate Change: A European Perspective* (pp. 65–105). Routledge, London.

Palmisano, J. (1996). *Air Permit Trading Paradigms for Greenhouse Gases: Why Allowances Won't Work and Credits Will*, Discussion Draft (Personal View). Enron Europe Ltd, London.

Parker, R. W. (1998). *Designs for Domestic Carbon Emissions Trading: Comments on WTO Aspects*, Discussion Draft 22 June. The H. John Heinz III Center for Science, Economics and the Environment, Washington, DC.

Paterson, M. (1997). Institutions for global environmental change. *Global Environmental Change*, **7**, 2, 175–177.
Pearce, D. (1995). Joint implementation: a general overview. In: C. J. Jepma (Ed.), *The Feasibility of Joint Implementation* (pp. 15–32). Kluwer, Dordrecht.
Pearce, D. W., et al. (1996). The social costs of climate change: greenhouse damage and the benefits of control. In: J. P. Bruce, H. Lee, & E. F. Haites (Eds), *Climate Change 1995: Economic and Social Dimensions of Climate Change* (Contribution of Working Group III to the Second Assessment Report of the IPCC, pp. 179–224). University Press, Cambridge.
Petsonk, A. (1999). *The Kyoto Protocol and the WTO: Integrating Greenhouse Gas Emissions Allowance Trading Into the Global Marketplace*, Draft 22 November, Relationship Between the Emissions Trading Systems of the Kyoto Protocol and the Rules of the World Trade Organization, Duke Environmental Law and Policy Forum.
Pew Center. (2002). *Pew Center Analysis of President Bush's February 14th Climate Change Plan*. The Pew Center on Global Climate Change, www.pewclimate.org/policy.
Pierson, P. (2000). Increasing returns, path dependence, and the study of politics. *American Political Science Review*, **94**, 2, 251–267.
Pitelis, C. (1993). Transaction costs, markets and hierarchies: the issues. In: C. Pitelis (Ed.), *Transaction Costs, Markets and Hierarchies* (pp. 7–19). Blackwell, Oxford.
Puhl, I. (1998). *Status of Research on Project Baselines Under the UNFCCC and the Kyoto Protocol*, Annex I Expert Group on the UNFCCC, Information Paper (Draft). Organization for Economic Co-operation and Development (OECD), Paris.
Richels, R., Edmonds, J., Gruenspecht, H., & Wigly, T. (1996). *The Berlin Mandate: The Design of Cost-Effective Mitigation Strategies (Draft)*, Energy Modeling Forum-14. Stanford University, Stanford.
Ringius, L. (1999). *The European Community and Climate Protection: What's Behind the "Empty Rhetoric"?*, CICERO Report 1999:8. Center for International Climate and Environmental Research (CICERO), Oslo.
Rizzo, M. J. (1994). Cost. In: P. J. Boettke (Ed.), *The Elgar Companion to Austrian Economics* (pp. 92–95). Edward Elgar, Cheltenham.
Rolfe, C., Michaelowa, A., & Dutschke, M. (1999). *Closing the Gap: A Comparison of Approaches to Encourage Early Greenhouse Gas Emission Reductions*, HWWA Report 199. Institut für Wirtschaftsforschung (HWWA), Hamburg.
Rolph, E. S. (1983). Government allocation of property rights: who gets what? *Journal of Policy Analysis and Management*, **3**, 1, 45–61.
Romstad, E. (1998). Environmental regulation and competitiveness. In: T. Barker, & J. Köhler (Eds), *International Competitiveness and Environmental Policies* (pp. 185–196). Edward Elgar, Cheltenham.
Rose, C. M. (1999). Expanding the choices for the global commons: comparing newfangled tradable allowance schemes to old-fashioned common property regimes. *Duke Environmental Law & Policy Forum*, **10**, 45, 45–72.
Rose, A., & Kverndokk, S. (1999). Equity in environmental policy with an application to global warming. In: J. C. J. M. van den Bergh (Ed.), *Handbook of Environmental and Resource Economics* (pp. 352–379). Edward Elgar, Cheltenham.
Rose, A., & Stevens, B. (1993). The efficiency and equity of marketable permits for CO_2 emissions. *Resource and Energy Economics*, **15**, 117–146.

Rose, A., & Stevens, B. (2001). An economic analysis of flexible permit trading in the Kyoto Protocol. *International Environmental Agreements: Politics, Law and Economics*, **1**, 219–242.

Rose, A., Bulte, E., & Folmer, H. (1999). Long-run implications for developing countries of joint implementation of greenhouse gas mitigation. *Environmental and Resource Economics*, **14**, 19–31.

Rosenzweig, R., Varilek, M., & Janssen, J. (2002). *The Emerging International Greenhouse Gas Market*. Pew Center on Global Climate Change, Arlington.

Rowlands, I. (1998). EU policy for ozone layer protection. In: J. Golub (Ed.), *Global Competition and EU Environmental Policy* (pp. 34–59). Routledge, London.

Ruiter, D. W. P. (2000). *Economic and Legal Institutionalism*, Paper Presented at the NIG Conference "From Government to Governance", 22–23 November. University of Twente, Enschede, The Netherlands.

Russell, C. S., & Powell, P. T. (1999). Practical considerations and comparison of instruments of environmental policy. In: J. C. J. M. van den Bergh (Ed.), *Handbook of Environmental and Resource Economics* (pp. 307–328). Edward Elgar, Cheltenham.

Rutgeerts, A. (1999). Trade and environment: reconciling the Montreal Protocol and the GATT. *Journal of World Trade*, **33**, 4, 61–86.

Rutherford, M. (1994). Austrian economics and American (old) institutionalism. In: P. J. Boettke (Ed.), *The Elgar Companion to Austrian Economics* (pp. 529–534). Edward Elgar, Cheltenham.

Sari, A. P. (1999). On equity and developing country participation. In: C. J. Jepma, & W. P. van der Gaast (Eds), *On the Compatibility of Flexible Instruments* (pp. 125–139). Kluwer, Dordrecht.

SB. (2000). *Activities Implemented Jointly Under the Pilot Phase: Fourth Synthesis Report*, Subsidiary Bodies (SB), Document FCCC/SB/2000/6, 3 August 2000, Lyon, France.

SBSTA/SBI. (1999). *Mechanisms Pursuant to Articles 6, 12 and 17 of the Kyoto Protocol*, Synthesis of Proposals by Parties on Principles, Modalities, Rules and Guidelines, FCCC/SB/1999/8, 28 September 1999. Subsidiary Body for Scientific and Technological Advice (SBSTA)/Subsidiary Body for Implementation (SBI), Bonn.

SBSTA/SBI. (2000). *Mechanisms Pursuant to Articles 6, 12 and 17 of the Kyoto Protocol: Text for Further Negotiation on Principles, Modalities, Rules and Guidelines*, SBSTA/SBI Twelfth Session, FCCC/SB/2000/3, 12–16 June 2000. Subsidiary Body for Scientific and Technological Advice (SBSTA)/Subsidiary Body for Implementation (SBI), Bonn.

Schmalensee, R., Joskow, P. L., Ellerman, A. D., Montero, J. P., & Bailey, E. M. (1998). An interim evaluation of sulfur dioxide emissions trading. *Journal of Economic Perspectives*, **12**, 3, 53–68.

Schwarze, R. (1998). *Activities Implemented Jointly: Another Look at the Facts*. University of Technology, Berlin.

SEVEn/JIN. (1997). *The Experience with Joint Implementation in Central and Eastern Europe during the AIJ Pilot Phase*. Energy Efficiency Center SEVEn/Joint Implementation Network (JIN), Prague/Groningen.

Sharkansky, I. (2002). *Politics and Policymaking: In Search of Simplicity, Explorations in Public Policy Series*. Lynne Rienner Publishers, London.

Shogren, J., & Toman, M. (2000). *Climate Change Policy*, Discussion Paper 00-22, May 2000. Resources for the Future (RFF), Washington.

Sijm, J. P. M., et al. (2000). *Kyoto Mechanisms: The Role of Joint Implementation, the Clean Development Mechanism and Emissions Trading in Reducing Greenhouse Gas Emissions*, Report No. ECN-C-00-026. Energy Research Centre of the Netherlands (ECN), Petten.

Simon, H. A. (1997). Models of Bounded Rationality: Empirically Grounded Economic Reason. MIT Press, Cambridge, vol. 3.

Sokona, Y., & Nanasta, D. (2000). The clean development mechanism: an African delusion? *Change: Research and Policy Newsletter on Global Change from the Netherlands*, **54**, 8–11.

Spencer, B. J. (1990). What should trade policy target? In: P. R. Krugman (Ed.), *Strategic Trade Policy and the New International Economics* (pp. 69–87). MIT Press, London.

Stavins, R. N. (1995). Transaction costs and tradeable permits. *Journal of Environmental Economics and Management*, **29**, 133–148.

Stavins, R. N. (2002). *Lessons from the American Experiment with Market-Based Environmental Policies*, Nota di Lavoro 30.2002. Fondazione Eni Enrico Mattei (FEEM), Milan.

Steenge, A. E. (1997). On background principles for environmental policy: "polluter pays", "user pays" or "victim pays"? In: P. B. Boorsma, K. Aarts, & A. E. Steenge (Eds), *Public Priority Setting: Rules and Costs* (pp. 121–137). Kluwer, Dordrecht.

Stewart, R., et al. (1999). *The Clean Development Mechanism: Building International Public–Private Partnership*, Unedited Draft October 1999, Ad Hoc Working Group on the CDM. United Nations, Geneva.

Sutter, C., Brodmann, U., & Lüchinger, A. (2001). International CO_2 abatement projects: experience from the Swiss AIJ Pilot Program. *Energy & Environment*, **12**, 5/6, 521–529.

Svendsen, G. T. (1998). *Tradable Permit Systems in the United States and CO_2 Taxation in Europe*. Edward Elgar, Cheltenham.

Svendsen, G. T., & Vesterdal, M. (2003). Potential gains from CO_2 trading in the EU. *European Environment*, **13**, 303–313.

Tietenberg, T. (1980). Transferable discharge permits and the control of stationary source air pollution: a survey and synthesis. *Land Economics*, **56**, 4, 391–416.

Tietenberg, T. (1992). *Relevant Experience with Tradeable Entitlements*, Combating Global Warming: Study on a Global System of Tradeable Carbon Emission Entitlements, Chapter IV, UNCTAD. United Nations, New York, pp. 37–54.

Tietenberg, T. (1999). Lessons from using transferable permits to control air pollution in the United States. In: J. C. J. M. van den Bergh (Ed.), *Handbook of Environmental and Resource Economics* (pp. 275–292). Edward Elgar, Cheltenham.

Tietenberg, T. (2002). *The Tradable Permits Approach to Protecting the Commons: What have We Learned?*, Nota di Lavoro 36.2002. Fondazione Eni Enrico Mattei (FEEM), Milan.

Tietenberg, T., & Victor, D. G. (1994). *Possible Administrative Structures and Procedures for Implementing a Tradeable Entitlement Approach to Controlling Global Warming*, Combating Global Warming: Possible Rules, Regulations and Administrative Arrangements for a Global Market in CO_2 Emission Entitlements. United Nations Conference on Trade and Development (UNCTAD), Geneva.

Tietenberg, T., Grubb, M., Michaelowa, A., Swift, B., & Zhang, Z. X. (1999). *International Rules for Greenhouse Gas Emissions Trading: Defining the Principles, Modalities, Rules and Guidelines for Verification, Reporting and Accountability*, UNCTAD/GDS/GFSB/Misc.6. United Nations Conference on Trade and Development (UNCTAD), Geneva.

Trexler, M. C., & Gibbons, R. (1999). *Advancing the Development of Forestry and Land-Use Based Project Baseline Methodologies for the Developing CDM Regime*, Paper Presented at the CDM Workshop on Baseline for CDM, 25–26 February, Tokyo, Japan.

Trexler, M. C., & Kosloff, L. H. (1998). The 1997 Kyoto Protocol: what does it mean for project-based climate change mitigation? *Mitigation and Adaptation Strategies for Global Change*, **3**, 1–58.

Ulph, A. M. (1999). Strategic environmental policy and foreign trade. In: J. C. J. M. van den Bergh (Ed.), *Handbook of Environmental and Resource Economics* (pp. 433–448). Edward Elgar, Cheltenham.

UNCTAD. (1995). *Controlling Carbon Dioxide Emissions: The Tradeable Permit System*. United Nations Conference on Trade and Development (UNCTAD), Geneva.

Unruh, G. C. (2000). Understanding carbon lock-in. *Energy Policy*, **28**, 817–830.

van den Bergh, J. C. J. M., & van Beers, C. (1997). An empirical multi-country analysis of the impact of environmental regulations on foreign trade. *Kyklos*, **50**, 29–46.

van der Laan, R., & Nentjes, A. (2001). Competitive distortions in EU environmental legislation: inefficiency versus inequity. *European Journal of Law and Economics*, **11**, 2, 131–152.

van der Wurff, R. (1997). *International Climate Change Politics: Interests and Perceptions*, Dissertation. University of Amsterdam, Amsterdam.

van Deth, J. W., & Scarbrough, E. (Eds) (1995a). *The Impact of Values*. Oxford University Press, Oxford.

van Deth, J. W., & Scarbrough, E. (1995b). The concept of values. In: J. W. van Deth, & E. Scarbrough (Eds), *The Impact of Values* (pp. 21–47). Oxford University Press, Oxford.

van Heukelen, M. (2000). *Greenhouse Gas Emission Trading within the EU*, Presentation on Behalf of the European Commission DG Environment at the Netherlands Ministry for Economic Affairs, June 30, The Hague, The Netherlands.

Varilek, M., & Marenzi, N. (2001). *Greenhouse Gas Price Scenarios for 2000–2012: Impact of Different Policy Regimes*. University of St. Gallen/Institute for Economy and the Environment (IWOe-HSG), St. Gallen.

Vaughan, S. (1999). Tradeable emissions permits and the WTO. In: J. Hacker, & A. Pelchen (Eds), *Goals and Economic Instruments for the Achievement of Global Warming Mitigation in Europe* (pp. 367–376). Kluwer, Dordrecht.

Vermeulen, W. J. V., & Kok, M. T. J. (2002). Contemporary practices: greenhouse scepticism? In: M. T. J. Kok, W. J. V. Vermeulen, A. P. C. Faaij, & D. de Jager (Eds), *The Climate Neutral Society: Visions, Dilemma's and Opportunities for Change*, Dutch National Research Programme (NRP) on Global Air Pollution and Climate Change (pp. 39–58). Earthscan, London.

Versteege, H. M. J., & Vos, J. B. (1995). Verhandelbare Emissierechten in het Nederlandse Verzuringsbeleid, Publication Series Air and Energy No. 116. VROM Ministry of the Environment, The Hague, in Dutch.

Victor, D. G., Nakicenovic, N., & Victor, N. (1998). *The Kyoto Protocol Carbon Bubble: Implications for Russia, Ukraine and Emission Trading*, Interim Report IR-98-094/October. International Institute for Applied Systems Analysis (IIASA), Laxenburg.

Viguier, L. (2000). *Fair Trade and Harmonization of Climate Change Policies in Europe*, MIT Joint Program on the Science and Policy of Global Change, Report No. 66. Massachusetts Institute of Technology (MIT), Cambridge, MA.

Vikhlyaev, A. (2001). *The Use of Trade Measures for Environmental Purposes — Globally and in the EU Context*, Nota di Lavoro 68.2001. Fondazione Eni Enrico Mattei (FEEM), Milan.

Vogler, J. (1999). The European Union as an actor in international environmental politics. *Environmental Politics*, **8**, 3, 24–48.

Vollebergh, H. R. J. (1994). *Environmental Taxes and Transaction Costs*. Tinbergen Institute, Erasmus University, Rotterdam.

Vrolijk, C., & Grubb, M. (2000). *Quantifying Kyoto: How Will CoP-6 Decisions Affect the Market?*, Report of the Quantifying Kyoto Workshop, 30–31 August 2000. The Royal Institute of International Affairs (RIIA), London.

VROM-Raad. (1998). *Transitie naar een Koolstofarme Energiehuishouding: Advies ten Behoeve van de Uitvoeringsnota Klimaatbeleid*, Advies 010. VROM-Raad, Den Haag, in Dutch.

Wallström, M. (1999). *Reasonable Limits on Emissions Trading, Financial Times*, Letter to the Editor from M. Wallström (European Union Environment Commissioner), 10 November.

Waltz, K. N. (1979). *Theory of International Politics*. McGraw-Hill, New York.

WCI. (2002). The emerging market in emissions trading. *Ecoal: The Newsletter of the World Coal Institute*, **42**, 4–6.

Weber, M. (1976). *Wirtschaft und Gesellschaft: Grundrisse der Verstehenden Soziologie*. Mohr/Siebeck, Tubingen.

Weiss, A., & Woodhouse, E. (1992). Reframing incrementalism: a constructive response to the critics. *Policy Sciences*, **25**, 255–273.

Welch, W. P. (1983). The political feasibility of full ownership property rights: the cases of pollution and fisheries. *Policy Sciences*, **16**, 165–180.

Werksman, J. (1999a). *Responding to Non-compliance Under the Climate Change Regime*, ENV/EPOC(99)21/Final. Organization for Economic Co-operation and Development (OECD), Paris.

Werksman, J. (1999b). Greenhouse gas emissions trading and the WTO. *Review of European Community and International Environmental Law (RECIEL)*, **8**, 3, 251–264.

Westskog, H. (1995). *Market Power in a System of Tradeable CO_2-Quotas*, Unpublished Paper, Oslo, Norway.

Wiener, J. B. (2000a). Designing markets for international greenhouse gas control. In: R. J. Kopp, & J. B. Thatcher (Eds), *The Weathervane Guide to Climate Policy: An RFF Reader* (pp. 34–41). Resources for the Future (RFF), Washington.

Wiener, J. B. (2000b). *Policy Design for International Greenhouse Gas Control*, Climate Issues Brief No. 6, Revised and Updated Version. Resources for the Future (RFF), Washington.

Williamson, O. E. (1975). *Markets and Hierarchies: Analysis and Antitrust Implication — A Study in the Economics of Internal Organization*. Free Press, New York.

Williamson, O. E. (1979). Transaction-cost economics: the governance of contractual relations. *The Journal of Law and Economics*, **20**, 233–261.

Williamson, O. E. (1993). Transaction costs economics and organization theory. *Industrial and Corporate Change*, **2**, 107–156.

Williamson, O. E. (1997a). Transaction cost economics and public administration. In: P. B. Boorsma, K. Aarts, & A. E. Steenge (Eds), *Public Priority Setting: Rules and Costs* (pp. 19–37). Kluwer, Dordrecht.

Williamson, O. E. (1997b). Hierarchies, markets and power in the economy: an economic perspective. In: C. Menard (Ed.), *Transaction Cost Economics: Recent Developments* (pp. 1–29). Edward Elgar, Cheltenham.

Williamson, O. E. (1999). Public and private bureaucracies: a transaction cost economics perspective. *Journal of Law, Economics and Organization*, **15**, 1, 306–342.
Williamson, O. E. (2000). The new institutional economics: taking stock, looking ahead. *Journal of Economic Literature*, **38**, 595–613.
Windrum, P. (1999). *Unlocking a Lock-In: Towards a Model of Technological Succession*. MERIT/University of Maastricht, Maastricht.
Woerdman, E. (1999). The (im)possibilities of an international market for emission entitlements. *Change: Research and Policy Newsletter on Global Change from the Netherlands*, **45**, 1–4.
Woerdman, E. (2000a). Competitive distortions in an international emissions trading market. *Mitigation and Adaptation Strategies for Global Change*, **5**, 4, 337–360.
Woerdman, E. (2000b). Organizing emissions trading: the barrier of domestic permit allocation. *Energy Policy*, **28**, 9, 613–623.
Woerdman, E. (2001a). The "third way" in combating climate change: on the political advantages of project-based emissions trading. In: J. Platje, & M. A. Giacomin (Eds), *Environmental Protection in the Baltic Region: Awareness Building and Greenhouse Gas Emission Reduction* (pp. 103–120). Gdansk University/Nordic Council of Ministers, Poland/Denmark.
Woerdman, E. (2001b). *Lock-In of Environmental Policy Instruments: An Evolutionary Economics Approach*, Paper Presented at the Fourteenth Workshop in Law and Economics, April 11–12, 2001. University of Erfurt (Germany), Erfurt.
Woerdman, E. (2001c). Emissions trading and transaction costs: analyzing the flaws in the discussion. *Ecological Economics*, **38**, 293–304.
Woerdman, E. (2002). Why did the EU propose to limit emissions trading? A theoretical and empirical analysis. In: J. Albrecht (Ed.), *Instruments for Climate Policy: Limited Versus Unlimited Flexibility* (pp. 63–95). Edward Elgar, Cheltenham.
Woerdman, E. (2003). Developing carbon trading in Europe: does grandfathering distort competition and lead to state aid? In: M. Faure, J. Gupta, & A. Nentjes (Eds), *Climate Change and the Kyoto Protocol: The Role of Institutions and Instruments to Control Global Change* (pp. 108–127). Edward Elgar, Cheltenham.
Woerdman, E. (2004a). Tradeable emission rights. In: J. G. Backhaus (Ed.), *Elgar Companion to Law and Economics*. Edward Elgar, Cheltenham, in press.
Woerdman, E. (2004b). Path-dependent climate policy: the history and future of emissions trading in Europe. *European Environment*, in press.
Woerdman, E., & van der Gaast, W. P. (2001). Project-based emissions trading: the impact of institutional arrangements on cost-effectiveness. *Mitigation and Adaptation Strategies for Global Change*, **6**, 2, 113–154.
Woerdman, E., Boom, J. T., & Nentjes, A. (2002). Economy versus environment? Design alternatives for emissions trading from a lock-in perspective. In: M. T. J. Kok, W. J. V. Vermeulen, A. P. C. Faaij, & D. de Jager (Eds), *The Climate Neutral Society: Visions, Dilemma's and Opportunities for Change* (pp. 160–178). Earthscan, London.
Woerdman, E., van der Gaast, W. P., Manders, T., & Nentjes, A. (2003). The Kyoto Mechanisms: economic potential, environmental problems and political barriers. In: J. Gupta, & E. C. van Ierland (Eds), *Options for International Climate Policies* (pp. 115–136). Edward Elgar, Cheltenham.
Worsley, R., & Freedman, R. (Eds) (2002). *'Europarl' Daily Notebook 10-10-2002*. European Parliament, Brussels.

Wortman, T. J. (1995). Unlocking lock-in: limited liability companies and the key to underutilization of close corporation statutes. *New York University Law Review*, **70**, 1362, 1362–1408.
Yamin, F., & Lefevere, J. (2000). *Designing Options for Implementing an Emissions Trading Regime for Greenhouse Gases in the EC*, Final Report to the European Commission DG Environment, 22 February 2000. Foundation for International Environmental Law and Development (FIELD), London.
Yamin, F., Burniaux, J. M., & Nentjes, A. (2000). *Kyoto Mechanisms: Key Issues for Policy-Makers for COP-6*, Draft Paper Presented at the European Forum on Integrated Environmental Assessment Second Climate Workshop, 17–18 April, Amsterdam, The Netherlands.
Yandle, B. (1998). *Bootleggers, Baptists, and Global Warming*, PERC Policy Series Issue Number PS-14, November 1998.
Yandle, B. (1999). Grasping for the heavens: 3-D property rights and the global commons. *Duke Environmental Law & Policy Forum*, **10**, 13, 13–44.
Ybema, J. R., Kram, T., & van Rooijen, S. N. M. (1999). *Consequences of Ceilings on the Use of Kyoto Mechanisms: A Tentative Analysis of Cost Effects for EU Member States*. Energy Research Centre of the Netherlands (ECN), Petten.
Zapfel, P., & Vainio, M. (2002). *Pathways to European Greenhouse Gas Emissions Trading: History and Misconceptions*, FEEM Nota di Lavoro 85.2002. European Commission, Brussels.
Zhang, Z. X. (1998a). *Towards a Successful International Greenhouse Gas Emissions Trading*, Paper (Short Version) Prepared for the ETC/JIN Experts Meeting on Dealing with Carbon Credits After Kyoto, 28–29 May, Callantsoog, The Netherlands.
Zhang, Z. X. (1998b). *Towards a Successful International Greenhouse Gas Emissions Trading*, Working Paper (Complete Version) July 1998. University of Groningen, Groningen.
Zhang, Z. X. (1998c). Greenhouse gas emissions trading and the world trading system. *Journal of World Trade*, **32**, 5, 1–22.
Zhang, Z. X. (1999a). International greenhouse gas emissions trading: who should be held liable for the non-compliance by sellers? *Ecological Economics*, **31**, 323–329.
Zhang, Z. X. (1999b). Should the rules of allocating emissions permits be harmonised? *Ecological Economics*, **31**, 11–18.
Zhang, Z. X. (2000a). The design and implementation of an international trading scheme for greenhouse gas emissions. *Environment and Planning C: Government and Policy*, **18**, 321–337.
Zhang, Z. X. (2000b). Estimating the size of the potential market for the Kyoto flexibility mechanisms. *Weltwirtschaftliches Archiv (Review of World Economics)*, **136**, 3, 491–521.
Zhang, Z. X. (2001). The liability rules under international GHG emissions trading. *Energy Policy*, **29**, 501–508.
Zhang, Z. X., & Nentjes, A. (1999). International tradeable carbon permits as a strong form of joint implementation. In: J. Skea, & S. Sorrell (Eds), *Pollution for Sale: Emissions Trading and Joint Implementation* (pp. 322–342). Edward Elgar, Cheltenham.

Subject Index

actionable subsidy, 141, 152, 153, 171, 273, 274
Activities Implemented Jointly, 16, 41, 129
adaptation tax, 14, 127
adaptive expectations, 58
additional cost, 34, 36
additionality period, 15
additionality problem, 99
administrative capacity, 106
administrative costs, 33, 36, 61, 62, 121, 130
administrative oversight, 120
adverse effect, 153, 161
afforestation, 6, 13
agenda formation, 158, 213
allocation problems, 217, 238, 243, 249, 275
allowances, 10, 28, 259
allowance price—see permit price
allowance trading—see permit trading
ambiguity, 122, 136, 155, 271
Americans, 1, 4, 9, 160, 161, 190, 206, 221, 230, 244, 253, 257, 277
Annex B Parties, 4, 211
antitrust policy, 147, 170
Appendix (Questionnaire), 281
approval costs, 14, 113
Asians, 9
assigned amount, 5, 99, 202
Assigned Amount Units, 5
attitude change, 168, 190
attitudes, 224, 229, 256, 258, 275, 277
auctioning, 32, 34, 44, 118, 121, 143, 144, 161, 169, 176, 272
authorization, 10
average costs, 260, 278
average transaction costs, 130

balance of interests, 74
banking, 6, 13, 91, 92
bankruptcy, 147, 169, 272
bargaining analysis, 231
bargaining behavior, 254, 276
bargaining chip, 244, 254, 277
bargaining model, 252
bargaining position, 244, 252, 277
bargaining power, 250, 254
bargaining space, 252
barrier to trade, 199
baseline, 45, 86, 98, 127
baseline inflation, 100, 107, 125, 217, 218, 271
baseline problem, 271
baseline standardization, 103, 125, 271
benchmarking approach to baseline standardization, 104, 107
benchmarks, 105
bilateral approach to project-based emissions trading, 40, 126
binding targets, 107
black box, 119
bottom-up, 6
bounded rationality, 67
branching process, 57
breakout, 22, 52, 56
broker, 18, 127, 257
brokerage fees, 123
bubble, 6, 129, 244
Buenos Aires Plan of Action, 159, 200
bureaucratic costs, 65
bureaucratic hierarchy, 233
bureaucrats, 50
business-as-usual, 86, 88, 94, 104, 185, 270
buy their way out, 205, 209, 240, 275, 276
buyer, 18, 59, 93, 94, 96, 208
buyer liability, 96, 118, 212
bygones, 63, 270

Subject Index

capacity building, 106, 125, 126
cap-and-trade—see permit trading
capital costs, 98
capital gift, 145–147, 163, 170, 173, 174, 177, 273
capital reserve, 170
car driver, 37, 38
carbon colonialism, 48
carbon leakage, 45
carbon sequestration, 13
carbon tax, 33
carbon trading schemes of Denmark and the United Kingdom, 184
cash outflow, 146, 150, 170
ceiling on emissions trading, 250
ceiling on trade, 117, 201, 204, 211, 243, 275
certification, 42
Certified Emission Reductions, 5
CERUPT, 17
chemical industry, 259
Chicago Climate Exchange, 17
chipcard, 38
civil servants, 233, 258
Clean Development Mechanism, 5, 6, 97, 124, 135, 201, 224, 231, 248, 267, 269, 271, 276
clearinghouse, 30, 120, 126, 127
climate change, 4, 9, 209, 278
climate institutions, 3, 267
climate leadership, 215, 243, 258
climate policy, 1, 256, 267
climate system, 4
clustering of hypotheses, 223, 247, 275, 276
co-decision procedure, 181, 189, 190, 258, 259, 277
command-and-control, 34, 216
command-without-control, 10
commitment period, 254
commitment period reserve, 118, 202, 230
common law, 80
common market, 175, 176
comparative disadvantages, 142
comparative institutional analysis, 124
compensation payment, 151
competition, 142
competition law, 175, 187, 192

competitive advantages, 11, 16, 93, 143, 183
competitive disadvantages, 16
competitive distortion, 141, 142, 169, 178, 217, 272
competitive distortion as inefficiency, 142
competitive distortion as inequity, 142
competitive edge, 143
competitive opportunities, 154
competitive relations, 150, 171, 273
competitiveness, 178, 218
complete analogies, 58
complete information, 70
complexity, 34, 79, 114, 119, 206, 212
compliance, 95, 118, 123, 210, 243, 244, 272
compliance costs, 19, 90, 168, 211, 218, 258
compliance culture, 96
concrete ceiling, 201, 238, 243, 250
conditions for equal competition, 180
conditions of competition, 154
consensus, 87, 230, 233, 271
consistency, 187, 191–193, 216, 240
content analysis, 231, 235, 249, 276
conversion measure, 18, 30
coordination advantages, 61
co-ordination benefits, 260
corporation laws, 80
cost advantage, 144, 169, 173, 177
cost barrier, 119
cost components, 131
cost concepts, 64
cost internalization, 172, 174
cost minimization, 133, 206
cost savings, 5, 46, 113, 115, 206, 220, 251, 257
cost-effectiveness, 5, 33, 43, 101, 117, 132
cost-raising institutions, 117
costs of adaptation, 155
Council of Ministers, 258
covenant, 28, 44, 45, 49, 106
credible leadership, 215
credible threat, 250, 258
credit sharing, 42, 43
credit trading, 28, 31, 40, 49, 51, 56, 135, 151, 186, 255, 259, 260, 268, 278
credit-based approaches, 124, 161, 249

credit-based instruments, 167
crisis, 72
criteria for state aid, 173
crowding out, 15, 16
cultural barrier, 199, 257
cultural change, 51, 74, 76, 229, 255, 256, 258, 276, 277
cultural compatibility, 64
cultural constraints, 57
cultural objections, 7, 249
cultural opposition, 74
cultural resistance, 63, 69, 75, 199, 270, 277, 278
cultural values, 275
culture, 79, 229, 268
"culture matters", 80

decision makers, 128
deep purse theory, 147
degrees of path dependence, 60
degressive aid, 177
demand, 28, 201, 204, 269, 278
design choices, 27
"design matters", 124
design options, 34
developed countries, 4, 44, 49, 209
developing countries, 4, 15, 18, 41, 49, 90, 93, 127, 159, 205, 209, 211, 222, 244, 251
development costs, 105, 113, 114
differential administrative costs, 61, 72
differential costs, 260
direct costs, 113
direct participants, 186
directional leader, 215
Directive on CO_2 permit trading, 259
Directive on GHG emissions trading EU, 180, 274, 277
Directive on GHG permit trading, 168, 188
distortion of efficient competition, 142
distortionary taxes, 144, 161, 176
distributional problems of permit trading, 158
domestic action, 5, 200, 202, 206, 243, 275
domestic constraints, 252
dominance, 57
dominant culture, 74
dominant institution, 8, 64

double counting, 34
downstream trading, 32, 116, 119, 213
dynamic baselines, 102
dynamic inconsistency problem, 210

economic benefits, 218
economic decline, 89
economic design, 189
economic growth, 28, 97, 102, 106, 151, 278
economic hierarchy, 21, 43, 48, 50, 134, 141, 269, 271
economic incentives, 270
economic instruments, 3, 255, 267
economic interests, 203, 231, 275
economic models, 18, 89, 128, 188, 203, 206
economic perspective, 243
economic theory, 55
economic value, 191, 259
economies in transition, 18, 127, 205, 251
economists, 3, 9, 10, 14, 21, 49, 112, 124, 129, 154, 158, 183, 203, 205, 208, 210, 220, 252, 259, 277
effectiveness, 21, 44, 67, 72, 85, 208, 257, 270
efficiency, 3, 5, 20, 21, 30, 33, 44, 111, 117, 134, 141, 144, 149, 167, 170, 178, 188, 190, 191, 207, 257, 268, 272, 274, 275, 278
efficiency loss, 122
electricity producers, 50, 120, 185
electricity sector, 8, 120
eligibility criteria, 30, 143, 162, 183
eligibility requirements, 92, 100, 125
eligibility to trade, 202
emerging market, 16
emission baseline, 86
emission ceiling, 4, 28, 45, 151, 185, 208, 267, 269, 270, 278
emission growth, 210
emission reduction commitments, 13, 253
Emission Reduction Units, 5
emission rights, 167, 191, 260, 272, 276, 278
emission target, 4, 30, 181, 182, 208
emissions trading, 48, 51, 88, 207, 233, 255, 257, 267, 277

energy sector, 130
energy use, 29, 86, 97, 100, 102
enforcement, 30, 38, 96, 116, 118, 132, 162, 212, 221
enforcement costs, 14, 113
engineers, 252
entrepreneurial policy arenas, 257
entry into force, 8
environmental adaptation, 155
environmental bureaucracy, 232, 251
environmental economics, 9, 22, 98, 108, 116, 128, 267
environmental economics in neoclassical tradition, 133
environmental economics in neo-institutional tradition, 133
environmental economists, 2
environmental effectiveness, 86
environmental integrity, 87, 101, 103, 117, 123, 125, 224
environmental ministers, 232
environmental performance, 86
environmental regulation, 100, 124, 271, 276, 278
environmental scarcity, 29, 151, 269, 278
environmentally "fit", 271
equal treatment, 274
equality principle, 188
equity, 3, 20, 21, 36, 117, 141, 148, 149, 157, 161, 163, 167, 171, 176, 178, 185, 189–191, 199, 209, 224, 229, 240, 244, 245, 247, 252, 267, 268, 272–276, 278
equity principle, 209
equity values, 275, 276
ERUPT, 17
ethical interpretation of environmental effectiveness, 86, 271, 208
European Commission, 7, 171, 173, 181, 184, 189, 233, 235, 238, 240, 243, 250, 258, 274, 276, 277
European Community (EC), 167
European Council, 177
European Court of Justice, 173
European Parliament, 181, 189, 258, 259
European Union (EU), 167
Europeans, 9, 160–162, 190, 199, 206, 229, 248, 253, 254, 258, 277

evolution to efficiency, 50, 80, 254, 269
evolution towards auctioning, 184
evolutionary perspective, 51, 79
evolutionary process, 3, 56
evolutionary setting, 269
evolutionary theory, 22, 23
ex ante perspective, 90
ex ante transaction costs, 113
ex ante uncertainty, 98
ex post baseline corrections, 101, 104
ex post perspective, 90
ex post transaction costs, 112
example-setting, 211, 244
excess emission reductions, 92
Executive Board, 100, 125
expectations, 61, 74
experiments, 71
export potential, 14
external cost, 9, 278
external pressure, 252
external shock, 70, 76, 190, 250, 256, 257, 270, 277
external threat, 190, 258
external trading, 123
externality, 9

fair competition, 148, 149, 159, 170, 177, 272–274
fair conditions of competition, 172
fair distribution, 217
fairness, 148, 188, 209, 259
fairness principle, 142
"fast" track for Joint Implementation, 101, 125, 128, 272
financial advantage, 146, 150, 163, 173, 174, 185, 217, 272, 273
financial effects, 167, 172
financial position, 144–146, 150, 177
financial resources, 150, 272, 273
firm advantage, 173
firm-to-firm trading—see permit trading
first-movers, 211
fixed costs, 61, 113
flexibility provisions, 5–7, 172
flexible instruments, 207, 233
foreign direct investment, 41
forest management, 6, 13
formal constraints, 19, 141, 190, 258, 268

Subject Index 319

formal institutions, 3, 57, 59, 267
formal interpretation of environmental effectiveness, 86, 208, 271
fossil fuels, 4, 32
fragmented carbon market, 119
Framework Convention on Climate Change (FCCC), 4, 209, 215
free lunch, 259
free trading, 160
free-rider problem, 9
free-riding, 101
fuel sales, 37
fuel use, 37
fungibility, 18

gainers, 204
gaming, 101, 103
gas- and electricity bill, 38
General Agreement on Tariffs and Trade (GATT), 151
General Agreement on Trade in Services (GATS), 151
generic allocation criterion, 37
global warming, 4
government, 62, 75, 279
government failure, 117
government revenue foregone, 153
government trading, 30, 46, 47, 51, 121, 222
government-issued licenses, 152
governments, 22, 47, 161, 274, 275
grandfathering, 32, 34, 44, 118, 143, 144, 147, 161, 167, 169, 176, 217, 272–274
gratis allocation, 191
gratis permits, 272
Green Paper on GHG emissions trading, 168
greenhouse effect, 4
"green" values, 224, 247, 275, 276
growth targets, 106
guidelines on State aid, 172, 175, 177

harmonization, 143, 161, 162, 172, 181, 182, 189, 191, 273, 274
high-level officials, 229, 231, 240, 250, 253
historical responsibility, 5, 209, 224
history, 19, 50, 51, 254, 268, 269

"history matters", 57, 79
hot air, 117, 201, 205, 208, 238, 243, 244, 249, 251, 252, 254, 276, 277
hot air trading, 88, 208, 253, 270, 275
households, 33, 37, 38
hybrid liability, 97
hybrid property right, 114
hybrid system, 120
hybrid trading, 32
hypotheses, 208, 223, 229, 276, 236, 256
hypothesis testing, 239

ideological views, 122
imperfect capital market, 146
imperfect competition, 144, 146, 163, 213
imperfect foresight, 147
imperfect knowledge, 65
imperfect market, 58, 146, 213, 243
imperfect product market, 146, 170
implementation, 7, 19, 51, 158, 192, 208, 268
implementation costs, 39, 113
importers of fuel, 37
incentives, 75
incompatibility, 159
incomplete analogy, 58
incomplete information, 66, 67, 87, 269
increasing returns, 22, 55, 56, 60, 255
incremental change, 120, 255, 269, 275
incremental policy changes, 224
incrementalism, 79, 119, 216, 238, 243, 249, 255, 268
independent verification, 47
indirect effects, 99, 172
industrialized countries, 1, 19, 49, 90, 205, 209, 211, 214, 244, 267, 275
industrial organization, 147
inefficiency, 145
inefficient firm, 146
inefficient regulation, 275
inequality, 148
inequity, 148
inertia, 70
inevitability, 70
informal constraints, 19, 199, 258, 268, 275
informal institutions, 3, 57, 229, 254, 267, 275, 276

information, 60, 63, 121, 257, 259, 270, 277, 278
information and negotiating costs, 121
information deficit, 31
innovation, 59
institutional analysis, 2
institutional arrangements, 57, 113, 118, 124
institutional barrier, 3, 7, 16, 20, 51, 61, 88, 91, 116, 122, 208, 267
institutional breakout, 58, 70, 76, 229, 256, 259, 260, 269, 276–278
institutional capacity, 41, 47, 126, 128, 135
institutional change, 56, 63, 70, 71, 75
institutional constraints, 127
institutional coordination advantages, 62
institutional design, 15
institutional differences, 5
institutional dynamics, 2, 16, 90
institutional economics, 1, 2, 19, 20, 55, 79, 133, 255, 267, 278
institutional enhancements, 128
institutional evolution, 56, 79, 231, 267
institutional experience, 132
institutional features, 11, 47
institutional fitness, 109
institutional framework, 2
institutional law and economics, 21
institutional lock-in, 51, 52, 57, 69, 117, 122, 125, 128, 155, 199, 255, 260, 269, 271, 274–276, 278, 279
institutional opportunity, 98, 101, 103, 116, 125, 128, 271
institutional path dependence, 22, 57, 79, 268
institutional persistency, 58
institutional requirements, 6
institutional reversibility, 73
institutional rigidity, 57
institutional safeguard, 100
institutional scale, 61, 66, 255, 260
institutional set-up, 97, 122
institutional shortcomings, 127
institutional vacuum, 133
institutional void, 57, 80, 256
institutionalist approach, 267
institutionally "fit", 271

institutions, 19, 55, 57, 117, 268
insurance costs, 14, 113, 132
interest foregone, 144, 169
interest group, 22, 43, 50, 64, 74, 161
Intergovernmental Panel on Climate Change (IPCC), 4, 267
internal market, 183
internal pressure, 252, 253, 256, 258, 277
internal trading, 123
international emissions trading, 5, 159, 267
international negotiations, 7, 159, 180, 229, 253
inter-source trading, 28
interview, 234, 240
investment costs, 39
investment risks, 11
issue linkages, 121

Joint Implementation, 5, 6, 31, 97, 100, 124, 135, 201, 224, 231, 238, 248, 267, 269, 271, 276
judicial decisions, 192
jurisprudence, 157
JUSCANZ, 201, 230, 244, 250
justice, 5

key officials, 231, 233, 239, 277
Kyoto Mechanisms, 1, 5, 7, 11, 134, 199, 207, 223, 225, 230, 233, 248, 267, 275, 276
Kyoto Protocol, 1, 4, 11, 88, 125, 129, 159, 162, 168, 190, 199, 209, 230, 253, 267, 275, 277

labor costs, 124
large projects, 126
law and economics, 3, 10, 21, 55, 142, 191, 268, 274
"law matters", 80
lawyers, 158, 192, 193, 252
leakage, 99, 101
learning, 51, 66, 72, 120, 125, 129, 131, 187, 255, 260, 278
legal ambiguities, 7, 160, 178, 190, 193, 273, 274
legal and cultural barriers, 136
legal barrier, 21, 273

legal compatibility, 63, 74, 191
legal complexity, 157
legal constraints, 57
legal entities, 11, 18, 32, 42, 46, 93, 135, 156
legal fitness, 163, 190
legal form, 189, 274
legal framework, 63, 74, 270
legal instruments, 28, 255
legal objections, 179
legal precedent, 184
legal problems, 69, 75
legal protocol, 49
legal regime, 160
legal-economic inconsistencies, 192
legally "fit", 274
legal-theoretic debate, 192
legitimacy, 183
level playing field, 148, 150, 154, 163, 171, 173, 180, 182, 273
liability, 96, 118, 243
liberal environmentalism, 259
lobby groups, 160
lobbying, 63, 74, 118, 270
local pollution, 218
lock-in, 3, 22, 52, 56, 136, 229, 267
losers, 204
low-cost potential, 127
low-hanging fruits, 43
lump sum subsidy, 145, 170, 193, 272, 273

macro-baseline, 86
macro-economic effects, 238, 243
macro-economic impacts, 218
macro-focus, 156
marginal abatement cost, 5, 18, 45, 205–207, 221
marginal change, 255
market design, 123
market imperfection, 243
market instruments, 277
market justice, 149
market position, 174
market power, 31, 33, 116, 121, 204, 213
market price, 37, 39, 92
market skepsis, 7
market transaction cost, 112, 125, 128, 133, 135, 271, 272

market value, 8, 18, 145, 170, 174
market-based climate policy, 1, 8, 19, 267
market-oriented institution, 8
market-skeptic governments, 208, 233
mark-up, 33, 36, 120
matrix approach to baseline standardization, 104, 126
maximum permit price, 211
mercantilist theory, 142
methodological problems, 115, 131
micro–macro mismatch, 100
micro-baseline, 86, 97
micro-focus, 156
minimum permit price, 211
mitigation costs, 206
mixed system, 120
mixed trading, 32
model versus muddle, 115, 124, 134
monitoring, 30, 33, 37, 72, 132, 162, 221
monitoring and enforcement costs, 120, 121
monitoring costs, 14, 39, 113
monopoly, 59, 147, 204
moral resistance, 7, 229, 254
most-favored nation principle, 151, 172
motorists, 33, 121
multilateral approach to project-based emissions trading, 40, 126
multilateral fund, 40, 42, 125, 127
multiple equilibria, 69, 76
multi-year commitment period, 6
mutual reinforcement, 61

national allocation plan, 181, 189, 190, 260, 259, 278
national interests, 183, 205
national sovereignty, 183
national treatment principle, 151, 154, 172
negotiating behavior, 229, 231
negotiating position, 253
negotiating power, 88, 208, 213, 214, 224, 244, 252, 275–277
negotiating sessions, 235
negotiation costs, 14, 47, 113, 132
neoclassical approach, 267
neoclassical economic theory, 28
neoclassical economics, 2, 48, 56, 85, 135, 160, 255, 269, 278

neoclassical economists, 115, 135, 209, 272
neoclassical interpretation, 272
neoclassical law and economics, 141, 169
(neoclassical) law and economics approach, 186
neoclassical models, 203
neo-institutional approach, 199
neo-institutional economic approach, 190
neo-institutional economics, 20, 113, 135, 224, 268
neo-institutional economists, 229, 247
neo-institutional interpretation, 272
neo-institutional law and economics, 141, 169
new entrants, 145, 185, 187
new institutional economics, 20, 112, 135, 268
new interests, 74, 257
"new" positive (international) trade theory, 146
newcomer, 29, 44
non-actionable subsidy, 155
non-Annex B countries, 201
non-Annex B parties, 14
non-compliance, 34, 95, 118, 187, 210, 212
non-degressive aid, 177
non-discrimination, 171
non-distortion of competition, 159
non-economic objectives, 188
non-incremental measures, 216
non-incremental policy, 276
non-paper, 188, 232, 237
non-response, 233, 234, 240, 250, 276

official documents, 237
officials, 90, 250
one-time allocation, 155
operating aid, 177
operating licences, 177
operational costs, 127
operational entities, 103, 125
opinion analysis, 231, 249, 276
opportunity cost argument, 173, 182
opportunity costs, 28, 113, 144, 162, 169, 193, 217, 272, 274
optimal choice, 70
overselling, 96, 212, 222

parallel institutionalization hypothesis, 79
path dependence, 3, 20–22, 51, 56, 79, 117, 136, 229, 238, 255, 260, 267, 279
perceived benefits, 71
perceived costs, 63, 255
perception, 13, 66, 74, 92, 97, 116, 121, 128, 129, 149, 158, 160, 161, 179, 183, 190, 207, 244, 252, 257, 258, 272, 274, 277
perfect competition, 116, 144, 204, 213
perfect market, 59
performance standard, 29
performance standard rate trading, 28
permanent losses, 146
permit account, 38, 39
permit allocation, 118, 143, 150, 163, 217, 224, 272
permit costs, 120
permit price, 36, 181, 204, 209, 272
permit trading, 21, 28, 48–51, 55, 87, 114, 131, 135, 151, 167, 186, 192, 193, 199, 213, 224, 229, 247, 249, 250, 255, 257, 259, 269, 271, 275–278
permit-versus-credit discussion, 259, 260
perverse effect, 206, 217
picking low hanging fruits, 205
pincard, 38
pioneering role, 277
pioneering stage, 78
policy change, 216, 255
policy community, 257
policy entrepreneurs, 259
policy instruments, 3, 21, 48, 57, 182, 213, 256, 268
policy paradigm, 255
policy pioneer, 257
policy targets, 57
policy transfer, 51, 255
policy-science debate, 158
political acceptability, 49, 88, 157, 271, 274
political acceptance, 154, 155, 159, 176
political advantage, 29
political agenda, 2, 7, 126, 180, 257
political arguments, 187
political attraction, 44, 278

political barrier, 218
political change, 58
political compromise, 190
political considerations, 177, 186
political culture, 3, 224, 229, 231, 250, 254, 267, 275, 276
political debate, 161
political desires, 187
political developments, 135
political divisions, 277
political economy, 43, 51
political hierarchy, 21, 44, 48, 50, 141, 179, 208, 224, 233, 247, 249, 269, 272, 276
political institution, 160
political interests, 252
political negotiations, 179
political opposition, 50, 93, 231, 276
political perspective, 243
political precedent, 184, 188, 274
political preferences, 187
political pressure, 160, 220, 221, 224, 244
political priority, 122
political process, 39, 118, 150, 158, 171
political science, 21, 55, 79, 224, 226
political scientists, 158, 224
political transaction costs, 2, 20, 21, 44, 51, 63, 76, 79, 112, 115, 118, 122, 128, 135, 159, 260, 267, 270–272, 278
political will, 106, 192
political-legal conflict, 192
politically acceptable, 21, 192
political-strategic considerations, 224, 250, 275
political-strategic reasons, 253, 258
politicians, 43, 48, 50, 72, 79, 87, 90, 98, 113, 115, 116, 118–121, 125, 135, 141, 155, 159, 167, 170, 193, 253, 258, 269, 271–273, 277, 278
politics, 48, 61, 79, 114, 149, 158, 179, 183, 190, 192, 200, 240, 247, 267, 271
polluter pays principle, 167, 172, 181, 190–193
positive feedbacks, 66, 269
positive network externalities, 61, 121
potential, 259
power, 160, 255
power sector, 184, 186

pre-approval, 47
predatory pricing, 146
price distortion, 144, 170
price estimates, 219
price signal, 121
price uncertainty, 47, 120, 219
price war, 147, 170
private entities, 19, 30, 258
private property right, 10
private sector, 28, 42
private trading, 28, 222
probability, 69, 76, 234
problem-solving capacity, 67, 72, 257, 259, 260, 269, 277, 278
producers of fuel, 37
product price, 29, 144, 151, 169, 272
production costs, 61
production volume, 61
project exchange, 127
project-based approach, 107
project-based emissions trading, 40, 97, 124
project-based trading, 268, 278
property rights, 9, 63, 112, 113, 135, 143, 268, 270, 271
proportionality-based fairness principle, 142
Prototype Carbon Fund, 17, 40, 127
public administration, 114
public choice, 22, 50, 161
public goods, 9
public law, 59
public pays, 191
public property rights, 10
pulling attitudes, 74
pulling values, 64
purchase price, 28
pushing attitudes, 74
pushing values, 64

qualified majority, 177, 258
quantitative ceiling, 250, 253
quantitative ceiling on trade, 202
quantitative restriction, 92, 117
"quasi-judicial" system, 160
questionnaire, 229, 231, 239, 241, 250, 276
QWERTY-keyboard, 22, 55, 78, 270

radical policy changes, 224
ranking of hypotheses, 246
ratification, 1, 91, 221, 258, 277
rational choice, 91
reasonable expectations, 157
reforestation, 6, 13, 254
regulatory change, 270
regulatory lock-in, 69
regulatory property rights, 10, 123
regulatory tradition, 257
regulatory uncertainties, 129, 133
relative fitness problem, 78, 87, 97, 117, 136, 155, 271
remediableness, 60, 70
renegotiation, 93, 214, 216, 221
reporting costs, 130
representativity, 234
reserve, 254
respondents, 241, 244
restriction on trade, 230
revelation of preferences, 214
revenue foregone, 144, 169, 185, 186, 273
revenue recycling, 144
risk, 31, 34, 102, 118, 147, 170, 187, 210, 219, 224, 238, 243, 260, 279
risk of distortion of competition, 174
routine behavior, 134
rules of the game, 57
rules of thumb, 77
running costs, 59, 62, 65, 72, 73, 260, 278
Russians, 1, 8, 89, 190, 201, 222, 244, 277

satisficing, 67, 260
scale economies, 61
search and bargaining costs, 120
search costs, 14, 113, 132
second commitment period, 95, 210, 243, 251
secondary benefits, 218, 238, 243, 275, 276
self-enforcing, 37, 57, 159
self-financing, 31, 41, 126, 128
self-fulfilling prophecy, 62
self-reinforcing dynamics, 57
self-reinforcing mechanisms, 55, 58, 61, 67, 71, 76, 260, 269
seller, 18, 59, 94, 96, 208
seller liability, 96, 212

sensitivity analyses, 129, 246, 248
serious prejudice, 153
set-up costs, 32, 39, 51, 62, 65, 68, 113, 118, 125, 128, 133, 213, 259, 270, 271, 278
share of proceeds, 14
shared competence scheme, 182
shared liability, 212
side effects, 219
side-payment, 64, 75, 90, 218
simulation projects, 132
Single European Act, 175
sinks, 13, 202, 222, 230, 253, 254, 277
"slow" track for Joint Implementation, 101, 125
small projects, 126
"snapback" price effects, 99
social costs, 29
societal ambiguity, 77, 87
sociology, 55
soft landing, 210
sovereign states, 7
sovereignty aspects, 163
specificity, 173
spillover effects, 224
stakeholders, 183
standardized matrix, 107
standards, 28, 30, 34, 44, 45, 49, 100, 106, 128, 216, 256, 257
starting conditions, 149
start-up "capital", 44
start-up costs, 127, 260, 278
state aid, 162, 167, 171, 273, 274
state aid directorate, 187
state aid exemption, 175, 185, 186
state origin, 173
state origin criterion, 185, 273, 274
state resources, 174
state sovereignty, 157, 160, 161, 163, 172, 183, 218, 273
static baselines, 101
status quo, 63, 64, 79, 118, 150
stranded costs, 35
strategic behavior, 218
strategic move, 214
sub-optimal arrangements, 43
sub-optimal design, 51, 56
sub-optimal institution, 64

sub-optimality, 57
subsidization, 162, 179, 217, 272
subsidy, 145, 153, 170
sunk costs, 51, 63–65, 72, 73, 255, 257, 260, 269, 270, 278
superior alternative, 28, 48, 56, 58–60, 77, 247, 256, 269, 271
superior technologies, 64
superiority, 77, 136, 270
supervising agencies, 30
supplementarity, 200
supplementarity proposal, 199, 229, 275
supply, 28, 29, 201, 204, 231, 269, 276, 278
survey, 250
sustainable development, 14, 172
sustainable energy use, 29
switching costs, 63–65, 73, 76, 157, 161, 190, 199, 229, 255, 257, 260, 269, 270, 274, 275, 277, 278
systematic error, 104
systems boundary issue, 99

tax exemptions, 145, 170, 177
taxation, 30, 34, 44, 46, 49, 106, 143, 174, 176, 216, 256, 258
technical assistance costs, 130
technological change, 211
technological innovation, 29, 102, 210, 218, 238, 244, 275
technological lock-in, 52
technology transfer, 14, 15, 93, 126
temporary losses, 146
tenders, 17
theoretical ambiguity, 77, 87, 270
theoretical and societal consensus, 117
theoretical consensus, 78
"thick" market, 120, 146, 170
"thin" market, 122
third degree path dependence, 69
third party assessment, 103
third party check, 104, 126
third party verification, 101
threat power, 258
time costs, 39
top-down approach, 107
top-down approach to baseline standardization, 104
trade effect, 173

trade liberalization, 151
trade restriction, 129, 133, 206, 217, 251, 275
trade rivals, 222
tradeable emission rights, 8, 28, 141, 259
tradeable quota systems, 10
trade-off, 2, 94, 101, 117, 186, 234
trading rules, 123
trading volume, 119, 129, 204
tragedy of the commons, 9
transaction cost economics, 20, 79, 112
transaction costs, 2, 14, 16, 20, 29, 44, 46, 79, 101, 103, 107, 112, 117, 126, 133, 204, 213, 268, 269, 271
transaction tax, 92, 117
transaction value, 47, 132
transaction volume, 123
transport sector, 259
Treaty of Rome, 175
tropical air, 90, 222
two-track approach, 100
two-track system, 125

uncertainty, 34, 47, 60, 70, 78, 79, 87, 89, 117, 119, 122, 136, 155, 159, 219, 224, 238, 243, 251, 253, 270, 271
unequal conditions of competition, 175
unequal cost increases, 149
unilateral approach to project-based emissions trading, 40, 126
unit participants, 186
unitary actor, 243
unlocking lock-in, 70
unofficial documents, 237
upfront costs, 106
upstream monitoring, 37
upstream system, 120
upstream trading, 32

values, 183, 199, 224, 229, 256
variable costs, 113
vested interests, 64, 74, 260, 278
Vienna Convention on the Law of Treaties, 152
voluntary agreements, 34, 44, 256, 257
voluntary coordination, 161
voters, 64

wealth transfer, 119, 145, 161, 170, 176, 185, 191, 274
weighted score, 248
willingness to comply, 96, 118
windfall profit, 146
window of opportunity, 258

World Trade Organization, 141
WTO Agreement on Subsidies and Countervailing Measures, 152, 273
WTO compatible, 160
WTO dispute, 159
WTO panel, 159